Technology and Applications of Polymers Derived from Biomass

PLASTICS DESIGN LIBRARY (PDL)
PDL HANDBOOK SERIES

Series Editor: Sina Ebnesajjad, PhD (sina@FluoroConsultants.com)
President, FluoroConsultants Group, LLC
Chadds Ford, PA, USA
www.FluoroConsultants.com

The **PDL Handbook Series** is aimed at a wide range of engineers and other professionals working in the plastics industry, and related sectors using plastics and adhesives.

PDL is a series of data books, reference works and practical guides covering plastics engineering, applications, processing, and manufacturing, and applied aspects of polymer science, elastomers and adhesives.

Recent titles in the series
Biopolymers: Processing and Products, Michael Niaounakis (ISBN: 9780323266987)
Biopolymers: Reuse, Recycling, and Disposal, Michael Niaounakis (ISBN: 9781455731459)
Carbon Nanotube Reinforced Composites, Marcio Loos (ISBN: 9781455731954)
Extrusion, 2e, John Wagner & Eldridge Mount (ISBN: 9781437734812)
Fluoroplastics, Volume 1, 2e, Sina Ebnesajjad (ISBN: 9781455731992)
Handbook of Biopolymers and Biodegradable Plastics, Sina Ebnesajjad (ISBN: 9781455728343)
Handbook of Molded Part Shrinkage and Warpage, Jerry Fischer (ISBN: 9781455725977)
Handbook of Polymer Applications in Medicine and Medical Devices, Kayvon Modjarrad & Sina Ebnesajjad (ISBN: 9780323228053)
Handbook of Thermoplastic Elastomers, Jiri G Drobny (ISBN: 9780323221368)
Handbook of Thermoset Plastics, 2e, Hanna Dodiuk & Sidney Goodman (ISBN: 9781455731077)
High Performance Polymers, 2e, Johannes Karl Fink (ISBN: 9780323312226)
Introduction to Fluoropolymers, Sina Ebnesajjad (ISBN: 9781455774425)
Ionizing Radiation and Polymers, Jiri G Drobny (ISBN: 9781455778812)
Manufacturing Flexible Packaging, Thomas Dunn (ISBN: 9780323264365)
Plastic Films in Food Packaging, Sina Ebnesajjad (ISBN: 9781455731121)
Plastics in Medical Devices, 2e, Vinny Sastri (ISBN: 9781455732012)
Polylactic Acid, Rahmat et. al. (ISBN: 9781437744590)
Polyvinyl Fluoride, Sina Ebnesajjad (ISBN: 9781455778850)
Reactive Polymers, 2e, Johannes Karl Fink (ISBN: 9781455731497)
The Effect of Creep and Other Time Related Factors on Plastics and Elastomers, 3e, Laurence McKeen (ISBN: 9780323353137)
The Effect of Long Term Thermal Exposure on Plastics and Elastomers, Laurence McKeen (ISBN: 9780323221085)
The Effect of Sterilization on Plastics and Elastomers, 3e, Laurence McKeen (ISBN: 9781455725984)
The Effect of Temperature and Other Factors on Plastics and Elastomers, 3e, Laurence McKeen (ISBN: 9780323310161)
The Effect of UV Light and Weather on Plastics and Elastomers, 3e, Laurence McKeen (ISBN: 9781455728510)
Thermoforming of Single and Multilayer Laminates, Ali Ashter (ISBN: 9781455731725)
Thermoplastics and Thermoplastic Composites, 2e, Michel Biron (ISBN: 9781455778980)
Thermosets and Composites, 2e, Michel Biron (ISBN: 9781455731244)

To submit a new book proposal for the series, or place an order, please contact Edward Payne, Acquisitions Editor at edward.payne@elsevier.com

Technology and Applications of Polymers Derived from Biomass

Syed Ali Ashter
Principal at Ashter Consultancy LLC.,
Massachusetts, USA

ELSEVIER
William Andrew
Applied Science Publishers

William Andrew is an imprint of Elsevier
The Boulevard, Langford Lane, Kidlington, Oxford, OX5 1GB, United Kingdom
50 Hampshire Street, 5th Floor, Cambridge, MA 02139, United States

Notices

Knowledge and best practice in this field are constantly changing. As new research and experience broaden our understanding, changes in research methods, professional practices, or medical treatment may become necessary.

Practitioners and researchers must always rely on their own experience and knowledge in evaluating and using any information, methods, compounds, or experiments described herein. In using such information or methods they should be mindful of their own safety and the safety of others, including parties for whom they have a professional responsibility.

To the fullest extent of the law, neither the Publisher nor the authors, contributors, or editors, assume any liability for any injury and/or damage to persons or property as a matter of products liability, negligence or otherwise, or from any use or operation of any methods, products, instructions, or ideas contained in the material herein.

Library of Congress Cataloging-in-Publication Data
A catalog record for this book is available from the Library of Congress

British Library Cataloguing-in-Publication Data
A catalogue record for this book is available from the British Library

ISBN: 978-0-323-51115-5

Publisher: Matthew Deans
Acquisition Editor: David Jackson
Editorial Project Manager: Lindsay Lawrence
Production Project Manager: Sruthi Satheesh
Cover Designer: Mark Rogers

Typeset by SPi Global, India

Dedication

Dedicated to my parents, Prof (Late) ***Syed Zahier Haider*** *and* ***Mrs. Naseem Haider***

Contents

Preface

Technology and Applications of Polymers Derived from Biomass is a one-stop practical guide consisting of 13 chapters that provides fundamental understanding of biomass-derived polymers, as well as information on the range of different possible routes from biomass to polymeric materials, discusses the value and limitations of using biomass in material applications, and compares petrochemical-derived polymers with bio-based polymers.

Chapter 1 provides information on application of biomass, global biomass market, current market trends, and limitation of biomass. Chapter 2 primarily focuses on biomass and its resources which includes understanding on different categories of biomass materials, types of biomass system, handling of biomass, calculating composition of biomass, and challenges of biomass commercialization.

Chapter 3 reviews chemistry of biomass. In this chapter, basic organic and carbohydrate chemistry is reviewed. Chemistry of cellulosic polymers such as cellulose esters, celluloid, lignin, hemicellulose, and lignocellulosic are discussed in Chapter 4. Chapter 5 provides a review on classification of raw biomass and different biomass conversion approaches. Manufacturing of different industrial chemicals from biomass is discussed in Chapter 6.

Chapter 7 discusses derivation of monomers from biomass using gasification and fermentations while Chapter 8 reviews polymerization of different biomass-based monomers. Chapter 9 provides information on direct derivatives of polymers from biomass. Some of the polymers include epoxy resin, C6 carbon compounds, 3-HPA, glucaric acid derivatives, and derivatives derived from microbial polymers.

Chapter 10 provides discussion on different techniques to process biomass-derived polymers, while Chapter 11 talks about material and energy balance required for conversion of biomass. Application of biomass-derived polymers and chemicals is discussed in Chapter 12. Some of the applications include automotive, packaging, horticulture, medical devices/biomedical, pharmaceutical/hygiene, and filtration/purification. Lastly, Chapter 13 talks about the environmental impact of biomass conversion. This chapter includes discussion on life cycle assessment, its benefit, and how to estimate carbon dioxide emissions.

Syed A. Ashter

Acknowledgment

I would like to thank my family especially my wife, Tahira, and my two kids, Zayn and Noor for showing tremendous patience and unconditional support. This work would have never been completed otherwise.

Introduction

1

1.1 Background

Biomass is a renewable and sustainable source of energy developed from mixture of organic materials [1]. This mixture of organic molecules contains hydrogen, oxygen, and nitrogen. Biomass is primarily categorized into five categories as shown in Fig. 1 [2–7]. They are as follows:

1. Wood
2. Energy crops
3. Agricultural residues
4. Food waste
5. Industrial waste and coproducts

1.2 Applications of biomass

Biomass can be as a replacement for products that can be made from fossil fuels. For example, fuels and power production. Overall considering the scope of the applications of biomass, they can be divided in two types: (1) energy and transportation, and (2) kind of products.

1.2.1 Energy and transportation

The main biomass feedstocks for power are paper mill residue, lumber mill scrap, and municipal waste. Many traditional food crops, such as corn, sugar, and vegetable oils, are also some of the most commonly used energy feedstocks. Furthermore, agricultural land may be shifted from producing food to the production of dedicated energy crops.

Fig. 2 shows commonly used biomass sources for energy and transportation applications [8]. Typically, there are six main sources of biomass:

1. Agricultural crops and residues
2. Sewage waste
3. Municipal solid waste
4. Animal residues
5. Industrial residues
6. Forestry crops and residues

These feedstocks are used to generate three different types of biofuels:

1. Solid biofuels—wood pellets, wood chips, briquettes, and firewood
2. Liquid biofuels—biodiesel, bioethanol, and advanced biofuels
3. Gaseous biofuels—biogas, biomethane, and syngas

Technology and Applications of Polymers Derived from Biomass. https://doi.org/10.1016/B978-0-323-51115-5.00001-3

Fig. 1 Examples of biomass: (A) wood, (B) energy crops, (C) agricultural residues, (D) food wastes, and (E) industrial waste and coproducts.
From (A) https://commons.wikimedia.org/wiki/File:Glycyrrhiza_glabra_(Pile_of_Spanish_wood_chips).jpg; (B) http://hoffmannursery.com/miscanthus; (C) http://www.inforse.org/europe/dieret/Biomass/biomass.html; (D) https://commons.wikimedia.org/wiki/File:GI_Market_food_waste.jpg; (E) https://en.wikipedia.org/wiki/Waste#/media/File:Shredded_solid_waste.jpg.

1.2.2 Products

For many industrial products made from fossil-based fuel, biomass can be a renewable substitute. Some of the examples include bio-based foams, plastics, fertilizers, lubricants, and industrial chemicals (Fig. 3) [9–13].

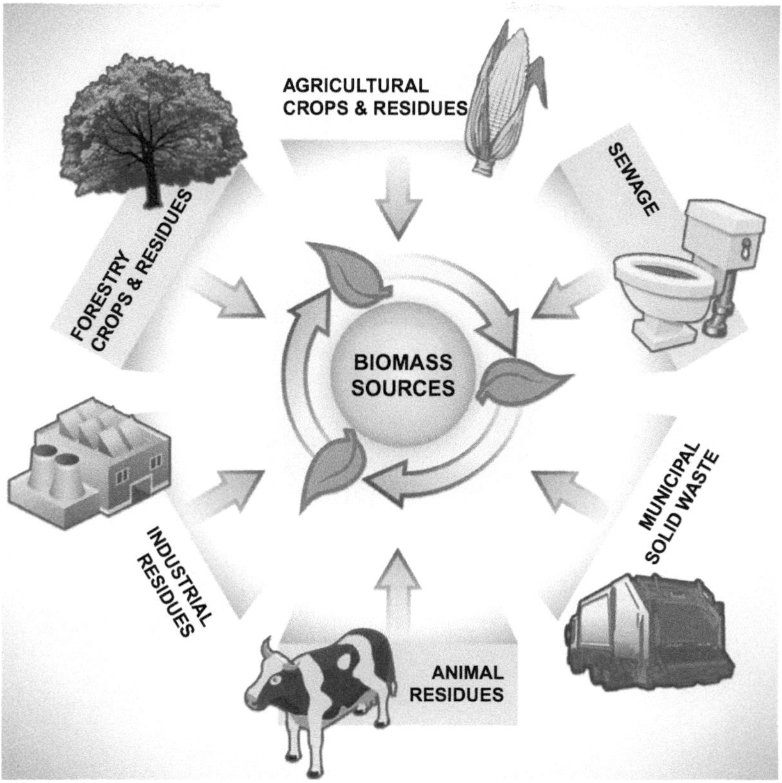

Fig. 2 Sources of biomass and their applications [8].
From http://all-mitsubishi.info/2017/bpics-biomass-energy.awp.

1.3 Understanding global markets

Biomass has major benefits over other renewable resources such as solar, wind, and hydropower because of its extensive natural resources. It provides practical solution to fossil-based fuels, especially in (a) wood bioenergy market, (b) power sector, and (c) gasification market.

1.3.1 Wood bioenergy markets

Wood bioenergy have shown significant growth in recent years mainly due to the timber-derived product market that is popular for residential and commercial applications. Other markets such as biomass pellets have also seen growth in the global market mainly due to increased government initiatives and stringent environmental regulations. Biomass pellets are commonly used as a heating fuel [14].

The global biomass pellet market will grow significantly between 2010 and 2025. Europe will account for the largest share of this global biomass pellet market growth

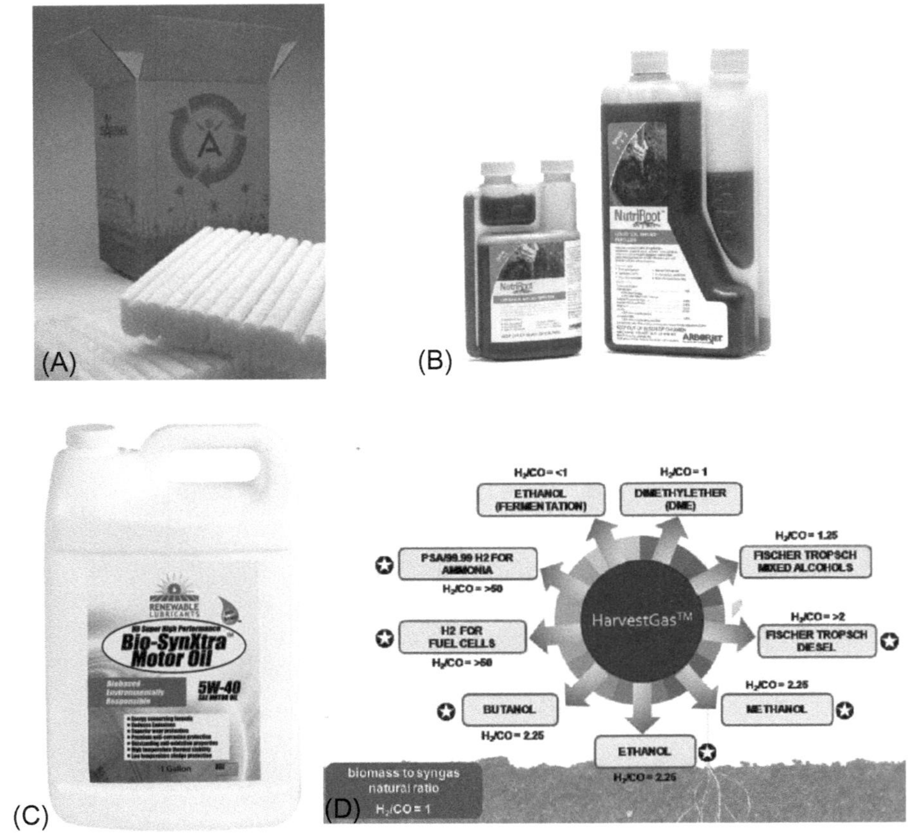

Fig. 3 Applications of biomass. (A) Bio-based foams, (B) bio-based plastics, (C) bio-based lubricants, and (D) industrial chemicals.
From (A) Greencell foam; (B) NutriRoot; (C) http://renewablelube.com/store/motor-oils?product_id=96; (D) http://www.chemicals-technology.com/projects/syngest-bioammonia/syngest-bioammonia2.html.

with 38 million metric tons consumed in 2025 [14,15]. North America will account for the second largest share in the global biomass pellet market primarily due to increase in demand [15].

1.3.2 Power sector

The global power sector market will observe the fastest growth during 2015–20 as shown in Fig. 4. Non-OECD-based countries will account for the largest share of the global power market with 600 TWh power consumed in 2020. China will account for the second largest share in the global power sector market with power consumption at 500 TWh followed by Asia in the third place [16].

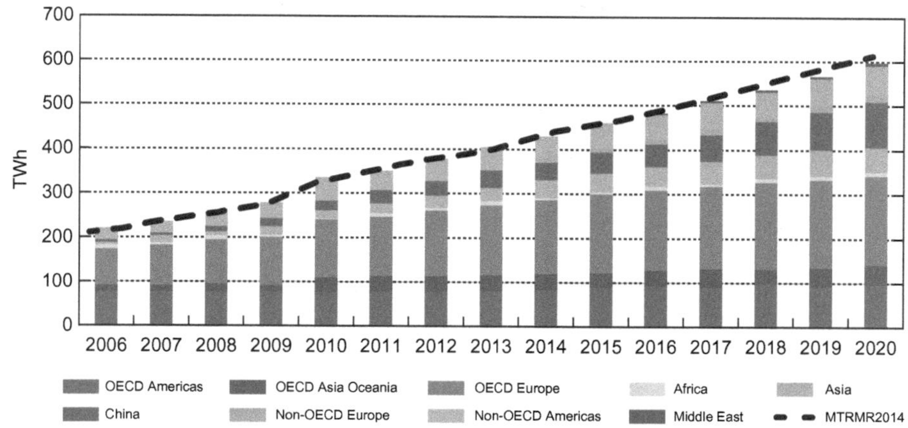

Fig. 4 Power sector market.
From https://www.iea.org/topics/renewables/subtopics/bioenergy/. OECD/IEA.

1.3.3 Gasification market

A variety of commodities can be produced from syngas, the primary product of gasification, providing for many potential markets. This product flexibility as well as the versatility and efficiency of gasification makes it an attractive technology now and for the future.

The four main markets for gasification are shown in Fig. 5 [17]

1. Chemicals
2. Liquid fuels

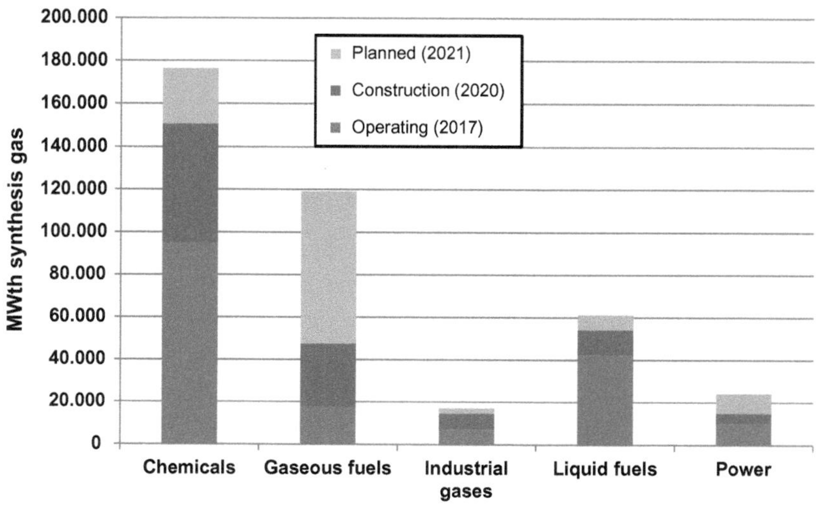

Fig. 5 Gasification market.
From https://www.netl.doe.gov/research/coal/energy-systems/gasification/gasifipedia/markets. Used with permission from Gasification-syngas, GSTC Database, 2017.

3. Power
4. Gaseous fuels

Globally, 60% of all the syngas produced is utilized in the production of chemicals and 24% in the production of liquid fuels. Eight percent of the total syngas produced will be used for producing power and gaseous fuels. China will account for the largest share of the chemicals market because of the increase in the capacity by 2019 [17]. The United States will account for the largest market share for liquid fuel application with significant growth of 19.9 million barrel per day (BPD) by 2035. Power and gaseous fuel accounts account for small share of the global gasification market because they are relatively new in the gasification market [17].

1.4 Current market trends

Fig. 6 shows the total energy demand supplied from biomass in the year 2015. More than any other renewable energy source, bioenergy has been significantly contributing to global energy supply. Bioenergy has been growing at around 2% per year since 2010 and its share in total global primary market has been relatively steady since 2005, at around 10%, despite a 24% increase in overall global energy demand between 2005 and 2015. Bioenergy plays a role in all three main energy-use sectors: heating, electricity, and transport [18].

Solid biomass represents the largest share of biomass that is used for heating and electricity generation, whereas liquid biofuel represents the largest source in the transport sector (Fig. 7) [18].

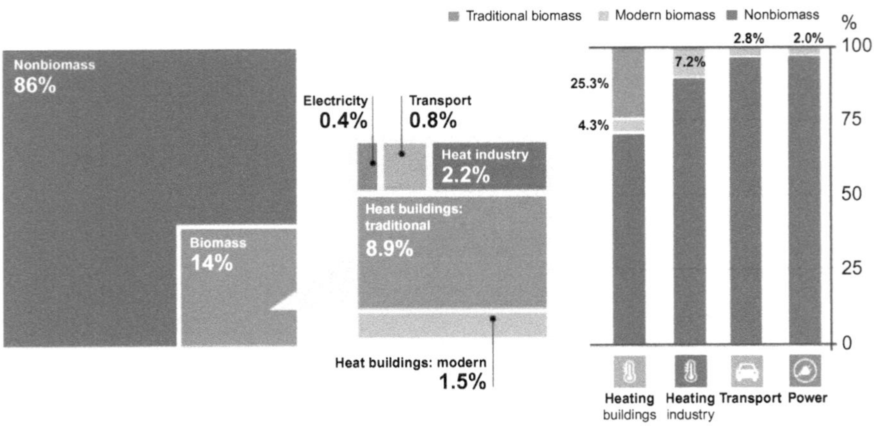

Fig. 6 Shares of biomass in total final energy consumption and in final energy consumption by end-use sector.
From http://www.ren21.net/wp-content/uploads/2016/06/GSR_2016_Full_Report.pdf, Original data—Table R10 derived from IEA, World Energy Outlook 2015, Energy Access Database, http://www.worldenergyoutlook.org/resources/energydevelopment/energyaccessdatabase, and submissions from report contributors.

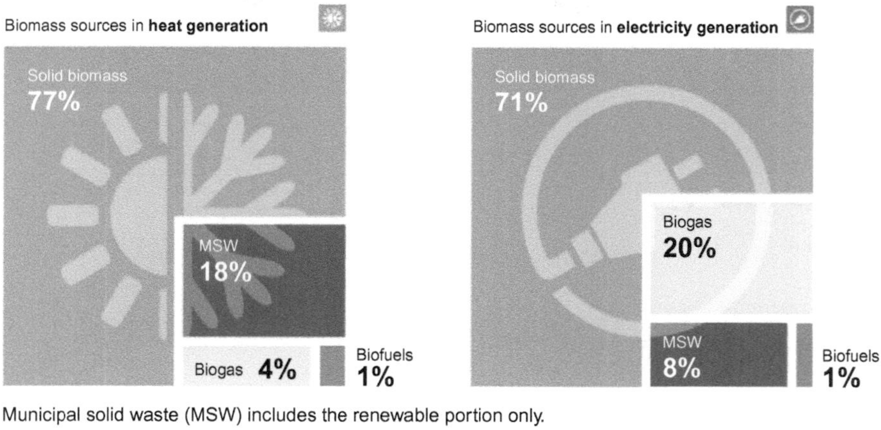

Fig. 7 Shares of biomass sources in global heat and electricity generation.
From http://www.ren21.net/wp-content/uploads/2016/06/GSR_2016_Full_Report.pdf.

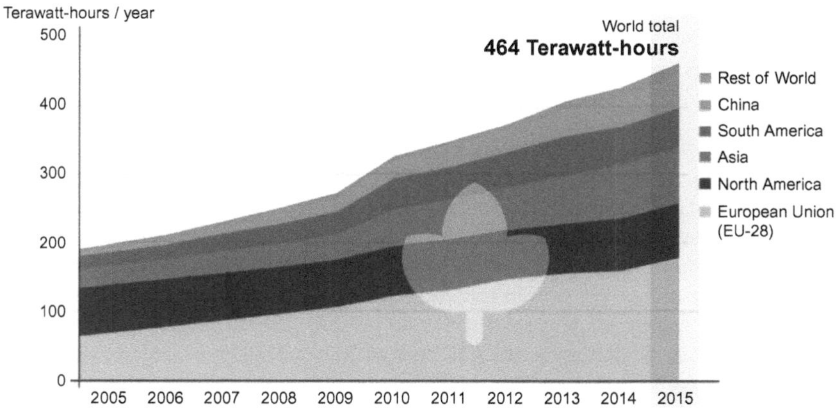

Fig. 8 Bio-power global generation, by country/region.
From http://www.ren21.net/wp-content/uploads/2016/06/GSR_2016_Full_Report.pdf.

Bio-power global generation per capita is shown in Fig. 8. The global bio-power generation increased by 8% to 464 TWh. Some of the leading countries involved in bioenergy production in 2015 were the United States (69 TWh), Germany (50 TWh), China (48 TWh), Brazil (40 TWh), and Japan (36 TWh) followed by the United Kingdom and India [18].

As shown in Fig. 9, United States and Brazil dominate biofuel global production market. Together, these two countries produce approximately 72% of all the biofuels followed by Germany, Argentina, and Indonesia. Among all the biofuels, 67% of the biofuel production was bioethanol while biodiesel was 33%. Hydrogenated vegetable oils and other advanced biofuels comprised of small fraction of the global biofuel production [18].

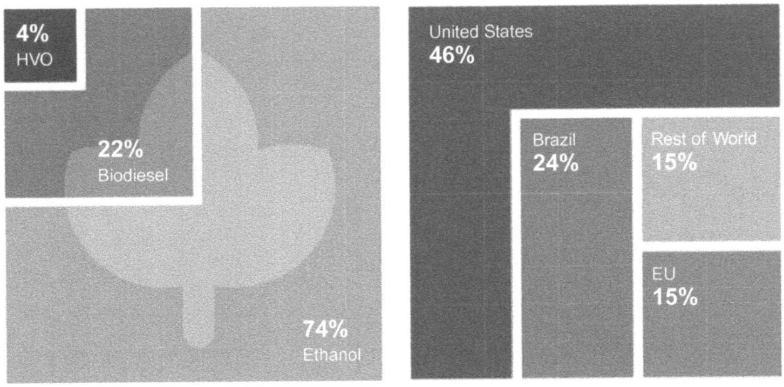

Fig. 9 Biofuels global production, shares by type and by country/region, 2015. From http://www.ren21.net/wp-content/uploads/2016/06/GSR_2016_Full_Report.pdf.

1.5 Limitations

Although fossil fuel such as wood has always served as the fuel source for heating and electricity, advancements in biomass technologies have made it as an alternative fuel source. However, production and the use of this renewable bioenergy has some limitations:

1. *Arable land requirement*: Some biomass energy crops require an arable land to produce the feedstock. With growing scarcities in necessities such as food and water, the use of the suitable arable land and other natural resources may need to be dedicated to producing sufficient food to feed the population [19,20].
2. *Sustainable behaviors of communities*: Some communities with a permanent processing facilities might discourage overall recycling participation rates and development of other forms of clean energy, such as solar and wind [19,20].
3. *Use of energy*: Some biomass feedstocks, such as corn and soybeans, require large amounts of energy to grow and process. This results in very little to no net gain in energy resources. These biomass feedstocks also require energy to transport to where they will be processed and made into fuel [19,20].
4. *Deforestation*: The use of trees as a biomass energy feedstock can lead to widespread deforestation. This is particularly concerning, as the need for forests and their important role in carbon sequestration in a world of climate change cannot be understated. We need to be planting forests, not cutting them down for fuel [19,20].
5. *Air pollution*: Some biomass energy resources can produce pollutants when burned, such as nitrous oxide, soot, ash, carbon monoxide, and carbon dioxide.
6. *Seasonal availability*: Some biomass energy feedstocks are seasonal.

References

[1] http://www.reenergyholdings.com/renewable-energy/what-is-biomass/&gt.
[2] http://www.biomassenergycentre.org.uk/portal/page?_pageid=76,15049&_dad=portal&gt.

[3] http://stobartwoodfuel.com/products/virgin-wood-biomass/&gt.
[4] http://hoffmannursery.com/miscanthus&gt.
[5] http://www.inforse.org/europe/dieret/Biomass/biomass.html&gt.
[6] http://biorefinerycentre.ifr.ac.uk&gt.
[7] http://www.energynext.in/mnre-encouraging-waste-to-energy-projects/&gt.
[8] http://www.geni.org/globalenergy/library/articles-renewable-energy-transmission/bio energy.shtml&gt.
[9] https://www.pinterest.com/maddy0499/bio/&gt.
[10] https://bestinpackaging.com/2009/11/30/the-potential-of-bio-based-plastics/&gt.
[11] http://www.greenindustrypros.com/news/12179608/liquid-fertilizer-nabs-certified-bio based-label.
[12] http://industrial-oil.net/bio-based-lubricants.html&gt.
[13] http://www.chemicals-technology.com/projects/syngest-bioammonia/syngest-bioammonia2.html&gt.
[14] http://www.prnewswire.com/news-releases/global-biomass-pellet-market-size-share-development-growth-and-demand-forecast-to-2020–industry-insights-by-application-heat-sector-power-sector-300217373.html.
[15] https://www.statista.com/statistics/243910/global-wood-pellet-consumption-outlook/-global wood pellet demand????&gt.
[16] https://www.iea.org/topics/renewables/subtopics/bioenergy&gt.
[17] https://www.netl.doe.gov/research/coal/energy-systems/gasification/gasifipedia/mar kets&gt.
[18] http://www.ren21.net/wp-content/uploads/2016/06/GSR_2016_Full_Report.pdf&gt.
[19] http://www.biomass.net/Biomass-Disadvantages.html&gt.
[20] http://greentumble.com/biomass-energy-advantages-and-disadvantages/&gt.

Biomass and its sources

2

2.1 Definition

Biomass is the collective term used to identify organic matter generated either in the form of waste such as solid municipal waste or from living plants such as trees, grass, and agricultural crops and residues. The composition of these organic matters will affect how they are handled and processed. Even varietal or hybrid differences within the same crop can also influence processing characteristics [1,2].

2.2 Difference between biomass and fossil-based fuels

Fossil-based fuels such as coal, oil, and natural gas have played a significant role in our day-to-day lives. In addition to producing electricity, and fuels for heating and transportation, fossil-based fuels are used in the production of industrial plastics [3]. However, there have been major concerns that have limited the use of fossil-based fuels and have started transitioning to green energy. These issues include

- Limited supply
- Environmental impacts—release of pollutants such as hydrocarbons, sulfur, smoke, ash, and large amount of carbon dioxide.

Severity of impact to the environment depends on the type of the fossil-fuel used. For example, when coal is burned, one carbon dioxide is emitted into the atmosphere while burning oil releases more energy per carbon dioxide emitted compared to coal. Similarly, natural gas is mainly methane which has four hydrogen atoms and one carbon, but still only emits one carbon dioxide. This makes natural gas as the best fossil-based fuel in terms of energy output per unit carbon dioxide emitted.

In comparison to fossil-based fuels, biomass-based fuel is an energy source that comes from sources such as wood and food crops like sugarcane and corn. It also includes by-products such as agricultural waste, for example, corn straw (stover) and rice husks and other food waste such as vegetable oil recycled from cooking processes. Biomass can be burned to generate electricity and heat, digested to produce biomethane, and converted into liquid biofuels, biochemicals, and bioplastics [3].

Biomass is considered a low carbon fuel because of its ability to absorb carbon dioxide from the atmosphere and store it. When it is burned, it releases carbon dioxide into the atmosphere. However, when the plant regrows, it undergoes process of photosynthesis as shown in Fig. 1 [3]. During this process, it reabsorbs the same amount of carbon dioxide. The carbon dioxide in burning biomass is emitted very quickly but only reabsorbed over the growth period of the biomass. The timescale of biomass carbon neutrality can be 20 or 30 years in the case of some trees. A significant advantage

Technology and Applications of Polymers Derived from Biomass. https://doi.org/10.1016/B978-0-323-51115-5.00002-5

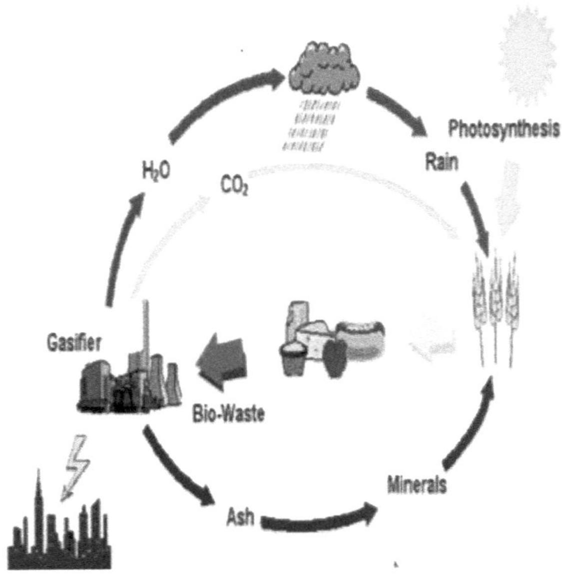

Fig. 1 Process of photosynthesis. From https://reich-chemistry.wikispaces.com/Michael+Gautreau+Brett+Fredericks+wiki+project2009.

of Giant King Grass is that it is very fast growing and can be harvested every 6 months or less. This is called a short rotation crop and the carbon dioxide emitted is reabsorbed very quickly [3].

2.3 Historical developments

History of biomass has evolved over the past decades. New innovations in technologies have created new ways to use biomass energy. For example, the biomass heater and a biomass pelletizer. Biomass heater produces constant flame which is used for residential use while the biomass pelletizer produces pellets that are efficient for pellet stoves and generate thermal energy for furnaces and boiler. Table 1 presents historical timeline of biomass [4].

2.4 Properties of biomass

It is the inherent properties of the biomass source that determines both the choice of conversion process and any subsequent processing difficulties that may arise. Equally, the choice of biomass source is influenced by the form in which the energy is required and it is the interplay between these two aspects that enables flexibility to be

Table 1 Historical timeline of biomass

	Timeline
1840	First commercially used biomass gasifier is built in France
1880	Henry Ford uses ethanol to fuel one of his first automobiles, the quadricycle
1910	Coal begins to replace the use of wood in city homes
1930s	Use of biomass heaters and stove was conceptualized. It used burning scrap and loose wood to produce heat
1950	Electricity and natural gas displace wood used in most homes
1970s	• There was an active search for alternative energy source during this period because of the uncontrollable increase in price for fossil fuels • Scientists were able to make car run using the by-product of degasification. Degasification is a process of converting biomass products to usable energy • The congress enacted Public Utility Regulatory Policies Act which will strengthen the development of alternative energy sources
1980s	• Infrastructures for generators which use wood fire were built • In 1982 and 1983, small plantations were made capable of producing electricity from biomass materials like wood waste; sawmill residues and by-products from agriculture, trimmings from yard and common household wood waste
1990	The clean air act controls the sale of oxygenated fuels in areas with higher levels of carbon monoxide
2000	• Programs were developed so that the fuel produced by biomass can be combined with nonrenewable resources coal to reduce the consumption of fossil fuels • There were also researches about crops that can be grown for power generation
2014	Industries increase biomass as a source of energy, while following environmental laws in order to protect the atmosphere

From Unfolding the improvements of biomass history, Alternative Energy Solutions Info. http://www.alternativeenergysourcesinfo.com/biomass-history.html.

introduced into the use of biomass as an energy source. Dependent on the energy conversion process selected, particular material properties become important during subsequent processing [5].

Three main biomass properties are as follows:

1. Physical
2. Chemical
3. Mechanical

Most biomass materials include [6]

- Agricultural residues (corn stover, wheat straw, rice straw, and soybean stalk)
- Forestry residues (saw dust, woody chips)
- Processing wastes (food processing wastes, animal wastes, and municipal solid wastes)
- Energy crops (switch grass, sweet sorghum, sugar canes, algae, cattail, and duckweed)

2.4.1 Physical properties

Physical properties of biomass include [6]

- Shape and size of biomass particles
- Aspect ratio of biomass particles
- Sphericity of particles
- Particle size distribution
- Density
- Porosity
- Densification of biomass

2.4.1.1 Shape and size of biomass particles

Biomass particles have different shapes and sizes: (a) flake like, (b) cylinder like, (c) near spherical, and (d) poplar samples. Shape and size of biomass particles are important because they affect mixing and fluidization, surface area for heat and mass transfer, angle of internal friction, and flow characteristics [6].

2.4.1.2 Aspect ratio of biomass particles

Aspect ratio of the particles can be defined as:

$$A = \frac{b}{l}$$

where b is the width and l is the length of the particle.

2.4.1.3 Sphericity of particles

Sphericity is a measure of roundness of an object. It is defined as the ratio of the volume of triaxial ellipsoid to the volume of a sphere with $D = D_{\max}$ with a power of one-third and is expressed by following equation [6]:

$$\text{Sphericity} = \left[\frac{\text{Volume of ellipsoid with equivalent diameters}}{\text{Volume of circumscribed sphere}}\right]^{1/3} = \left[\frac{(D_{\text{major}} D_{\text{inter}} D_{\text{minor}})^{1/3}}{D_{\text{major}}}\right]$$

where D_{major}, D_{minor}, and D_{inter} are the major, minor, and intermediate diameters, respectively.

It is important to note that an assumption that the volume of the object is equal to that of a triaxial ellipsoid with diameters equal to major, minor, and intermediate diameter.

Table 2 Different sizes of sieves

Sieve no.	Opening (mm)
4	4.76
6	3.36
8	2.38

From L. Wang, A. Shahbazi, Biomass properties, BEEMS Module A2, USDA Higher Education Challenger Program 2009-38411-19761, North Carolina Agricultural and Technical State University, 2009.

2.4.1.4 Particle size distribution

The particle size distribution can be characterized by using a set of sieves with different opening sizes. The opening size of sieves decreases from top to the bottom as shown in Table 2. Biomass is added to the most top sieve on the set of sieves. These sieves are then loaded in an automatic shaker which retains a mass fraction on each sieve. Large size particles are retained on the top sieve while the smaller particles are collected on the bottom sieves [6].

2.4.1.5 Density of biomass

Density of biomass is:

$$\frac{\text{Mass of biomass}}{\text{Volume}}$$

The density of biomass can be divided into three types:

1. Solid density: It is defined as mass per unit volume of the solid material and excludes any interior pores which are filled with air. It is important to note that if the volume of the internal pores is negligible, then the solid density is equal to the bulk density.
2. Particle density: It is the mass per unit volume of an individual particle including internal pores.
3. Bulk density: It is the mass per total volume occupied by a large quantity of particles including the pores space among particles.

2.4.1.6 Porosity

Porosity is defined as percentage of the total volume occupied by the air and is expressed by the following equations [6]:

$$\text{Porosity}\,(\%) = \frac{(\text{Total volume} - \text{Particle volume})}{\text{Particle volume}} \times 100\%$$

$$= \frac{\left(\dfrac{\text{Mass}}{\text{Bulk density}} - \dfrac{\text{Mass}}{\text{Particle density}}\right)}{\left(\dfrac{\text{Particle volume}}{\text{Bulk density}}\right)} \times 100\%$$

$$= \left(1 - \frac{\text{Bulk density}}{\text{Particle density}}\right) \times 100\%$$

2.4.1.7 Densification of biomass

One of the major barriers to their widespread use is that biomass has a lower energy content than traditional fossils fuels, which means that more fuel is required to get the same amount of energy. When combined low energy content with low density, the volume of biomass handled increases enormously (Fig. 2). Compaction or densification is one way to increase the energy density and overcome handling difficulties [7].

The low density of biomass materials poses a challenge for the handling, transportation, storage, and combustion processes. These problems may be addressed through densification, a process that produces either liquid or solid fuel with denser and more uniform properties than the raw biomass.

The main advantages of biomass densification for combustion are as follows:

- Simplified mechanical handling and feeding
- Uniform combustion in boilers
- Reduced dust production
- Reduced possibility of spontaneous combustion in storage
- Simplified storage and handling infrastructure, lowering capital requirements at the combustion plant
- Reduced cost of transportation due to increased energy density

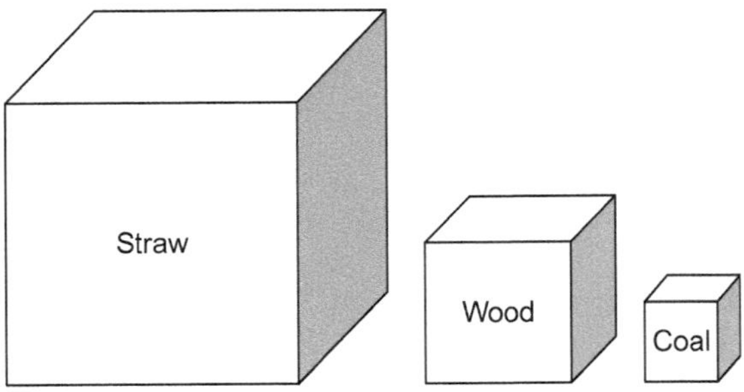

Fig. 2 Equivalent energy content by volume of unprocessed materials.
Original from Preto, F. (2007). Strategies and techniques for combustion of agricultural biomass fuels. Growing the Margins Energy Conference.

2.4.2 Chemical properties

Chemical properties of biomass include

- Chemical composition
- Elemental composition
- Equilibrium moisture content

2.4.2.1 Chemical composition

The chemical composition of biomass comprises of five main components:

1. Water
2. Protein
3. Lipid
4. Carbohydrate
5. Ash

Composition of selected biomass is shown in Table 3 [8–10].

The chemical composition of plant biomass varies among species. Plants are made of approximately 30% lignin and 70% carbohydrates or sugars. The carbohydrate fraction consists of many sugar molecules linked together in long chains or polymers. Two categories are distinguished: cellulose (glucose) and hemicellulose (xylose). The lignin fraction consists of nonsugar type molecules that act as a glue holding together the cellulose fibers [11].

2.4.2.2 Elemental composition

Biomass contains a large number of complex compounds, moisture, and a small amount of inorganic impurities known as ash. The organic compounds comprise four principal elements: carbon (C), hydrogen (H), oxygen (O), and nitrogen (N). Biomass may also have small amounts of chlorine (Cl) and sulfur (S). The composition can be obtained by performing ultimate analysis of a biomass.

Table 3 Composition of selected biomass

	Composition (% dry mass)			
Biomass	**Protein**	**Lipid**	**Carbohydrate**	**Ash**
Corn	10.36	3.29	84.96	1.39
Corn stover	7.82	1.57	86.9	3.9
Switchgrass	3.2	1.0	80.9	5.7
Microalgae	50.8	20.7	12.9	–

From L. Wang, A. Shahbazi, Biomass properties, BEEMS Module A2, USDA Higher Education Challenger Program 2009-38411-19761, North Carolina Agricultural and Technical State University, 2009; S.M. Drury, J. Agric. Food Chem. 56 (2008) 4623–4630; B.S. Dien, Biomass Bioenergy. 30 (2006) 880–891; S.M. Renaud, Aquaculture 211 (2002) 195–214.

Wang et al. derived chemical formulas by statistically analyzing the data of the ultimate analysis of different biomass materials [12].

Grasses	$CH_{1.488}O_{0.666}N_{0.016}S_{0.001}$
Crop residues	$CH_{1.482}O_{0.656}N_{0.021}S_{0.001}$
Woody biomass	$CH_{1.428}O_{0.593}N_{0.007}S_{0.001}$

These chemical formulas can be used if the ultimate analysis of specific biomass materials is not available.

2.4.2.3 Equilibrium moisture content

When dealing with biomass, there is always a water vapor exchange between moist air and biomass. The moisture content at which equilibrium between biomass and air occurs is called the equilibrium moisture content [6].

Equilibrium moisture content depends on two factors:

1. Relative humidity of air
2. Storage temperature

In order to evaluate equilibrium moisture content of a biomass, percent relative humidity of air is plotted as a function of percent moisture at a constant temperature. Set of equations can be used to predict equilibrium moisture content. These equations are [6,12]:

Modified Henderson equation	$M_{\mathrm{dd}} = \left[\dfrac{\ln(1-\mathrm{RH})}{-C_1(T+C_2)}\right]^{1/C_3}$
Modified Halsey equation	$M_{\mathrm{dd}} = \left[\dfrac{-\exp(C_1+C_2T)}{\ln(\mathrm{RH})}\right]^{1/C_3}$
Modified Oswin equation	$M_{\mathrm{dd}} = \left(C_1 + {}^{C_2}/_{T}\right)\left(\dfrac{\mathrm{RH}}{1-\mathrm{RH}}\right)^{1/C_3}$
Modified Chung-Pfost equation	$M_{\mathrm{dd}} = \frac{1}{C_1}\ln\left[\left(\frac{C_2-T}{C_3}\right)\ln(\mathrm{RH})\right]$

Where C_1, C_2, and C_3 are coefficients.

Arabhosseini et al. have calculated coefficients for the predictions of equilibrium moisture content of Miscanthus and is shown in Table 4 [6,12]

2.4.3 Mechanical properties

One of the major barriers against the use of these bulky residues as feedstocks is on their collection, handling, transportation, and storage [13,14]. Therefore these bulky residues can be densified into pellets. Pelletizing is a method of increasing the bulk density of biomass by mechanical pressure [15].

Uniaxial compression test is used to evaluate compression properties of a biomass. The biomass is first densified in pellet form. These pellets are then cut into cylindrical or rectangular shapes and loaded between a pair of compression plate (top and

Table 4 Coefficients for the predictions of equilibrium moisture content of miscanthus

Biomass	Henderson	Halsey	Oswin	Chung-Pfost
Stems				
C_1	0.222	−4.438	0.132	0.156
C_2	1.123	−0.0284	−0.011	−0.026
C_3	17.011	2.104	2.406	1.147
Leaves				
C_1	0.435	−4.491	0.155	0.1654
C_2	1.349	−0.036	−0.00137	−9.008
C_3	2.166	2.361	2.729	117.24

From L. Wang, A. Shahbazi, Biomass properties, BEEMS Module A2, USDA Higher Education Challenger Program 2009-38411-19761, North Carolina Agricultural and Technical State University, 2009; L.J. Wang, A. Shahbazi, M.A. Hanna, Biomass Bioenergy, 35 (2011) 171–178.

Table 5 Compressive strength of biomass pellets

Biomass material	Forming pressure (MPa)	Compressive strength of pellets (MPa)
Oak sawdust	34	25
	69	28
	103	45
	138	49
Pine sawdust	34	25
	69	35
	103	44
	138	45

From L. Wang, A. Shahbazi, Biomass properties, BEEMS Module A2, USDA Higher Education Challenger Program 2009-38411-19761, North Carolina Agricultural and Technical State University, 2009; L. J. Wang, A. Shahbazi, M.A. Hanna, Biomass Bioenergy, 35 (2011) 171–178.

bottom). These plates hold the samples and a good contact between the plate and the pellets is essential. The load cell is attached to the bottom plate which measures amount of force acting on the sample. A force is then applied on the top plate which moves the platen downward and thereby compresses the pellets. As the top plate moves downward, load cell records the force with change in length. This information can be used to calculate compressive strength (Table 5).

2.5 Categories of biomass materials

There are three categories of biomass resources based on their source: (1) waste biomass including organic waste such as livestock excrement and food residues discharged from households and food-related industries; (2) unused biomass composed of inedible parts of agricultural products such as logging residues, rice straw, and rice husk; and (3) resource crops (dedicated energy crops) that can be used to produce bioethanol and biodiesel fuel, such as rapeseed, corn, and rice. The biomass is produced in agriculture as well as forestry and fishery industries as either main crops or waste and unused materials [16].

Biomass resources are converted into the following six types of end products:

1. Fertilizers and feed
2. Raw materials of chemical products such as amino acids and useful chemical substances
3. Materials such as plastics and resins
4. Fuel such as ethanol, diesel, wood pellets, biogas, and solid fuel
5. Thermal energy
6. Electric energy

These products are generally used in five categories: (1) recycled use, (2) use of products, (3) power generation, (4) use of heat, and (5) use of fuel.

2.6 Sources of biomass

There are many potential sources of biomass. One of the economical sources of biomass is spent biomass from fermentation process for manufacturing of biochemical products such as antibiotics, enzymes, organic acids, and possibly vitamins. These materials contain extraneous insoluble matter such as medium components or product recovery chemicals [17].

An abundant source of mixed microbial biomass is the activated sludge wastewater treatment process. In a sewage treatment facility, metals such as cadmium, copper, chromium, magnesium, and lead are removed largely by filtration and adsorption [17].

Another inexpensive source of biomass is marine algae. Cultured fresh water algae serves as a continuous biomass source for the biosorbents family. Algae have been known to concentrate heavy metals, including radioactive metals like radium. Another natural source of biomass is fungi. They are nonphotosynthetic group of microorganisms relying on organic substances as their sole source of carbon and energy for growth. The principal active species in the fungal cell wall are polysaccharides. Two such molecules are chitin and chitosan. Uptake of metals such as iron, zinc, copper, lead, and mercury is taken up by chitin by coordination binding with the nitrogen atom of amine and amide groups [17].

Fig. 3 shows major sources of plant material and possible conversion technologies [18].

2.7 Types of biomass system

There are two types of biomass systems, closed-loop and open-loop.

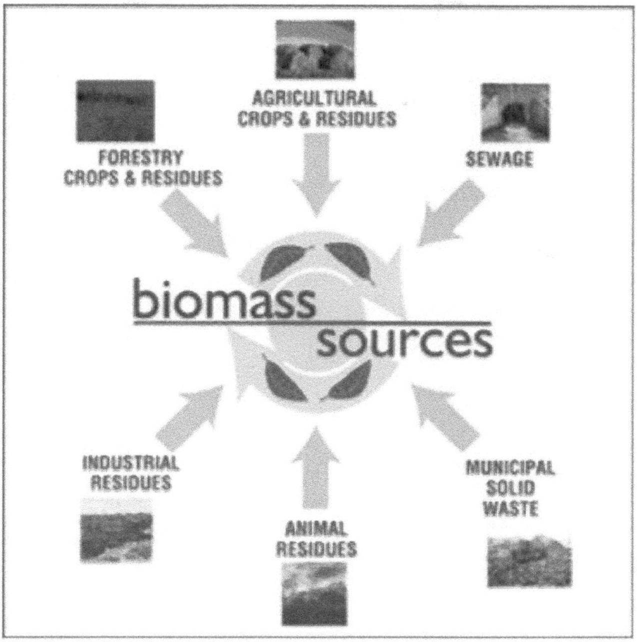

Fig. 3 Sources of biomass, possible conversion technologies, and fuel products. From https://reich-chemistry.wikispaces.com/Gray.Garanito.wiki.spring.2011.

2.7.1 Closed-loop biomass system [19]

In a closed-loop biomass system, the material to be burned in the plant has been planted and harvested specifically for that purpose. E3 biofuels were the first to start closed-loop ethanol plant fueled mainly by biogas from animal waste instead of coal or natural gas. Their unique technology brought together three main components into one, closed-loop system [19]:

1. A large cattle feedlot or dairy that produces large quantities of cow manure needling treatment
2. An anaerobic digester that transform the cow manure into biogas
3. An ethanol plant that runs on the biogas instead of natural gas or coal, and whose leftover wet grain is fed back to the cattle

Ethanol production is a closed-loop system that produces energy by combining manure, collected from an adjacent 28,000-head cattle feedlot, with thin stillage, a cellulose by-product of ethanol refining. The hot liquid mixture is decomposed inside an anaerobic digester, where bacteria extract methane-rich biogas that is used for fire the plant's ethanol boilers [19].

The left over grain is fed wet to the cattle at the nearby feedlot, whose manure in turn powers the plant and creates high-quality fertilizer as a by-product. The on-site cows are treated to fresh wet cake from the nearby plant, thus avoiding both the cost and fossil-fuel pollution of drying and transporting the grain [19].

Greenhouse gases are 23 times more potent than carbon dioxide in causing global warming in comparison to the gases released by the decomposing manure to the atmosphere if it were not used in the process. Greenhouse gases can be reduced by capturing the methane gas [19].

2.7.2 Open-loop biomass system

Open-loop systems derive biomass from the waste stream of some other primary process not intended for use as fuel source such as municipal solid waste, agricultural residues, or forestry residues. In these cases, the biorefinery directly integrates into, and rearranges, established resource flows and land-use practices. All existing cellulosic-ethanol facilities operating in agricultural centers are examples of open-loop systems.

2.8 Handling of biomass

The physical handling of biofuels during collection or at a processing plant possesses a significant challenge. Biofuels tend to vary with density, moisture content, and particle size and can also be corrosive. Therefore biomass fuel handling equipment is often a difficult part of a plant to adequately design, maintain, and operate [20].

It is important to understand properties of feedstock prior to characterization. Some of the properties include moisture, bulk density, particle size and range, seasonal properties, abrasiveness, and contaminants. For example, a woody biomass may differ from a switchgrass and corn stover in material properties so that correct equipment and processing methods can be implemented [20].

The characterization results will dictate how the biomass is received, stored, and transported. If the blending of two different feedstocks is required, system should be capable of handling incoming material separately. For example, miscanthus bails cannot be processed or stored the same as a truckload of wood chips, so blending the materials together requires additional equipment and expenses [20].

A simple conveying operation of biomass can sometimes be a problem in biomass processing operations. Results from characterization will dictate the type of conveyor style and maintenance, conveyor belt type, belt travel distance, and angle of travel. It is important to perform a detailed analysis to design with respect to the feedstock. This will be critical in achieving operation throughput requirements [20].

2.9 Advantages of biomass

There are several advantages of biomass over fossil-based fuels [21]:

- Renewable energy
- Carbon neutral
- Local availability
- Cheaper
- Less dependency

- Reduction in waste
- Variety of applications

1. Renewable energy
   Energy produced from biomass is considered as renewable because it is created from organic matter which is continuously produced. The organic matter includes wood, crop waste, garbage, sewage sludge, and manure.
2. Carbon neutral
   Biomass is a natural part of the carbon cycle as the only carbon emitted from biomass fuels to the environment is the amount that was absorbed by plants during their life cycle. This develops and maintains neutrality that witnesses no new carbon generated. This aspect renders biomass uniquely clean.
3. Local availability
   Most biomass types are characterized by local availability. Since the material involved is usually of low density and of low value, it is not suitable for transportation over larger distances. The value created by biomass is returned to a large extent to the local communities, and the energy conversion facilities will probably be situated close to the biomass production area, therefore creating more local job opportunities. Finally, the potential of cogeneration or tri-generation of energy from biomass means that local communities may benefit from receiving heat or cooling from the biomass energy conversion facility at a competitive price and with higher reliability [22].
4. Cheaper
   Unlike production of fossil fuels, producing biomass energy does not involve high upfront capital costs such as drilling, constructing pipelines, and collection of biomass fuel. This low-cost resulting from the production of biomass fuel is passed on to customers. This means that customer's energy bills will not depend on aspects such as availability. Low biomass cost makes this form of energy attractive to manufacturers and producers since they are able to generate higher profits from extremely low output.
5. Less dependency
   Virtually anyone can produce biomass energy because the raw materials are available everywhere. Traditional forms of electricity can sometimes be unreliable due to power outages. Biomass energy produced at home can be a good backup for traditional electricity.
6. Reduction in waste
   Most waste produced in homes is either plant matter or biodegradable. This kind of waste can be channeled to more profitable use. Biomass energy generation utilizes any waste that would have otherwise found way into landfills. This minimizes the impacts of waste in landfills to the environment.
7. Variety of applications
   Biomass energy is also versatile, as different forms of organic matter can be used to create different products. For example, bioethanol can be made from corn and other crops.

2.10 Composition of biomass

Plant cell wall is comprised of three main components: (1) cellulose, (2) hemicellulose, and (3) lignin. These components vary in proportion in different groups of plants as well as in the same group. For example, if the concentration of sugar is high, the biomass will be sugary, e.g., sugarcane and sugar beet. Similarly, high starch content

Table 6 **Chemical composition of lignocellulosic material [23]**

	Cellulose	Hemicellulose	Lignin	Protein
Biomass source	% dry weight			
Birch angiosperm	44.9	32.7	19.3	0.5
Spruce gymnosperm	46.1	24.6	26.3	0.2
Crop residues	30–45	16–27	3–13	3.6–7.2
Wood residues	45–56	10–25	18–30	1.3

From H. Ando, T. Sakaki, T. Kokusho, M. Shibata, Y. Uemura, Y. Hatate, Decomposition behaviour of plant biomass in hot compressed water, Ind. Eng. Chem. Res. 39 (2000) 3688–3693.

will yield starchy biomass, e.g., potato and tapioca. Variation in chemical constituents (cellulose, hemicellulose, and lignin) in some plants is shown in Table 6 [23].

The structural differences between cellulose, hemicellulose, and lignin lead to different chemical reactivities, making the relative composition of biomass a crucial factor for the efficiency of a biomass conversion process [24–28]. There are many wet chemical analytical methods that are based on fractionation of biomass samples to isolate purified fractions and further quantify using analytical equipment [28].

2.10.1 Thermogravimetric analysis

Thermogravimetric analysis (TGA) of plant biomass is used for determining thermal decomposition mechanisms [28]. TGA provides a mass change profile as a function of the sample temperature or analysis time; thus the thermal decomposition processes can be monitored. The TGA curves and the derivative thermogravimetric curves give information about the thermal stability of the sample and the overall kinetic progress of the thermal decomposition.

Fig. 4 shows thermogravimetric analyzer within in-line mass spectrometer to analyze biomass sample. The in-line mass spectrometer analyzes volatiles generated as a result of sample decomposition. Samples weighing 2–4 mg are placed into the platinum sample pan and the furnace is flushed with the carrier gas. Linear heating rate of 10°C/min or 20°C/min is applied. When undergoing thermal decomposition, argon gas is used for flushing [29]. The pyrolytic decomposition of woody plant tissues in inert atmospheres occurs at mild temperatures for hemicelluloses (250–300°C) followed by cellulose (300–350°C) and finally lignin (300–500°C) [28].

Table 7 summarizes the major thermogravimetric parameters of several biomass samples as well as cellulose, hemicellulose, and lignin [29].

Fig. 5 shows a typical thermogravimetric curve [29]. $T_{1\%}$ and T_{hc} are the temperatures used for the characterization of the beginning of the thermal decomposition. $T_{1\%}$ marks the temperature belonging to 1% weight loss of the dry sample, and T_{hc} denotes the extrapolated temperature of the beginning of decomposition. Large amounts of extractable materials start to evaporate or decompose at lower temperatures and with low-mass low rate. T_c stands for the extrapolated temperature of the

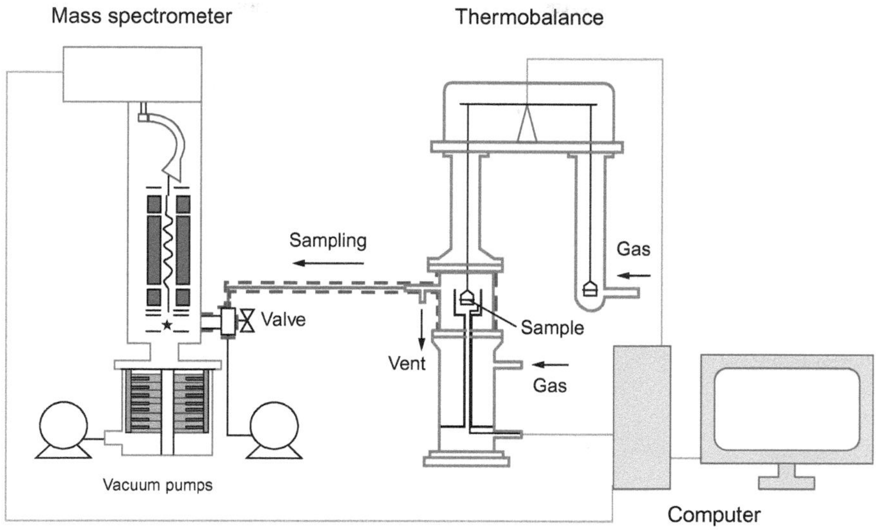

Fig. 4 Thermogravimetric analyzer with an in-line mass spectrometer.
From A. Pandey, T. Bhaskar, M. Stocker, R.K. Sukumaran, Analytical techniques as a tool to understand the reaction mechanism, Recent Advances in Thermochemical Conversion of Biomass, Elsevier, 2015, pp. 78–81.

end of cellulose decomposition, in the case of the cellulosic samples. T_{peak} denotes the temperature of the maximal decomposition rate (DTG_{max}).

2.10.2 Ultimate analysis

The ultimate analysis generally reports the elemental carbon (C), hydrogen (H), nitrogen (N), sulfur (S), and oxygen (O) composition. Ultimate analysis is performed and reported on a dry basis in order to avoid confusion and provide good representation of the fuel. Biomass materials that grow fast such as straw, the determination of chlorine is important, because it represents a possible pollutant and corrosive agent in pyrolysis, gasification, and combustion systems [30].

The ultimate analysis of different biomass is shown in Table 8 [31–38].

2.10.3 Proximate analysis

The proximate analysis classifies the fuel in terms of its moisture (M), volatile matter (V), fixed carbon (FC), and ash. The thermal behavior of biomass is studied by measuring the rate of weight loss of the sample as a function of time and temperature using TGA. The observed rates are not only functions of time and temperature but also of the size and the density of the sample [30].

Table 7 Thermogravimetric parameters of different biomass

Sample	T_{hc} (extrapol) (°C)	$T_{1\%}$ (°C)	T_{peak} (°C)	T_c (°C)	DTG_{max} (%/s)	Char yield (%)
Avicel cellulose	300	290	352	382	0.755	4.1
Whatman cellulose	330	300	379	410	0.708	6.4
Hemicellulose (xylan)	230	221	272	320	0.487	19
Cotton wool	325	277	379	408	0.661	8.3
Flax fiber	240	260	371	400	0.641	10.5
Hemp	220	252	365	397	0.582	12.8
Wheat straw	190	213	328	368	0.303	26.0
Rape straw	190	199	333	376	0.253	13.7
Eucalypt (wood and leaves)	180	207	352	378	0.203	22.8
Olive tree (wood and leaves)	170	161	350	398	0.170	22.4
Pine	240	218	379	413	0.306	16.1
Birch	250	237	392	422	0.321	10.3
Willow	190	207	377	408	0.272	16.8
Poplar	190	210	370	402	0.281	17.3
Black locust	190	195	369	403	0.281	16.7
Beech wood	250	247	381	411	0.349	13.4
MWL beech	210	229	364	560	0.146	27.2
MWL spruce	190	198	416	590	0.109	35.6

From A. Pandey, T. Bhaskar, M. Stocker, R.K. Sukumaran, Analytical techniques as a tool to understand the reaction mechanism, Recent Advances in Thermochemical Conversion of Biomass, Elsevier, 2015, pp. 78–81.

A typical thermogram of straw is divided into two main sections and is shown in Fig. 6. The first part of the thermogram is exposed to nitrogen atmosphere and shows the loss of moisture and volatiles. This moisture represents physically bound water only—water released by chemical reactions during slow pyrolysis. Fast pyrolysis techniques employing very rapid heating rates might yield more volatile matter. The second part of the thermogram is performed in air. The ash content is determined by combustion of the fixed carbon. The resulting ash fraction is not representative of the original ash due to the oxidation process employed in its determination [30].

The proximate analysis of different biomass is shown in Table 9 [31,32,34–38].

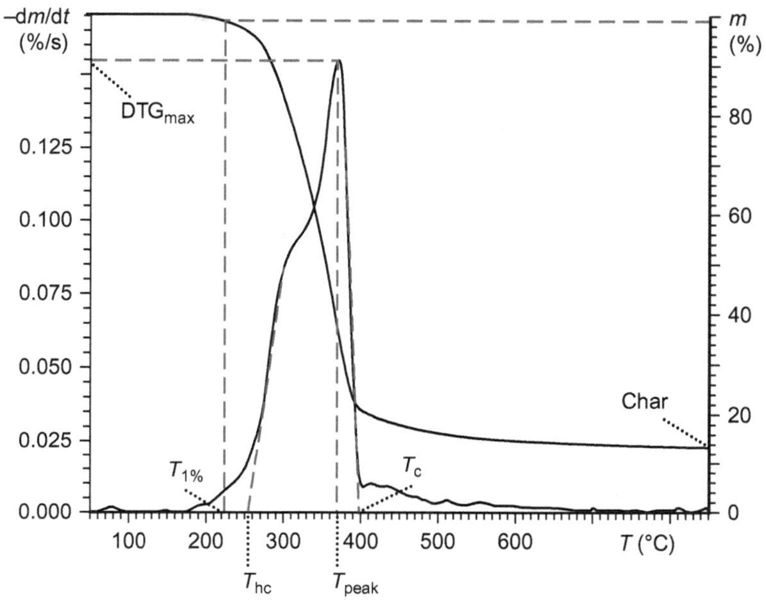

Fig. 5 A typical thermogravimetric curve.
From A. Pandey, T. Bhaskar, M. Stocker, R.K. Sukumaran, Analytical techniques as a tool to understand the reaction mechanism, Recent Advances in Thermochemical Conversion of Biomass, Elsevier, 2015, pp. 78–81.

Table 8 Ultimate analysis of different biomass sources

Biomass	C	H	O	N	S	Ash
Sugarcane bagasse	44.8	5.3	39.6	0.38	0.01	9.8
Barley straw	45.7	6.1	38.3	0.4	0.1	6.0
Cotton stalk	43.6	5.8	43.9	–	–	6.7
Corn grain	44.0	6.1	47.2	1.2	0.14	1.3
Corn stover	43.7	5.6	43.3	0.61	0.01	6.3
Pine bark	52.3	5.8	38.8	0.2	–	2.9
Poplar	48.5	5.9	43.7	0.47	0.01	1.4
Redwood	53.5	5.9	40.3	0.1	–	0.2
Rice straw	41.8	4.6	36.6	0.7	0.08	15.9
Switchgrass	47.5	5.8	42.4	0.74	0.08	3.5
Wheat straw	43.2	5.0	39.4	0.61	0.11	11.4

From J. Cheng, Thermochemical conversion of biomass to power and fuels, Biomass to Renewable Energy Processes, CRC Press, 2010, p. 443.

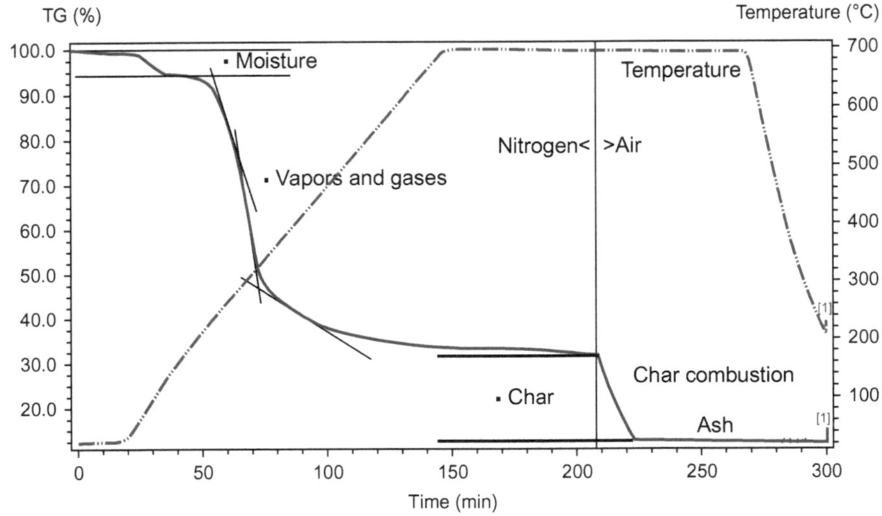

Fig. 6 Thermogram showing a slow pyrolysis of wheat straw.
From Definition of a standard biomass, Integrated project sustainable energy systems, SES6-CT-2003-502705, Forschungszentzrum Karlsruhe GmbH, 2004.

Table 9 Proximate analysis of different biomass sources

Biomass	Volatiles (% wt)	Ash (% wt)	Fixed carbon (% wt)
Sugarcane bagasse	74	11	15
Barley straw	46	6	18
Cotton stalk	29	6	31
Corn grain	87	1	12
Corn stover	75	6	19
Pine needles	72	2	26
Poplar	82	1	16
Redwood	80	0.4	20
Rice straw	69	13	17
Switchgrass	81	4	15
Wheat straw	59	4	21

From J. Cheng, Thermochemical Conversion of Biomass to Power and Fuels, Biomass to Renewable Energy Processes, CRC Press, 2010, p. 443.

2.10.4 Basis of expressing biomass composition

The composition of a fuel is often expressed on different bases depending on the situation. The following four basis of analysis are commonly used [39]:

1. As-received
2. Air-dry
3. Total dry
4. Dry and ash free

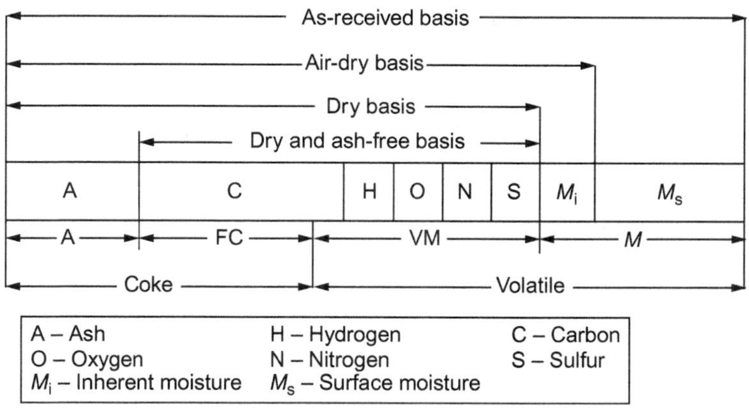

Fig. 7 Basis of expressing fuel composition.
From P. Basu, Composition of biomass, Biomass Gasification, Pyrolysis, and Torrefaction: Practical Design and Theory, Elsevier, 2013, pp. 78–80.

A comparison of this basis is shown in Fig. 7.

2.10.4.1 *As-received basis*

When using the as-received basis, the results of ultimate and proximate analysis can be expressed as:

$$\text{Ultimate analysis: } \mathrm{C+H+O+N+S+ASH}+M=100\%$$

$$\text{Proximate analysis: } \mathrm{VM+FC+ASH}+M=100\%$$

where VM, FC, M, and ASH are weight percentages of volatile matter, fixed carbon, moisture, and ash, respectively, measured by proximate analysis, and C, H, O, N, and S represent the weight percentages of carbon, hydrogen, oxygen, nitrogen, and sulfur, respectively, as measured by ultimate analysis. The ash and moisture content of the fuel is the same in both analyses [39].

2.10.4.2 *Air-dry basis*

When the fuel is dried in air, its surface moisture is removed while its inherent moisture is retained. The constituent on an air-dry basis can be expressed by following equation:

$$\mathrm{C_{Air\,dry}}=\frac{100\mathrm{C}}{(100-M_{\mathrm{a}})}$$

where M_a is the mass of surface moisture removed from 100 kg of moist fuel after drying in air. Other constituents of the fuel can be expressed similarly [39].

2.10.4.3 Total dry basis

When the fuel is completely free from moisture, it is important to make allowance for surface as well as inherent moisture, M_i. The constituent on total dry basis can be expressed as [39]:

$$C_{\text{total dry}} = \frac{100C}{(100 - M_i)}$$

where M is the total moisture in the fuel: $M = M_a + M_i$

2.10.4.4 Dry-ash free basis

Ash is another component that at times is eliminated along with moisture. The constituent on a dry-ash free basis can be expressed as [39]:

$$C_{\text{dry ash free}} = \frac{100C}{(100 - M - \text{ASH})}$$

where $(100 - M - \text{ASH})$ is the mass of biomass without moisture and ash.

The percentage of all constituents on any basis equals 100. Therefore

$$C_{\text{dry ash free}} + O_{\text{dry ash free}} + H_{\text{dry ash free}} + N_{\text{dry ash free}} + S_{\text{dry ash free}} = 100\%$$

2.10.5 Heating value of fuel

The heating value of biomass fuel is divided into two types [39]:

1. Higher heating value (HHV)
2. Lower heating value (LHV)

2.10.5.1 Higher heating value

HHV is the amount of heat released by the unit mass of volume of fuel once it is combusted and the products have returned to a temperature of 25°C. It includes the latent heat of vaporization of water. Table 10 shows HHVs of some selected biomass [39].

2.10.5.2 Lower heating value

The LHV, also known as the net calorific value, is the amount of heat released by fully combusting a specific quantity less the heat of vaporization of the water in the combustion product [39].

Table 10 **Higher heating values (HHV) of selected biomass**

Material	HHV (kJ/kg)
Bituminous coal	17,000–23,250
Douglas fir	21,000
Douglas fir bark	22,000
Eucalyptus grandis	19,400
Sugar cane bagasse	17,300
Wheat straw	17,500
Poplar	20,700
Rice hulls	15,300
Rice straw	15,800

From Biomass: higher heating value, The Engineering Toolbox. http://www.engineeringtoolbox.com/biomass-fuels-hhv-d_1818.html.

The relationship between HHV and LHV is given by:

$$\mathrm{LHV} = \mathrm{HHV} - h_{\mathrm{g}}\left(\frac{9H}{100} - \frac{M}{100}\right)$$

where LHV, HHV, H, M, and h_{g} are LHV, HHV, hydrogen percentage, moisture percentage, and the latent heat of steam.

2.10.6 Stoichiometric calculations for complete combustion

Complete oxidation of simple hydrocarbon fuels leads to the formation of carbon dioxide (CO_2) and water (H_2O). A representative equation for a hydrocarbon fuel with the general composition C_nH_m is shown as follows [40]:

$$C_nH_m + \left(n + \frac{m}{4}\right)O_2 \rightarrow nCO_2 + \frac{m}{2}H_2O$$

In order to have an ideal case of complete combustion, it is important to assume that:

- Nitrogen in the air may participate in the combustion process to produce nitrogen oxide.
- Fuels may contain elements other than carbon, and these elements may be transformed during combustion.
- Combustion is not complete, and the effluent gases contain unburned and partially burned products in addition to CO_2 and H_2O.

Some of the major elements that comprise air on a mole basis are nitrogen (~79.1%) and oxygen (~20.9%). Nitrogen (3.78 mol) is required for every mol of oxygen to initiate combustion. The stoichiometric relationship for complete oxidation of a hydrocarbon fuel C_nH_m is shown by the following equation:

$$C_nH_m + \left(n + \frac{m}{4}\right)(O_2 + 3.78N_2) \rightarrow nCO_2 + \frac{m}{2}H_2O + 3.78\left(n + \frac{m}{4}\right)N_2$$

Thus for every mole of fuel burned, $4.78(n + m/4)$ mol of air is required and $4.78(n + m/4) + m/4$ mol of combustion products are generated.

The molar fuel/air ratio for stoichiometric combustion is:

$$1/[4.78(n + m/4)]$$

Gas compositions are generally reported in terms of mole fractions. The product mole fractions for complete combustion of a hydrocarbon fuel can be represented by the following equations:

$$y_{CO_2} = \frac{n}{4.78(n + m/4) + m/4}$$

$$y_{H_2O} = \frac{m/2}{4.78(n + m/4) + m/4}$$

$$y_{N_2} = \frac{3.78(n + m/4)}{4.78(n + m/4) + m/4}$$

2.10.7 Composition of the product gas of gasification

The solid fuel is heated by combustion of a part of the fuel. The combustion gases are then reduced by being passed through a bed of fuel at high temperature. In complete combustion, carbon dioxide is obtained from the carbon and water from the hydrogen. Oxygen from the fuel will of course be incorporated in the combustion products, thereby decreasing the amount of combustion air needed [41].

Some of the major chemical reactions occurring in a gasifier include:

(a) Gasification with oxygen:

$$C + {}^{1}/{}_{2}O_2 \leftrightarrow CO$$

(b) Combustion with oxygen:

$$C + O_2 \leftrightarrow CO_2$$

(c) Gasification with steam:

$$C + H_2O \leftrightarrow CO + H_2$$

(d) Gasification with hydrogen:

$$C + 2H_2 \leftrightarrow CH_4$$

Table 11 Gasifier gas composition in Vol%

Gasifier gas	Composition (Vol%)
H_2	25–30
CO	30–60
CO_2	5–15
H_2O	2–30
CH_4	0–5
H_2S	0.2–1
COS	0–0.1
N_2	0.5–4
Ar	0.2–1
NH_3+HCN	0–0.3

From U.S. Department of Energy, Gasification Introduction, National Energy Technology Laboratory. https:/www.netl.doe.gov/research/coal/energy-systems/gasification/gasifipedia/reaction-transformations#figure.

(e) Water gas shift:

$$CO + H_2O \leftrightarrow H_2 + CO_2$$

(f) Methanation

$$CO + 3H_2 \leftrightarrow CH_4 + H_2O$$

The product gas of gasification is generally a mixture of several gases as shown in Table 11 [42].

2.11 Challenges of biomass commercialization

Although, cost is the key barrier affecting biomass usage, each conversion method has its own limitations. Different biomass conversion processes and their challenges toward commercialization are summarized in Table 12 [43].

Table 12 Biomass conversion processes and their challenges toward commercialization

Processes	
Biochemical conversion	
Pretreatment	• Lack of complete access to the cellulose • Generation of inhibitors that reduce the yield of fermentable sugars and that make downstream processing more difficult

Continued

Table 12 Continued

Processes	
Hydrolysis	Slow rate of reaction and inadequate yield of product hydrolyzed by current enzymes
Biochemical fuel synthesis	• Toxicity of by pretreated byproducts and fuel synthesis products to the microbes used for synthesizing fuel • Inability of existing microbes to process all the sugars in lignocelluloses efficiently • Lack of natural processes to produce hydrocarbon fuels (green gasoline, diesel, and jet fuel) • Lack of value-added coproducts
Thermochemical conversion methods	
Gasification	• Lack of low-cost production of syngas from a broad range of feedstocks • Insufficiently scalable gasification facilities • Insufficient lifetime of cleanup catalysts at high conversion efficiencies • High cost of catalysts and ash accumulation in catalyst bed • Low conversion capacities of syngas fermentation microbes • Low syngas bioreactor efficiencies • Low catalyst selectivity, free-pass yield, and lifetimes • Lack of cost-effective, small-scale catalytic facilities
Pyrolysis	• High oxygen and acid content of product bio-oil • Instability of product bio-oil • Expense of construction materials • Lack of low-cost hydrogen reactant for hydrotreating • Lack of low-cost, corrosion-resistant construction materials • Inadequate yield, selectivity, and stability of catalysts

From Biomass conversion: challenges for federal research and commercialization, a report by biomass conversion interagency working group, Biomass Research and Development Board, February 2011. https:/biomassboard.gov/pdfs/biomass_conversion_report.pdf.

References

[1] M. Montross, C. Crofcheck, Energy crops for the production of biofuels, in: Thermochemical Conversion of Biomass to Liquid Fuels and Chemicals, RSC Publishing, Cambridge, UK, 2010, p. 26.

[2] S. Tojo, T. Hirasawa, Biomass as local resource, in: Research Approaches to Sustainable Biomass System, Elsevier, Waltham, MA, USA, 2014, p. 8.

[3] Biomass compared to fossil fuels, solar and wind, Viaspace. http://www.viaspace.com/biomass_versus_alternatives.php&gt.

[4] Unfolding the improvements of biomass history, Alternative Energy Solutions Info. http://www.alternativeenergysourcesinfo.com/biomass-history.html&gt.

[5] Bioenergizeme virtual science fair: biomass history a timeline, Office of Energy Efficiency and Renewable Energy. https:/energy.gov/eere/bioenergy/bioenergizeme-virtual-science-fair-biomass-history-timeline.

[6] L. Wang, A. Shahbazi, Biomass properties, BEEMS Module A2, USDA Higher Education Challenger Program 2009-38411-19761, North Carolina Agricultural and Technical State University, 2009.
[7] Biomass densification for energy production, Ministry of Agriculture, Food and Rural Affairs, Ontario, Canada. http://www.omafra.gov.on.ca/english/engineer/facts/11-035.htm#1&gt.
[8] S.M. Drury, T.L. Reynolds, W.P. Ridley, N. Bogdanova, S. Riordan, M.A. Nemeth, R. Sorbet, W.A. Trujillo, M.L. Breeze, Composition of forage and grain from second-generation insect-protected corn MON 89034 is equivalent to that of conventional corn (Zea mays L.), J. Agric. Food Chem. 56 (2008) 4623–4630.
[9] B.S. Dien, H.J.G. Jung, K.P. Vogel, M.D. Casler, J.F.S. Lamb, L. Iten, R.B. Mitchell, G. Sarath, Chemical composition and response to dilute-acid pretreatment and enzymatic saccharification of alfalfa, reed canarygrass, and switchgrass, Biomass Bioenergy 30 (2006) 880–891.
[10] S.M. Renaud, L.V. Thinh, G. Lambrinidis, D.L. Parry, Effect of temperature on growth, chemical composition and fatty acid composition of tropical Australian microalgae grown in batch cultures, Aquaculture 211 (2002) 195–214.
[11] Biomass characteristics, European Biomass Industry Association. http://www.eubia.org/cms/wiki-biomass/biomass-characteristics/&gt.
[12] A. Arabhosseini, W. Huisman, J. Muller, Modeling of the equilibrium moisture content (EMC) of Miscanthus (Miscanthus x giganteus), Biomass Bioenergy 34 (2010) 411–416.
[13] A. Demirbas, Biomass resource facilities and biomass conversion processing for fuels and chemicals, Energy Convers. Manag. 42 (2001) 1357–1378.
[14] A.K. Tripathi, P.V.R. Iyer, T.C. Kandpal, A techno-economic evaluation of biomass briquetting in India, Biomass Bioenergy 14 (1998) 479–488.
[15] S. Mani, L.G. Tabil, S. Sokhansanj, Effects of compressive force, particle size and moisture content on mechanical properties of biomass pellets from grasses, Biomass Bioenergy 30 (2006) 648–654.
[16] S. Tojo, T. Hirasawa, Biomass as local resource, in: Research Approaches to Sustainable Biomass System, Elsevier Publishers, Waltham, MA, USA, 2014, p. 9.
[17] S.R.R. Rao, Biotechnological processes, in: Resource Recovery and Recycling From Metallurgical Wastes, Elsevier Publishers, Oxford, UK, 2006, pp. 109–110.
[18] Watt Committee on Energy Publications, The prospect of a biological-photochemical approach for the utilization of solar energy, in: Energy From the Biomass, Taylor and Francis, London, UK, 2004, p. 9. Watt Committee: Report Number 5.
[19] E# Biofuels efficient closed-loop ethanol plant in operation, The Energy Blog. http://thefraserdomain.typepad.com/energy/2007/07/e3-biofuels-eff.html&gt.
[20] Overview of biomass handling equipment, Bioenergy Consult. http://www.bioenergyconsult.com/biomass-handling-equipments/&gt.
[21] Rinkesh, What is biomass energy. http://www.conserve-energy-future.com/advantages_disadvantages_biomassenergy.php&gt.
[22] A.A. Rentizelas, Biomass supply chain, Biomass Combustion Science, Technology, and Engineering, Elsevier, Cambridge, UK, 2013.
[23] Biomass: a renewable source of energy. http://www.eplantscience.com/index/biotechnology/biotechnology_and_environment/biomass_a_renewable_source_of_energy/biotech_biomass_composition_of_biomass.php&gt.
[24] O. Bobleter, Hydrothermal degradation of polymers derived from plants, Prog. Polym. Sci. 19 (1994) 797–841.
[25] H. Ando, T. Sakaki, T. Kokusho, M. Shibata, Y. Uemura, Y. Hatate, Decomposition behaviour of plant biomass in hot compressed water, Ind. Eng. Chem. Res. 39 (2000) 3688–3693.

[26] J. Yanik, S. Ebale, A. Kruse, M. Saglam, M. Yuksel, Biomass gasification in supercritical water: II effect of catalyst, Fuel 86 (2007) 2410–2415.
[27] A. Loppinet-Serani, C. Aymonier, F. Cansell, Current and foreseeable applications of supercritical water energy and the environment, ChemSusChem 1 (2008) 486–503.
[28] M. Carrier, A.L.- Serani, D. Denux, J.-M. Lasnier, F.H.- Pichavant, F. Cansell, C. Aymonier, Thermogravimetric analysis as a new method to determine the lignocellulosic composition of biomass, Biomass Bioenergy 35 (2011) 298–307.
[29] A. Pandey, T. Bhaskar, M. Stocker, R.K. Sukumaran, Analytical techniques as a tool to understand the reaction mechanism, in: Recent Advances in Thermochemical Conversion of Biomass, Elsevier Publishers, Cambridge, UK, 2015, pp. 78–81.
[30] Definition of a standard biomass, Integrated project sustainable energy systems, SES6-CT-2003-502705, Forschungszentzrum Karlsruhe GmbH, 2004.
[31] R. Brown, Biorenewable resources, in: Engineering New Products from Agriculture, Blackwell Publishing, Ames, IA, USA, 2003.
[32] S. Guar, T. Reed, Thermal Data for Natural and Synthetic Fuels, Marcel Dekker, New York, NY, USA, 1998.
[33] E.A. Hiler, B.A. Stout, Biomass Energy: A Monograph, first ed., University Press, College Station, TX, USA, 1985.
[34] P. McKendry, Energy production from biomass (part 1): overview of biomass, Bioresour. Technol. 83 (2002) 37.
[35] P. McKendry, Energy production from biomass (part 1): overview of biomass, Bioresour. Technol. 83 (2002) 47.
[36] P. McKendry, Energy production from biomass (part 1): overview of biomass, Bioresour. Technol. 83 (2002) 55.
[37] T.B. Reed, Encyclopedia of biomass thermal conversion, in: The Principles and Technology of Pyrolysis, Gasification and Combustion, third ed., Biomass Energy Foundation Press, Golden, CO, USA, 2002.
[38] J. Cheng, Thermochemical conversion of biomass to power and fuels, in: Biomass to Renewable Energy Processes, CRC Press, Boca Raton, FL, USA, 2010, p. 443.
[39] P. Basu, Composition of biomass, in: Biomass Gasification, Pyrolysis, and Torrefaction: Practical Design and Theory, Elsevier, Amsterdam, 2013, pp. 78–80.
[40] R.C. Flagan, J.H. Seinfeld, Combustion fundamentals, in: Fundamentals of Air Pollution Engineering, Prentice-Hall, Englewood Cliffs, 1988, p. 63.
[41] Theory of gasification, FAO corporate document repository, Forest Department. http://www.fao.org/docrep/t0512e/T0512e09.htm>.
[42] Gasification introduction, National Energy Technology Laboratory, U.S. Department of Energy. https://www.netl.doe.gov/research/coal/energy-systems/gasification/gasifipedia/reaction-transformations#figure.
[43] Biomass conversion: challenges for federal research and commercialization, a report by biomass conversion interagency working group, Biomass Research and Development Board, February 2011. https://biomassboard.gov/pdfs/biomass_conversion_report.pdf.

Further reading

[1] Biomass: higher heating value, The Engineering Toolbox. http://www.engineeringtoolbox.com/biomass-fuels-hhv-d_1818.html>.

Chemistry of biomass

3

3.1 Introduction

Manufacturing of organic chemicals initially started with coal and alcohol from fermentation industry, but later moved on to petroleum and natural gas. Currently, more than 90% of organic chemicals are produced from petroleum and natural gas. However, rising cost of fossil fuels and continuous decrease in the reserves has pushed the chemical industry to look for an alternative feedstock [1].

Biomass is a naturally abundant resource that is not only cheap but also sustainable. It presents an environmentally friendly solution to the petroleum industry [1]. Conversion of biomass to energy typically involves the synthesis or breakdown of organic compounds that constitute different types of biomass. These organic compounds range in their complexity and contain numerous functional groups that influence the ultimate structure and chemistry of biomass. Understanding these organic compounds and their functional groups is important in developing an understanding on biological and chemical processes that convert biomass into renewable energy [1].

This chapter summarizes relevant concepts from organic and carbohydrate chemistry followed by a discussion of the structure and chemistry of different types of biomass [2].

3.2 Review of chemistry

3.2.1 Organic chemistry

Organic chemistry had a rapid expansion during the 20th century initially with coal, then with oil, as the understanding of the mechanism of its reactions improved rather than the huge quantities of the new compounds that were synthesized. The method to prepare the various products of organic chemistry uses the controlled or catalytic oxidization of these simple hydrocarbons to produce carbonyl molecules, which are the more important class of organic compounds [3].

Current biofuels, such as ethanol, have a lower energy content compared with conventional hydrocarbon fuels, petroleum, and natural gas. The aim is to produce fuels that have high carbon content and therefore have a higher volumetric energy density. This can be achieved by chemical reactions that remove oxygen atoms from organic compounds. The use of organic chemistry to convert platform molecules like levulinic acid [4] and furfural [5] into potential advanced biofuels is important because of the changing parts of the molecules that are responsible for their structure and function [6].

Technology and Applications of Polymers Derived from Biomass. https://doi.org/10.1016/B978-0-323-51115-5.00003-7

There are seven main types of organic compounds:

1. Ethers
2. Compounds with carbonyl groups
3. Alkanes
4. Alkenes
5. Alkynes
6. Aromatic
7. Alcohols

3.2.1.1 Ethers

Ethers are organic compounds that are formed when two alcohol molecules react under proper conditions. Ether is generally defined by $R_1—O—R_2$, where R_1 and R_2 can be aliphatic or aromatic group. Ethers can be classified into three types:

1. Symmetric
2. Asymmetric
3. Aromatic

When ether has two identical R groups attached to the oxygen atom, it is termed as symmetrical. However, asymmetrical ethers have different R groups attached to the oxygen atoms. Some ether structures are shown in Fig. 1 [7].

In comparison with similar molecular weight alcohols, ethers have lower boiling points because ether molecule is unable to engage in hydrogen bonding with each other. For example, the boiling point of diethyl ether is significantly lower than the boiling point of 1-butanol. Table 1 highlights physical properties of different ethers [8].

3.2.1.2 Organic compounds containing carbonyl groups

A carbonyl group is a functional group that contains a carbon atom double bonded to an oxygen atom (C=O). Aldehydes and ketones are the two simplest carbonyl groups. Other carbonyl groups include carboxylic acids and esters. Examples of compounds containing carbonyl groups are shown in Fig. 2 [8–13].

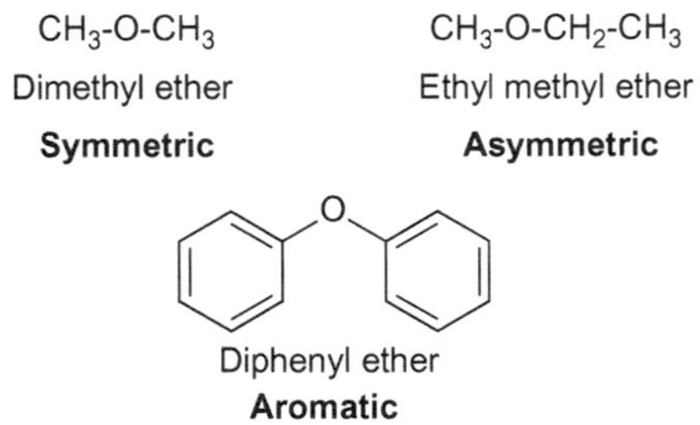

Fig. 1 Examples of ethers.
From http://www.3rd1000.com/chem301/chem301w.htm.

Table 1 **Physical properties of ethers**

Name	Type	Structure	Boiling point (°C)
Dimethyl ether	Symmetric	CH_3-O-CH_3	−25
Ethyl methyl ether	Asymmetric	CH_3-O-CH_2-CH_3	8
Diethyl ether	Symmetric	CH_3CH_2-O-CH_2CH_3	35
Dipropyl ether	Symmetric	$CH_3CH_2CH_2$-O-$CH_2CH_2CH_3$	91
Diphenyl ether	Aromatic	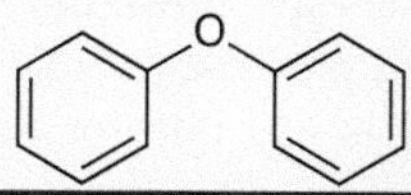	259

From https:/www.britannica.com/science/ether-chemical-compound.

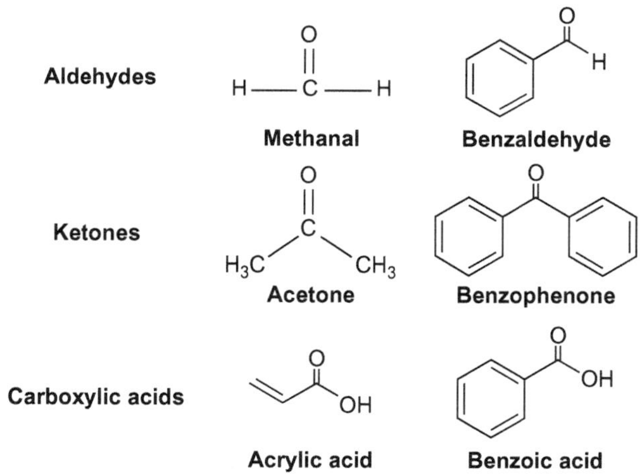

Fig. 2 Examples of compounds containing carbonyl groups.
From https://www.britannica.com/science/ether-chemical-compound; http://www.ck12.org/book/ck-12-chemistry-second-edition/r15/section/25.4/; http://www.sigmaaldrich.com/catalog/product/aldrich/418099?lang=en&region=US; http://badacetone.weebly.com/chemistry.html; http://www.chemicalbook.com/ChemicalProductProperty_EN_CB7307797.htm; http://web.chem.ucla.edu/~harding/IGOC/B/benzoic_acid.html.

The general structure for aldehyde is R—(CO)—H, where R is a substituent group, which can be a hydrogen atom, aliphatic or aromatic, with at least one hydrogen atom of the terminal carbonyl group is connected to the carbon atom. The general formula of ketone is R_1—(CO)—R_2, where R_1 and R_2 can be aliphatic or aromatic groups with two carbon atoms of the carbonyl group connected to the carbon atom. The general formula for carboxylic acid is R—COOH if an ester contains a carbonyl group that is bonded to a hydroxyl group, where R can be an aliphatic or aromatic group.

However, the general formula changes to R_1—(COOH)—R_2 if an ester contains a carbonyl group bonded to an oxygen atom. R_1 can be a hydrogen atom, aliphatic or aromatic group and R_2 can be an aliphatic or aromatic group.

3.2.1.3 Alkanes

Hydrocarbons that are made up of single bonds between carbon and hydrogen are called alkanes and represented by the general formula C_nH_{2n+2}. Due to the absence of double or triple bonds, they are referred to as saturated hydrocarbons. Examples of different variations of alkanes are shown in Fig. 3. The simplest alkanes are methane ($n=1$), ethane ($n=2$), and propane ($n=3$).

More than one compound is possible for the same formula when $n \geq 4$. These compounds are termed as structural isomers. Two distinct compounds (n-butane and isobutane) with the empirical formula C_4H_{10} are formed when $n=4$. The structural isomers of butane are shown in Fig. 4. Similarly, three structural isomers with the empirical formula of C_5H_{12} are formed when $n=5$. Formation of structural isomers significantly increases with increase in the number of carbon atoms. For example, C_7H_{16} has nine structural isomers and $C_{10}H_{22}$ has 75 structural isomers.

Fig. 5 shows the simplified representations of the structures of n-butane and isobutene, along with the three structural isomers of pentane ($n=5$). Representation of organic compounds is often simplified since including every carbon and hydrogen atom in structures can be cumbersome.

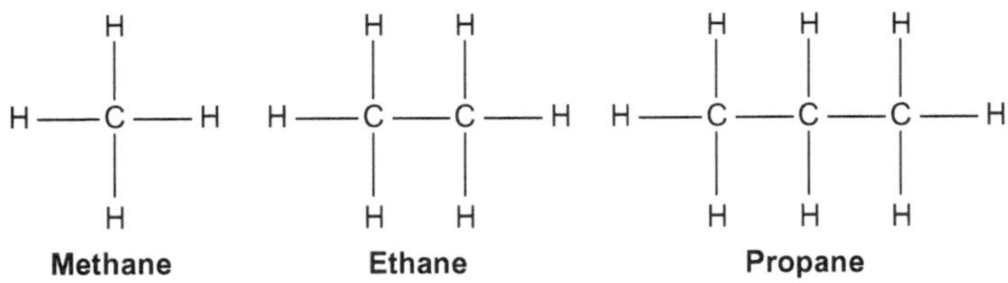

Fig. 3 Structures of simple alkanes.
From D.R. Keshwani, Biomass chemistry, in: J. Cheng (Ed.), Biomass to Renewable Energy Processes, CRC Publishers, 2009 (Chapter 2).

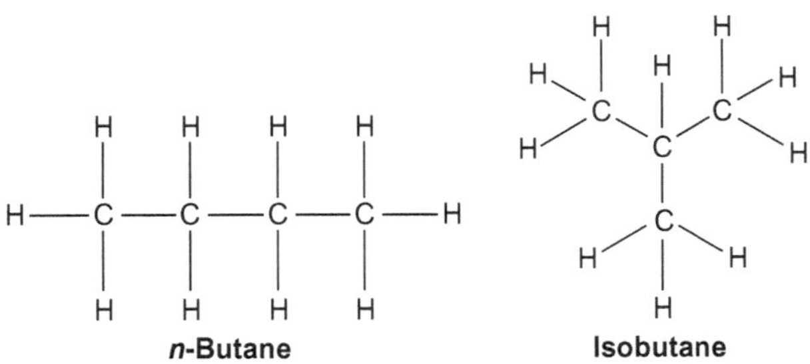

Fig. 4 Structural isomers of butane (C_4H_{10}).
From D.R. Keshwani, Biomass chemistry, in: J. Cheng (Ed.), Biomass to Renewable Energy Processes, CRC Publishers, 2009 (Chapter 2).

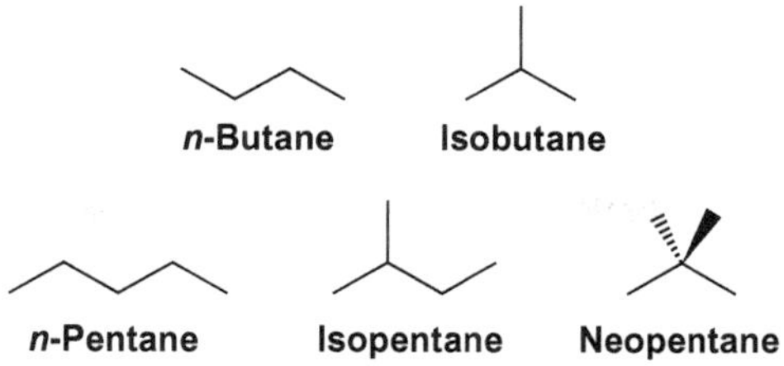

Fig. 5 Simplified representation of different isomers of butane and pentane.
From D.R. Keshwani, Biomass chemistry, in: J. Cheng (Ed.), Biomass to Renewable Energy Processes, CRC Publishers, 2009 (Chapter 2).

3.2.1.4 Alkenes

Hydrocarbons that are made up of at least one double bond between adjacent carbon atoms are called alkenes and represented by the empirical formula C_nH_{2n}. The simplest acyclic alkenes (ethene and propene) contain only one such double bond and are shown in Fig. 6.

In alkenes, structural isomers with empirical formula C_4H_8 can be produced either by changing the position of the double bonds or by changing the way the carbon atoms are joined to each other. In addition, the presence of these double bonds can lead to geometric isomerism as long as each carbon atom involved in the double bond gets attached to different functional group. Fig. 7 shows the structural and geometric isomers of butene [2].

$H_2C{=}CH_2$ $\quad$ $H_2C{=}CH{-}CH_3$

Ethene $\quad$ **Propene**

Fig. 6 Structures of simple alkenes.
From D.R. Keshwani, Biomass chemistry, in: J. Cheng (Ed.), Biomass to Renewable Energy Processes, CRC Publishers, 2009 (Chapter 2).

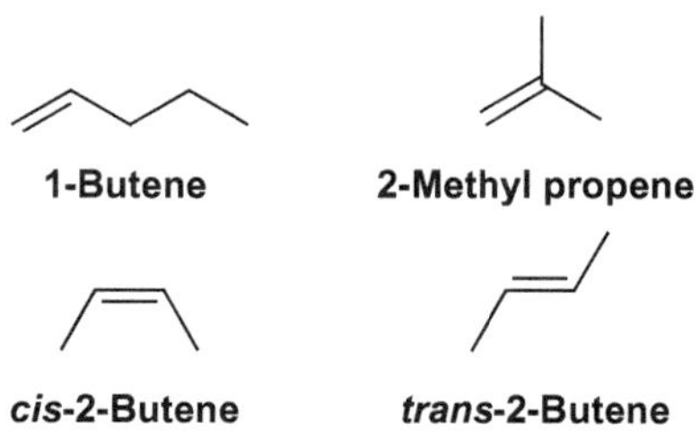

Fig. 7 Structural and geometric isomers of butene.
From D.R. Keshwani, Biomass chemistry, in: J. Cheng (Ed.), Biomass to Renewable Energy Processes, CRC Publishers, 2009 (Chapter 2).

3.2.1.5 Alkynes

Hydrocarbons that are made up of at least one triple bond between adjacent carbon atoms are called alkynes and represented by the general formula C_nH_{2n-2}. As shown in Fig. 8, the simplest examples of such alkynes are ethyne and propyne. Certain alkynes such as butyne and pentyne can have structural isomers depending on the position of the triple bond. However, alkynes do not exhibit geometric isomerism because of the linear nature of the carbon-carbon bond [2].

3.2.1.6 Aromatic compounds

In aliphatic compounds like alkanes, alkenes, and alkynes, electron pairs in the molecules either belong to a single atom or are shared between a pair of atoms. However, when some electron pairs cannot be assigned to a specific atom or pair of atoms they are called aromatic. Aromatic compounds contain a conjugated ring of alternating single and double bonds. Benzene is the most commonly discussed aromatic compound with an empirical formula of C_6H_6.

Fig. 9 shows the resonance structure of benzene. Benzene is represented by a six-carbon ring with a circle inside to indicate the equal distribution of the double

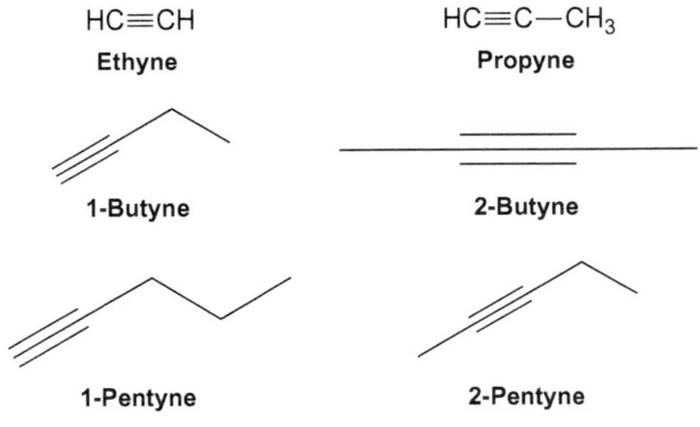

Fig. 8 Structures of ethyne, propyne, and isomers of butyne and pentyne. From D.R. Keshwani, Biomass chemistry, in: J. Cheng (Ed.), Biomass to Renewable Energy Processes, CRC Publishers, 2009 (Chapter 2).

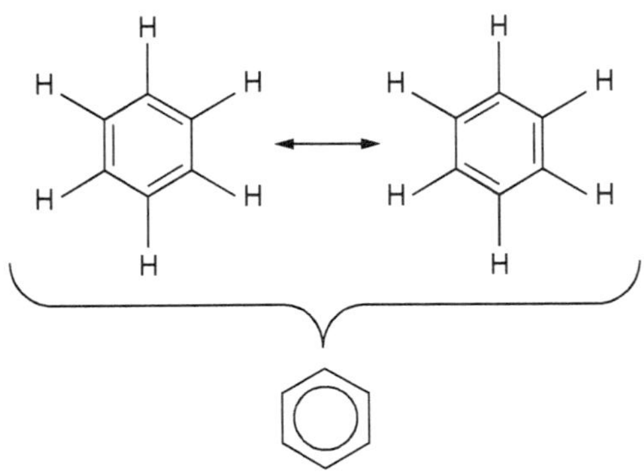

Fig. 9 Resonance structure of benzene. From D.R. Keshwani, Biomass chemistry, in: J. Cheng (Ed.), Biomass to Renewable Energy Processes, CRC Publishers, 2009 (Chapter 2).

bonds between all the carbon atoms. Depending on the location of the double bonds, it can have two structures and neither structure is an adequate representation since the carbon-carbon bonds are neither single nor double. The ability to exist in two structures is termed as resonance. Aromatic compounds can be heterocyclic, polycyclic, or substituted. Examples of each type of aromatic compounds are shown in Fig. 10.

3.2.1.7 Alcohols

When a compound has a hydroxyl (OH) group that is attached to at least one carbon atom, it is termed as an alcohol. Depending on the number of carbon atoms connected to the hydroxyl-bearing carbon atom, alcohols can be defined as primary, secondary, and tertiary. If one carbon is connected to the hydroxyl-bearing carbon atom, it is termed as primary. An alcohol is termed secondary if two carbon atoms are connected to the hydroxyl-bearing carbon atom while tertiary if three carbon atoms are connected to the hydroxyl-bearing carbon atom. Structures of primary and secondary alcohols are shown in Fig. 11 [2].

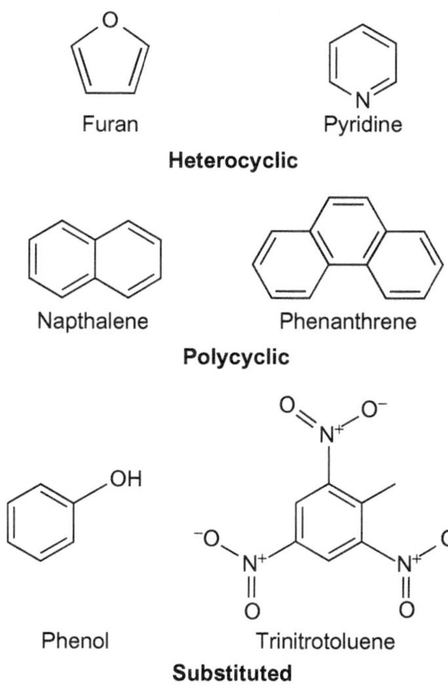

Fig. 10 Examples of heterocyclic, polycyclic, and substituted aromatic compounds. From D.R. Keshwani, Biomass chemistry, in: J. Cheng (Ed.), Biomass to Renewable Energy Processes, CRC Publishers, 2009 (Chapter 2).

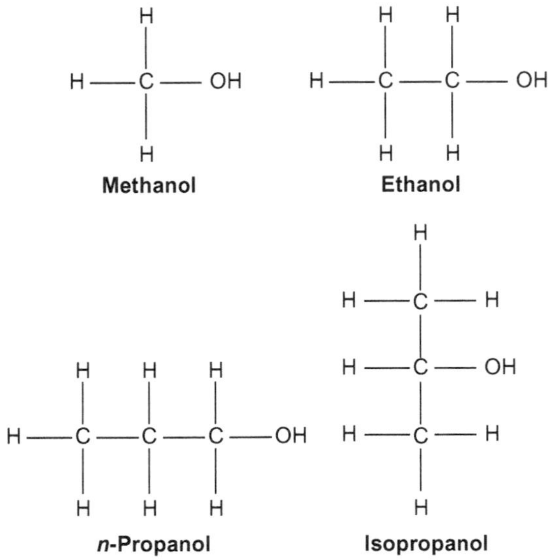

Fig. 11 Structures of primary and secondary alcohols.
From D.R. Keshwani, Biomass chemistry, in: J. Cheng (Ed.), Biomass to Renewable Energy Processes, CRC Publishers, 2009 (Chapter 2).

3.2.2 Carbohydrate chemistry

3.2.2.1 Monosaccharides

Compounds that contain carbonyl functional groups with numerous hydroxyl groups attached to the noncarbonyl carbon atoms are defined as carbohydrates. Simplest form of carbohydrate is monosaccharides and is represented by the general formula $C_n(H_2O)_n$. Carbohydrates can be classified in two ways, either based on the number of carbon atoms or on the nature of the carbonyl group. Fig. 12 shows some examples of monosaccharides structures [2].

Monosaccharides form ring structures when in solution and are in equilibrium with the chain structures. In general, aldehydes and ketones react with alcohols to form hemiacetals and hemiketals. Since, monosaccharides are comprised of several hydroxyl groups and with one aldehyde group within the same molecule, they can form cyclic hemiacetals. However, when several hydroxyl groups interact with one ketone group within the same molecule, they form cyclic hemiketals. Most monosaccharides form five-member rings called furanoses or six-member rings called pyranoses.

Since the mechanism for ring formation is similar for most sugars, the ring structure of monosaccharides is favored in solution. Fig. 13 shows formation of hemiacetal in D-glucose producing α and β anomers of D-glucopyranose. This is obtained when the oxygen of the hydroxyl group attached to the C5 carbon in the chain forms links with the carbonyl carbon. In the process, the hydrogen atom from the OH group of the C5 carbon is transferred to the carbonyl carbon [2].

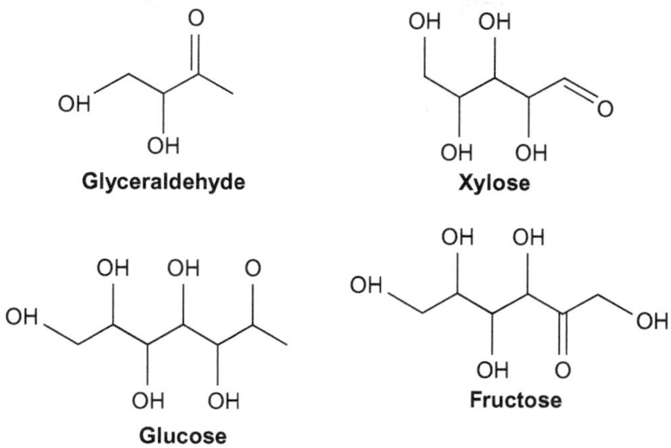

Fig. 12 Examples of monosaccharides.
From D.R. Keshwani, Biomass chemistry, in: J. Cheng (Ed.), Biomass to Renewable Energy Processes, CRC Publishers, 2009 (Chapter 2).

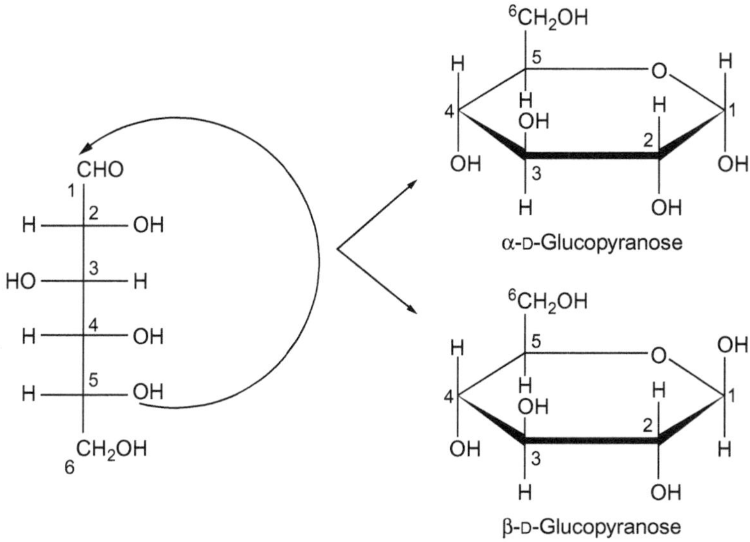

Fig. 13 Hemiacetal formation in D-glucose.
From D.R. Keshwani, Biomass chemistry, in: J. Cheng (Ed.), Biomass to Renewable Energy Processes, CRC Publishers, 2009 (Chapter 2).

3.2.2.2 Disaccharides and glycosidic bonds

Most common examples of disaccharides include maltose, sucrose, lactose, and cellobiose. Disaccharides consist of two simple sugars linked together through a glycosidic bond. This bond is formed as a result of the reaction between anomeric carbon of a monosaccharide and the oxygen atom from the OH group of another monosaccharide. The reaction is called as dehydration or condensation reaction as it is accompanied by the loss of one water molecule for every glycosidic bond.

Glycosidic bonds between monosaccharides are defined based on the stereochemistry of the OH group on the anomeric carbon and the specific carbon atoms involved in the bond formation. Fig. 14 shows formation of α-1,4 glycosidic bond resulting in maltose [2]. If the glycosidic bond involves a hydroxyl group oriented in alpha position, it is called alpha bond.

Fig. 15 shows formation of β-1,4 glycosidic bonds. Cellobiose consist of two β-D-glucopyranose molecules linked via a beta bond between the C1 carbon of one molecule and the oxygen attached to the C4 carbon of the other molecule.

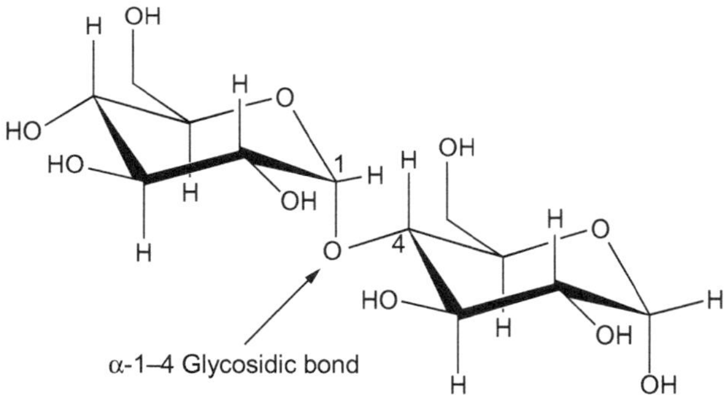

Fig. 14 α-1,4 Glycosidic bond formation in maltose.
From D.R. Keshwani, Biomass chemistry, in: J. Cheng (Ed.), Biomass to Renewable Energy Processes, CRC Publishers, 2009 (Chapter 2).

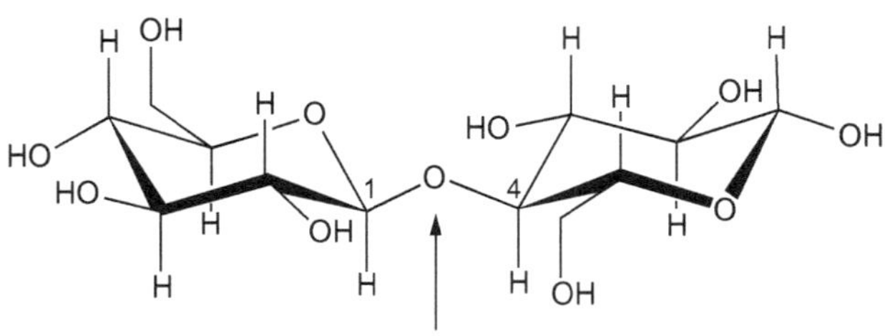

Fig. 15 β-1,4 Glycosidic bond formation in cellobiose.
From D.R. Keshwani, Biomass chemistry, in: J. Cheng (Ed.), Biomass to Renewable Energy Processes, CRC Publishers, 2009 (Chapter 2).

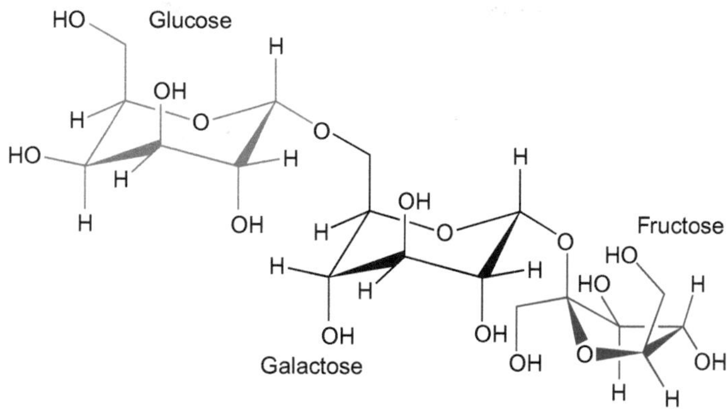

Fig. 16 Structure of raffinose.
From D.R. Keshwani, Biomass chemistry, in: J. Cheng (Ed.), Biomass to Renewable Energy Processes, CRC Publishers, 2009 (Chapter 2).

Disaccharides are part of a larger group of oligosaccharides, which consist of 2–10 monosaccharide units connected through glycosidic bonds. Fig. 16 shows structure of raffinose, which is a trisaccharide, composed of galactose, fructose, and glucose units.

3.3 Starch

Starch is primarily made up of two polysaccharides [14]: amylose, a mostly linear α-D (1,4′)-glucan and branched amylopectin, having the same backbone structure as amylose but with many α-1,6′-linked branch points as shown in Fig. 17 [15]. The starch chains have a lot hydroxyl groups, two secondary hydroxyl groups at C2 and C3 of each glucose residue, as well as one primary hydroxyl group at C-6 when it is not linked. The available hydroxyl groups on the starch chains can be oxidized and reduced, and can help in the formation of hydrogen bonds, ethers, and esters [16].

Starch comprises of 10%–20% amylose and 80%–90% amylopectin depending on the source [17]. Amylose is soluble in water and forms a helical structure [18]. Short branched amylopectin chains can form helical structures which crystallize resulting in starch to exist as discrete granules. Starch granules exhibit hydrophilic properties and strong intermolecular association via hydrogen bonding formed by the hydroxyl groups on the granule surface. In amylose, α and β amylases can hydrolyze amylose; however, in amylopectin, α and β amylases can hydrolyze α-1,4 glycosidic linkage but cannot hydrolyze α-1,6 glycosidic linkage at branch points of amylopectin.

Starch is typically stored as granules in roots, tubers, fruits, and seeds. While the granular nature is a universal occurrence, starches can occasionally be produced in amorphous forms, as is the case in certain lichens. While starch granules primarily consist of carbohydrates, minor constituents such as proteins, lipids, phosphate groups, and other plant cell wall materials are also present.

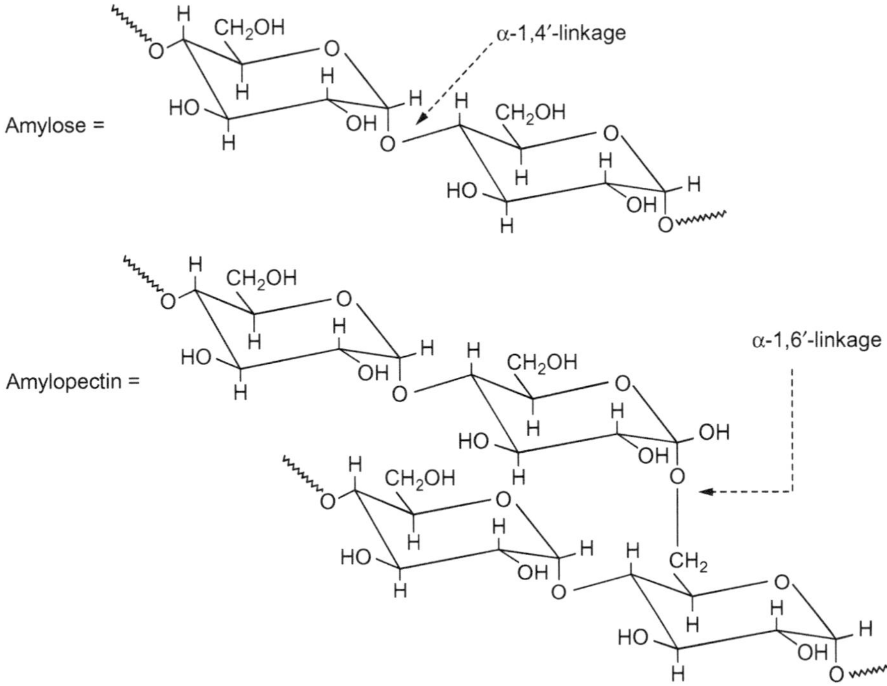

Fig. 17 Molecular structure of starch.
From D.R. Lu, C.M. Xiao, S.J. Xu, Starch-based completely biodegradable polymer materials, eXPRESS Polym. Lett. 3 (6) (2009) 366–375.

3.4 Cellulose

Cellulose is a natural polymer made from long chains that are linked together by smaller molecules. These links in the cellulose chains consist of β-D-glucose. These sugar units are linked when water is eliminated by combining the H and hydroxyl group as shown in Fig. 18 [19].

When two of these sugars are linked together, they produce a disaccharide called as cellobiose. The structure of cellobiose is shown in Fig. 19 [20,21]. In this structure, the spatial arrangement of the acetal linkages is very important.

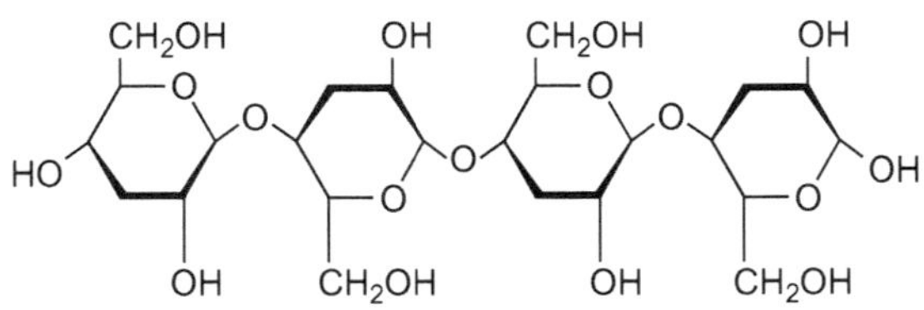

Fig. 18 Molecular structure of starch.
From Sketch the cellulose structure. http://www.biochemcs.com/cellulose_structure.htm.

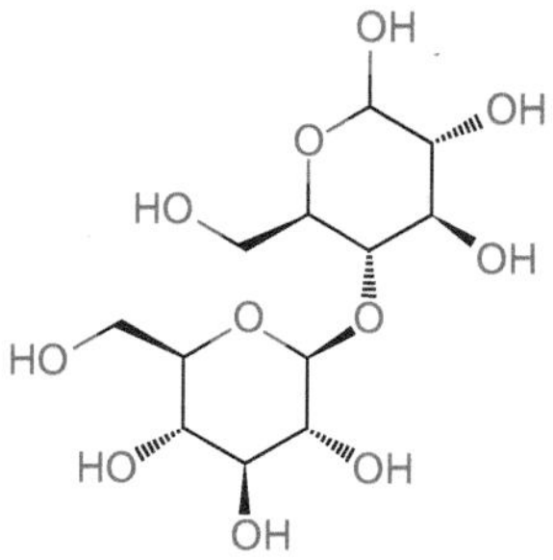

Fig. 19 Molecular structure of starch.
From Cellobiose, European Molecular Biology Laboratory. http://www.ebi.ac.uk/chebi/searchId.do?chebiId=CHEBI:17057; The Merck Index, eigth ed., Merck & Co, Rahway, NJ, USA, 1968.

Fig. 20 shows two ring structures of the cellulose molecule. These structures are called the pyranose rings. In this ring structure, all the groups are larger than hydrogen which sticks out from the periphery of the rings [22]. The spatial arrangement of carbon atom in the glucose molecule at position 2, 3, 4, and 5 is fixed; however, in a ring structure such as pyranose, the hydroxyl group at C4 can approach the carbonyl at C1 from either side, thus resulting in two different spatial arrangements at carbon position 1. When the hydroxyl group at C1 is on the same side of the ring as the C6 carbon, it is said to be in α configuration. In cellulose, the C1 oxygen is in the opposite or β configuration [23].

Because of the alignment of the hydroxyl groups on the cellulose chain, they tend to stick out laterally along the extended molecule and are readily available for hydrogen bonding. The chains are grouped together in a highly ordered structure by means of hydrogen bonding. Due to long length of the chains, they tend to pass through several crystalline regions with some areas of disorder in the middle. These disordered middle areas are termed as fringed-micelle regions [24,25]. Because of the strong hydrogen bonding in the crystalline regions, cellulose has good strength and resistance toward solvents. This also prevents them from melting. In comparison, less-ordered regions, more hydrogen is available to bond with other molecules such as water because the

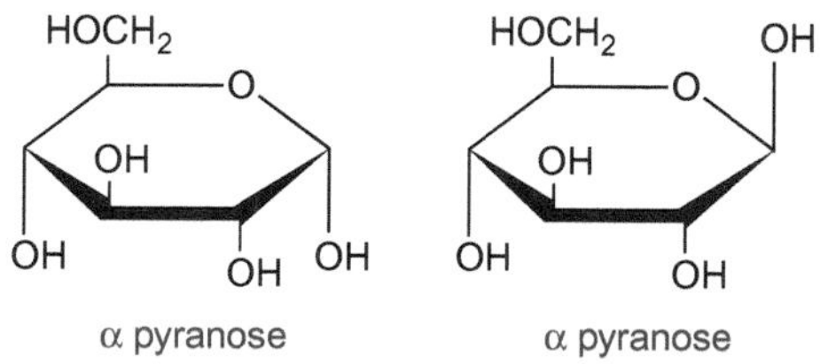

Fig. 20 Molecular structure of starch.
From http://science.uvu.edu/ochem/index.php/alphabetical/o-p/pyranose/.

chains are located far apart from one another. Most cellulose structures can absorb large quantities of water and hence they are hygroscopic. This makes cellulose to swell but does not dissolve in water [23].

3.5 Vegetable oil

Vegetable oils are lipid materials found in plants. Although, vegetable oils and fats can be obtained from animal sources, approximately 71% of the total edible oil production is from plant sources. Vegetable oils comprise 96%–98% of fatty acids triglycerides, and 2%–4% of major and minor nontriglycerides. Major nontriglycerides include free fatty acids, diglycerides, monoglycerides, and phospholipids. Minor nontriglycerides are present in very small amounts and include trace metals, tocopherols, sterols, and pigments [2,26,27].

Fig. 21 shows formation of mono-, di-, and triglycerides via esterification between glycerol and fatty acid [2]. Monoglycerides are formed when only one

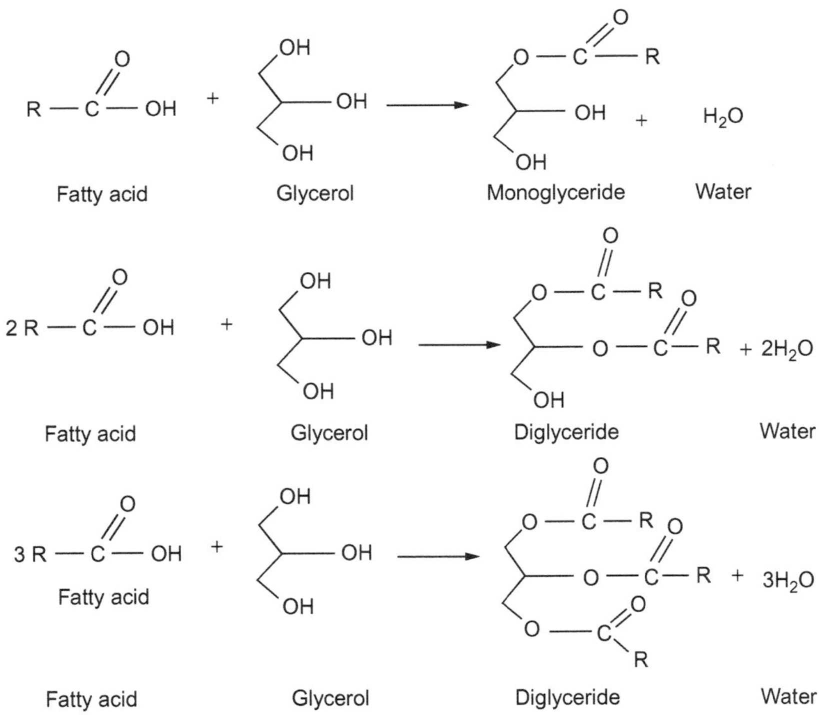

Fig. 21 Formation of mono-, di-, and triglycerides.
From D.R. Keshwani, Biomass chemistry, in: J. Cheng (Ed.), Biomass to Renewable Energy Processes, CRC Publishers, 2009 (Chapter 2); http://scialert.net/fulltext/?doi=jas.2010.3199.3206.

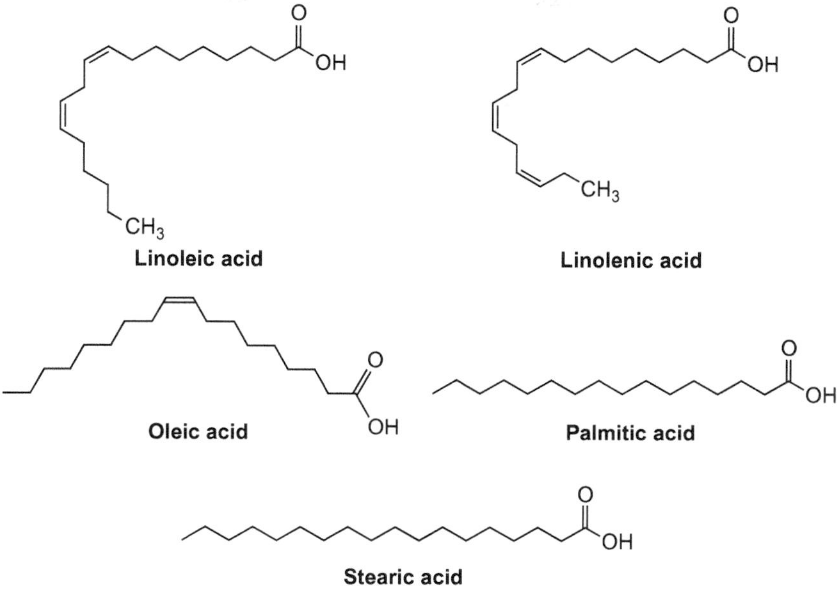

Fig. 22 Examples of fatty acids in vegetable oils.
From D.R. Keshwani, Biomass chemistry, in: J. Cheng (Ed.), Biomass to Renewable Energy Processes, CRC Publishers, 2009 (Chapter 2); http://chemistry.about.com/od/factsstructures/ig/Chemical-Structures—L/Linoleic-Acid.htm; http://chemistry.about.com/od/factsstructures/ig/Chemical-Structures—L/Alpha-Linolenic-Acid.htm; http://www.scienceofcooking.com/oleic_acid.htm; http://www.pixell.club/palmitic-acid-structure/; http://chemistry.about.com/od/factsstructures/ig/Chemical-Structures—S/Stearic-Acid-Octadecanoic-Acid.htm.

hydroxyl group of glycerol form ester bond while diglycerides are formed when two hydroxyl groups of glycerol are esterified. Triglycerides are formed when all the three hydroxyl groups of glycerol form ester bonds with three fatty acid molecules. Since, esterification process is a condensation reaction, it releases water molecules in the process. Monoacid triglycerides are relatively rare [28].

Fatty acids are aliphatic compounds with a carboxylic acid functional group. Most natural fatty acids contain 4–28 carbon atoms. They are classified either by their chain length or by the presence or absence of double bonds. When classifying via chain length, fatty acid can be short chain (4–8 carbon atoms), medium chain (8–14 carbon atoms), and long chain (more than 14 carbon atoms). Saturated fatty acids contain no double bonds, monounsaturated fatty acids contain one double bond, and polyunsaturated fatty acids contain two or more double bonds. Fig. 22 shows the structure of the fatty acids typically seen in vegetable oils. Typical fatty acids in vegetable oils are linoleic acid, linolenic acid, oleic acid, palmitic acid, and stearic acid [2].

3.6 Extractives

The term extractives are used to describe a wide range of compounds present in hardwoods, softwoods, and grasses that can be extracted using polar and nonpolar solvent. Most woody biomass, grasses, and agricultural residues contain 2%–10% extractives by weight. The three main groups of extractives are terpenes and terpenoids, fats and waxes, and phenolic compounds.

Fig. 23 shows the structures of terpenes and terpenoids present in extractives [2,29–31]. Terpenes and their derivatives are a large group of compounds that are characterized by the presence of two or more isoprene units in their structures. They have a general formula of $(C_{10}H_{16})_n$ and are characterized as monoterpenes ($n=1$), sesquiterpenes ($n=1.5$), diterpenes ($n=4$), and polyterpenes ($n>4$).

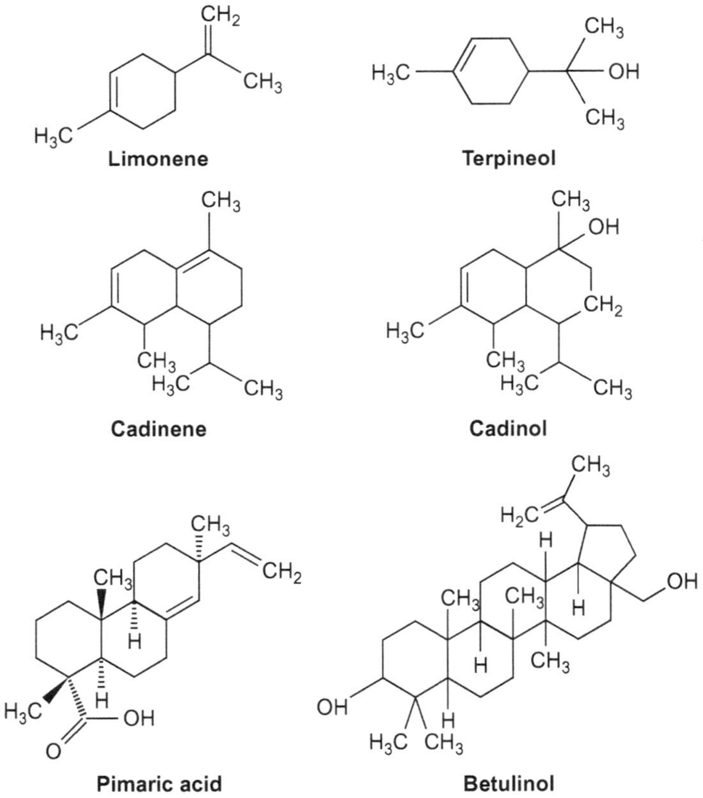

Fig. 23 Examples of terpenes and terpenoids in extractives.
From D.R. Keshwani, Biomass chemistry, in: J. Cheng (Ed.), Biomass to Renewable Energy Processes, CRC Publishers, 2009 (Chapter 2); http://www.chegg.com/homework-help/questions-and-answers/limonene-one-compounds-give-lemons-tangy-odor-draw-structure-product-formed-limonene-treat-q6527793; http://www.sigmaaldrich.com/catalog/product/aldrich/432628?lang=en&region=US; http://www.coompo.com/compounds/Pimaric+acid+C1647.html.

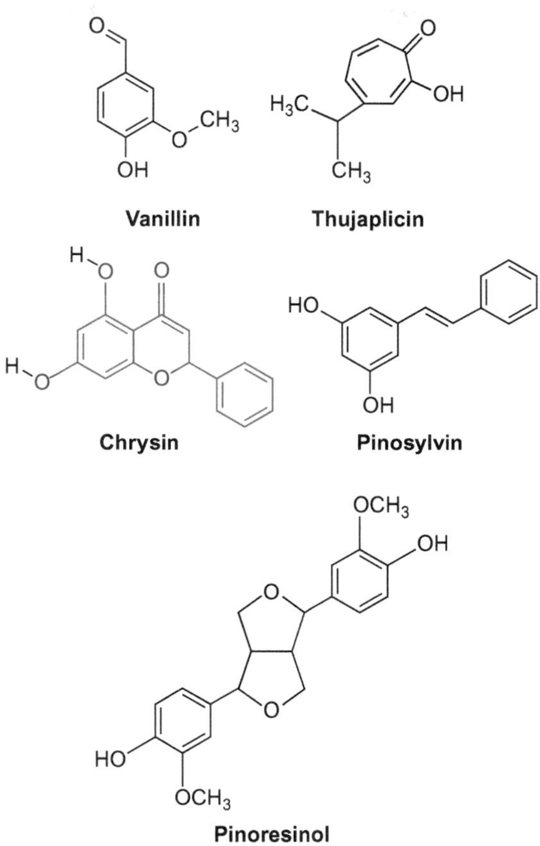

Fig. 24 Examples of phenolic compounds in extractives.
From http://chemistry.about.com/od/factsstructures/ig/Chemical-Structures—V/Vanillin.htm; http://www.sigmaaldrich.com/catalog/product/aldrich/469521?lang=en&region=US; http://wildflowerfinder.org.uk/Flowers/P/Passionflower(Blue)/Passionflower(Blue).htm; http://www.enzymetek.com/product-Pinosylvin.html; http://www.bifcpresidency.tn.gov.in/Pinoresinol.html; https://pubchem.ncbi.nlm.nih.gov/compound/chrysin#section=Top.

Phenolic extractives are a heterogeneous group of compounds that are characterized by the presence of phenolic units in their structures. Some of the examples of phenolic compounds found in extractives are vanillin, thujaplicin, chrysin, pinosylvin, and pinoresinol and are shown in Fig. 24 [32–37].

References

[1] L.A. Lucia, D.S. Argyropolulos, L. Adamopoulos, A.R. Gaspar, Chemicals and energy from biomass, Can. J. Chem. 84 (2006) 960–970.

[2] D.R. Keshwani, Biomass chemistry, in: J. Cheng (Ed.), Biomass to Renewable Energy Processes, CRC Publishers, Boca Raton, FL, USA, 2009 (Chapter 2).

[3] http://www.biomass-chemistry.com/biomass-chemistry/biomass-chemistry.html&gt.
[4] http://pubs.acs.org/doi/abs/10.1021/acs.energyfuels.6b01207&gt.
[5] https://www.ncbi.nlm.nih.gov/labs/articles/25873007/&gt.
[6] http://www.rsc.org/Membership/Networking/InterestGroups/OrganicDivision/organic-chemistry-case-studies/organic-chemistry-biofuels.asp&gt.
[7] http://www.3rd1000.com/chem301/chem301w.htm&gt.
[8] https://www.britannica.com/science/ether-chemical-compound&gt.
[9] http://www.ck12.org/book/ck-12-chemistry-second-edition/r15/section/25.4/&gt.
[10] http://www.sigmaaldrich.com/catalog/product/aldrich/418099?lang=en&region=US&gt.
[11] http://badacetone.weebly.com/chemistry.html&gt.
[12] http://www.chemicalbook.com/ChemicalProductProperty_EN_CB7307797.htm&gt.
[13] http://web.chem.ucla.edu/~harding/IGOC/B/benzoic_acid.html&gt.
[14] R. Pareta, M.J. Edirisinghe, A novel method for the preparation of starch films and coatings, Carbohydr. Polym. 63 (2006) 425–431.
[15] D.R. Lu, C.M. Xiao, S.J. Xu, Starch-based completely biodegradable polymer materials, eXPRESS Polym. Lett. 3 (6) (2009) 366–375.
[16] P. Tomasik, C.H. Schilling, Chemical modification of starch, Adv. Carbohydr. Chem. Biochem. 59 (2004) 175–403.
[17] M. Ramesh, J.R. Mitchell, S.E. Harding, Amylose content of rice starch, Starch 51 (1999) 311–313.
[18] R.A. Wallace, J.L. King, G.P. Sanders, Biology—The Science of Life, Goodyear Publishing Company, California, 1981.
[19] Sketch the cellulose structure. http://www.biochemcs.com/cellulose_structure.htm&gt.
[20] Cellobiose, European Molecular Biology Laboratory. http://www.ebi.ac.uk/chebi/searchId.do?chebiId=CHEBI:17057&gt.
[21] The Merck Index, eigth ed., Merck & Co, Rahway, NJ, 1968.
[22] Pyranose. http://science.uvu.edu/ochem/index.php/alphabetical/o-p/pyranose/&gt.
[23] Cellulose. http://www.fibersource.com/f-tutor/cellulose.htm&gt.
[24] P.M. Visakh, S. Thomas, Preparation of bionanomaterials and their polymer nanocomposites from waste and biomass, Waste Biomass Valoriz. 1 (1) (2010) 121–134.
[25] Y. Sawai, T. Nozoe, Y. Yoshitani, D. Tsukada, Development of new cellulose based polymers with excellent melt processability, Fujifilm Res. Dev. 57 (2012).
[26] https://secondaryscience4all.files.wordpress.com/2013/12/bbc-gcse-bitesize-vegetable-oils.pdf&gt.
[27] http://www.fediol.eu/web/chemical+composition/1011306087/list1187970121/f1.html&gt.
[28] http://scialert.net/fulltext/?doi=jas.2010.3199.3206&gt.
[29] http://www.chegg.com/homework-help/questions-and-answers/limonene-one-compounds-give-lemons-tangy-odor-draw-structure-product-formed-limonene-treat-q6527793&gt.
[30] http://www.sigmaaldrich.com/catalog/product/aldrich/432628?lang=en&region=US&gt.
[31] http://www.coompo.com/compounds/Pimaric+acid+C1647.html&gt.
[32] http://chemistry.about.com/od/factsstructures/ig/Chemical-Structures—V/Vanillin.htm&gt.
[33] http://www.sigmaaldrich.com/catalog/product/aldrich/469521?lang=en&region=US&gt.
[34] http://wildflowerfinder.org.uk/Flowers/P/Passionflower(Blue)/Passionflower(Blue).htm&gt.
[35] http://www.enzymetek.com/product-Pinosylvin.html&gt.
[36] http://www.bifcpresidency.tn.gov.in/Pinoresinol.html&gt.
[37] https://pubchem.ncbi.nlm.nih.gov/compound/chrysin#section=Top&gt.

Further reading

[1] http://chemistry.about.com/od/factsstructures/ig/Chemical-Structures—L/Linoleic-Acid.htm&gt.
[2] http://chemistry.about.com/od/factsstructures/ig/Chemical-Structures—L/Alpha-Linolenic-Acid.htm&gt.
[3] http://www.scienceofcooking.com/oleic_acid.htm&gt.
[4] http://www.pixell.club/palmitic-acid-structure/&gt.
[5] http://chemistry.about.com/od/factsstructures/ig/Chemical-Structures—S/Stearic-Acid-Octadecanoic-Acid.htm&gt.

Chemistry of cellulosic polymers 4

4.1 Introduction

Biopolymers represent a growing field and a vast number of biodegradable polymers (e.g., cellulose, chitin, starch, polyhydroxyalkanoates, polylactide, polycaprolactone, collagen, and other polypeptides) have been synthesized or are formed in natural environment during the growth cycles of organisms. Some microorganisms and enzymes capable of degrading such polymers have been identified.

Biopolymers are classified according to their origin into three classes: naturally produced renewable polymers, synthetic polymers derived from renewable resources, and synthetic polymers derived from petroleum-based resources.

Natural biopolymers are produced in nature by all living organisms. Biodegradation reactions are typically enzyme catalyzed and occur in aqueous media. Natural macromolecules containing hydrolyzable linkages, such as protein, cellulose, and starch are generally susceptible to biodegradation by the hydrolytic enzymes of microorganisms. Cellulose, lignin, and hemicellulose are some of the biopolymers naturally occurring in plants or animals.

This chapter will review chemistry of different forms of cellulose esters and provide discussion on other form of natural biodegradable polymers such as lignin, hemicellulose, and lignocellulosic.

4.2 Review of chemistry

4.2.1 Cellulose esters

Cellulose esters are part of a large family of cellulose derivatives that have found use in pharmaceutical and other applications. Cellulose ester is divided into two categories—enteric and nonenteric. Enteric esters are those which are relatively insoluble in acid solutions but soluble in mildly acidic to slightly alkaline solutions such as cellulose acetate phthalate (CAP). However, nonenteric esters are not dependent on pH solubility characteristics. They are mostly insoluble in water except for cellulose acetates with low level of acetyl [1].

Acid catalysis of an acid and alcohol also known as Fischer esterification process is the most important method for preparing an ester. In comparison, traditional esterification method uses controlled acid catalysis of an acid or acid anhydride and the hydroxyl groups attached to cellulose to make cellulose acetate. Because this is a reversible reaction, the reaction requires complete removal of water for completion. Degree of substitution (DS) is determined by the amount of water. The general reaction to make cellulose ester is shown as illustrated reaction:

Technology and Applications of Polymers Derived from Biomass. https://doi.org/10.1016/B978-0-323-51115-5.00004-9

$$\text{Cellulose} + \text{acid} \xrightarrow{\text{Catalyst}} \text{cellulose ester} + \text{water}$$

Cellulose esters are commercially available in a variety of applications. Some of the applications include solvent-borne coatings, separation, medical and controlled release, composites and laminates, and plastic applications.

The three most common types of cellulose esters are as follows:

1. Cellulose acetate (CA)
2. Cellulose acetate propionate (CAP)
3. Cellulose acetate butyrate (CAB)

These three cellulose esters are thermoplastic materials that are produced through esterification of cellulose. To make powdered cellulose esters biopolymer, different raw materials such as cotton, recycled paper, wood cellulose, and sugarcane are needed. For example, Bioceta, plasticized cellulose acetate, is prepared from cotton flakes and wood pulp through an esterification process with acetic anhydride. Cellulose acetate propionate (CAP) and cellulose acetate butyrate (CAB) are mixed esters produced by treating cellulose with appropriate acids and anhydrides in the presence of sulfuric acid [2]. Table 1 shows melting points of cellulose esters [3].

4.2.1.1 Cellulose acetate (CA)

Cellulose acetate (CA) is the first organic ester of cellulosic family. It was originally discovered by Schutzenberger in 1865 and later developed by Franchimont in 1879 and Miles in 1903. By 1929, commercial grades of cellulose acetate in the form of molding powder, fibers, sheets, and tubes were being produced in the United States [4].

Fig. 1 shows general mechanism to prepare cellulose acetate [5,6]. Cellulose acetate is prepared by mixing cellulose with acetic anhydride using acetic acid as solvent

Table 1 Melting points of cellulose esters

Cellulose ester	Melting point (°C)
Cellulose	450
Acetate	306
Butyrate	183
Caproate	94
Caprylate	86
Caprate	88
Heptylate	88
Laurate	91
Myristate	106
Palmitate	106
Propionate	234
Valerate	122

From O. Olabisi, K. Adewale, Handbook of Thermoplastics, Marcel Dekker, 1997, p. 340.

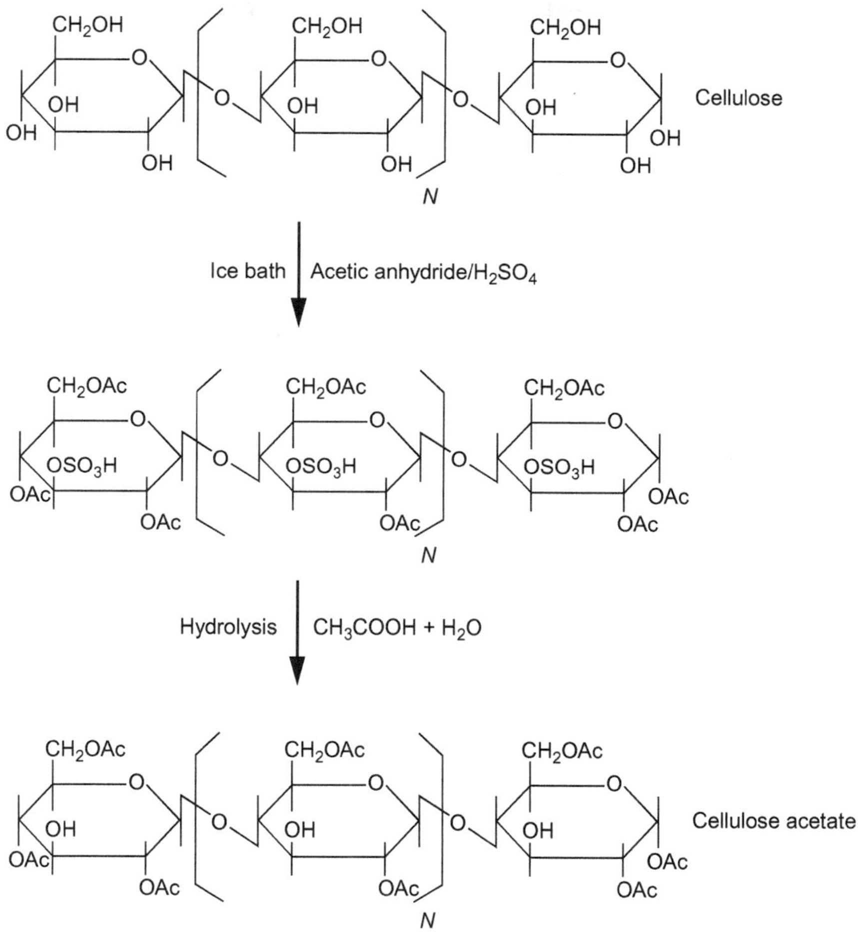

Fig. 1 General mechanism to prepare cellulose acetate.
From http://wpage.unina.it/avitabil/testi/AcetilCellulosa.pdf; http://textilelibrary.weebly.com/acetate-rayon-mfg-process.html.

and sulfuric acid as catalyst. Sulfuric acid reacts with acetic anhydride to form acetyl sulfuric acid. During the acetylation process, both sulfuric acid and acetyl sulfuric acid react with cellulose to form cellulose sulfate acid ester.

Most of the sulfate groups are replaced by the acetyl groups during the acetylation process. This mixture reacts until the three hydroxyl groups per anhydroglucose have been replaced with the acetyl groups. In the next step, hydrolysis is performed to remove acetyl and sulfate groups under controlled conditions of time, temperature, and acidity. This helps in improving the physical properties of cellulose acetate. Hydrolysis process also removes the combined acid sulfate ester and thereby improving the thermal stability of cellulose acetate [5,6].

A pretreatment of cellulose with water or acetic acid is performed to make more rapid and uniform ester. This pretreatment swells the fiber and thereby reducing intramolecular hydrogen bonding. The rate of acetylation of preswollen cellulose is about three times higher than that of nonswollen cellulose. It also makes them accessible to the esterifying mixture, and hence increases their reactivity.

Table 2 shows some properties of cellulose acetate. Cellulose acetate is inexpensive material and is either in transparent or colored. They exhibit poor thermal, electrical and weathering properties, and are not resistant to chemicals. They are typically used in films and sheets in packaging and display industries. Cellulose acetate is easily processed and have been used in brush handles, combs, and spectacle frames. They are also used in vacuum-formed display containers for hardware or food products, and

Table 2 Properties of cellulose acetate

Property	Value
Molding qualities	Excellent
Relative density	1.22–1.34
Tensile strength (MPa)	13–62
Compressive strength (MPa)	14–248
Impact strength, Izod (J/mm)	0.02–0.26
Hardness, Rockwell	R34–R125
Thermal expansion (10^{-4}/°C)	20–46
Resistance to heat (°C)	60–105
Dielectric strength (V/mm)	9,840–23,620
Dielectric constant (at 60 Hz)	3.5–7.5
Dissipation factor (at 60 Hz)	0.01–0.06
Arc resistance (s)	50–310
Water absorption, 24 h (%)	1.7–6.5
Burning rate (mm/min)	Slow to self-extinguishing
Effect of sunlight	Slight
Effect of acids	Decomposes
Effect of alkalis	Decomposes
Effect of solvents	Soluble in ketones, esters, chlorinated hydrocarbons, aromatic hydrocarbons
Machining qualities	Excellent
Optical qualities	Transparent to opaque

From https://www.ncsu.edu/bioresources/BioRes_07/BioRes_07_1_1068_Loo_HP_Recycling_Paper_Dust_Cellulose_Acetate_Pretreat_2309.pdf.

barrier films. Cellulose acetate coated films are used in magnetic recording tapes and photographic film. Cellulose acetate plastics are made into fibers for use in textiles [7].

The degree of substitution (DS) of acetyl groups per anhydro-D-glucose unit of cellulose can be calculated from the acetyl content as follows [7]:

$$DS = \frac{3.86 \times \%\,\text{acetyl}}{102.4 - \%\,\text{acetyl}}$$

Because the formula weight is 60 for CH_3COOH and 43 for CH_3CO, the ratio between these two is 1.395. Hence, the relationship between acetyl content and combined acetic acid can be obtained by using the following equation:

$$\text{Combined acetic acid} = \text{acetyl content} \times 1.395$$

Table 3 shows relationship of cellulose acetate degree of substitution (DS) to acetyl content and combined acetic acid. The thermoplastic characteristics of cellulose acetate are greatly improved as the acetyl content is increased from around 20% (DS = 1) to around 39% (DS = 2.4). Higher acetyl content gives poorer flow but better moisture resistance; lower acetyl content gives better impact strength. The melting point of cellulose acetate also drops from 300°C to approximately to 230°C with decrease in acetyl content. The high melting point can potentially lead to thermal degradation of cellulose ester. To prevent this degradation during processing, plasticizers are added which helps in lowering the melting point [7].

4.2.1.2 Cellulose acetate butyrate (CAB)

Hercules Powder Company and Eastman Chemical jointly developed cellulose acetate butyrate (CAB) during the mid-1930s. Fig. 2 shows a general structure of cellulose acetate butyrate (CAB) [8]. CAB is produced when cellulose is reacted with a mixture of sulfuric and acetic acids followed by esterification process. During the

Table 3 Relationship of cellulose acetate DS to acetyl content and combined acetic acid

DS	Acetyl content	Combined acetic acid
0.50	11.7	16.3
0.75	16.7	23.3
1.0	21.1	29.4
1.5	28.7	40.0
2.0	35.0	48.8
2.5	40.3	56.2
3.0	44.8	62.5

From https://www.ncsu.edu/bioresources/BioRes_07/BioRes_07_1_1068_Loo_HP_Recycling_Paper_Dust_Cellulose_Acetate_Pretreat_2309.pdf.

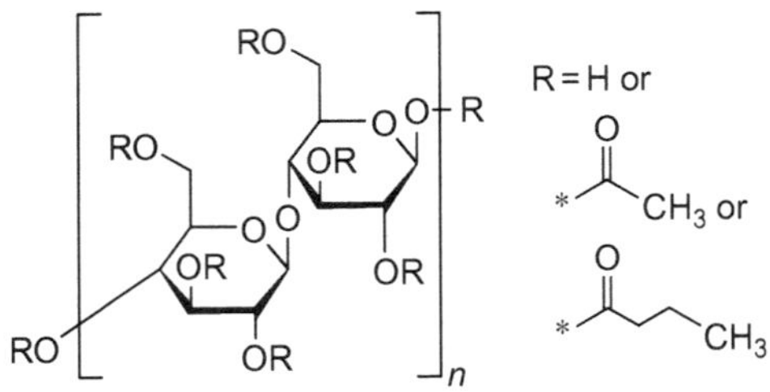

Fig. 2 General structure of cellulose acetate butyrate (CAB).
From http://www.sigmaaldrich.com/catalog/product/aldrich/419044?lang=en&region=US.

esterification process cellulose is reacted with butyric acid and acetic anhydride. This reaction is similar to the ones used in producing cellulose acetate except that butyrate is also used. The end product has acetyl groups (CH_3CO) and butyl groups ($CH_3CH_2CH_2CH$) in the repeating cellulose unit.

Cellulose acetate butyrate (CAB) has acetyl content in the range of 13–15 wt%, a butyryl content of 34–39 wt%, and a free hydroxyl group of 1–2 wt%. Cellulose acetate butyrate products have good dimensional stability among family of other acetates. They have excellent toughness, moisture resistance, and are available in fine colors. They are typical extruded and have found their use for automobile steering wheels, knobs, tool handles, packaging blisters, illuminated advertising signs, machine hoods, lamp covers, and dome lights [8].

4.2.1.3 Cellulose acetate propionate (CAP)

Cellulose acetate propionate (CAP) was originally developed by the Celanese Plastics Company in 1931. Like other acetates, it is made with the addition of propionic acid (CH_3CH_2COOH) in place of acetic anhydride. The general structure is shown in Fig. 3 [9]. Plastic grade cellulose acetate propionate (CAP) has an acetyl content in the range of 1.5–7 wt% and a propionyl content of about 39–42 wt%. Because of its high transparency, low-level light scattering and good impact resistance, CAP is used in high quality frames for sun glasses, personal protective equipment, and sport goggles.

Fig. 3 General structure of cellulose acetate propionate.
From http://www.sigmaaldrich.com/catalog/product/aldrich/330183?lang=en&region=US.

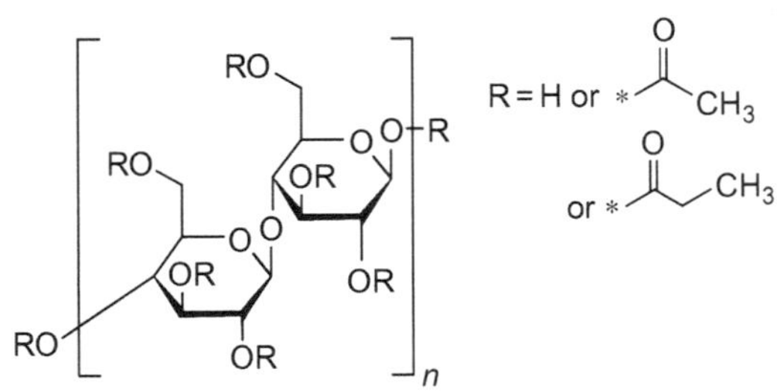

4.2.2 Celluloid

In the middle of the 19th century during the times of early experimentation and synthesis, celluloid was invented. It all started when cellulose was treated with strong nitric acid to form nitrocellulose, which found some use as an explosive. During 1860–70, English chemists tried to stabilize the process with different solvent. They were somewhat successful however the process proved to be unprofitable. Later, an American scientist developed a material which was produced by mixing camphor with nitrocellulose. The camphor dissolved the nitrocellulose, but also affected the structure of the product and performed as a plasticizer. This development was ultimately successful and became trademarked as celluloid [10].

Fig. 4 shows the flow chart of celluloid manufacture. The raw materials required to synthesize celluloid include cellulose fibers, nitric acid, sulfuric acid, water, alcohol, camphor, colorants, and fillers [11,12].

Cellulose, the starting raw material is obtained from a variety of sources, including cotton, ramie, flax, jute, and wood pulp. The structure of cellulose is shown in Fig. 5 [13]. In the early days, the main commercial source of cellulose for cellulose nitrate synthesis in the United States was cotton mill waste, including cloth scraps from clothing production and waste fibers from spinning plants.

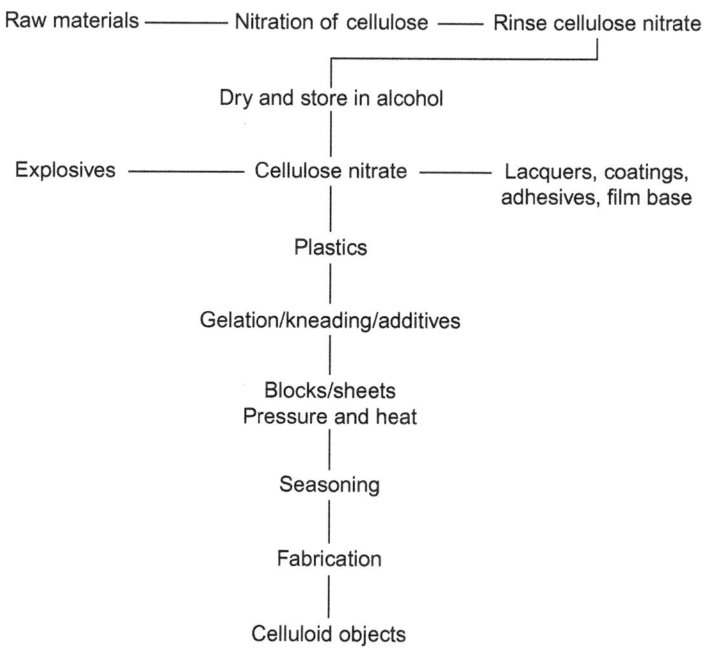

Fig. 4 Flowchart of celluloid manufacture.
From J.A. Reilly, Celluloid objects: their chemistry and preservation, J. Am. Inst. Conserv. 30 (2) (1991) pp. 145–162. http://cool.conservation-us.org/jaic/articles/jaic30-02-003.html; http://www.ottawavalleybuttonclub.ca/Articles/tabid/544/articleType/ArticleView/articleId/409/The-History-and-Properties-of-Celluloid.aspx.

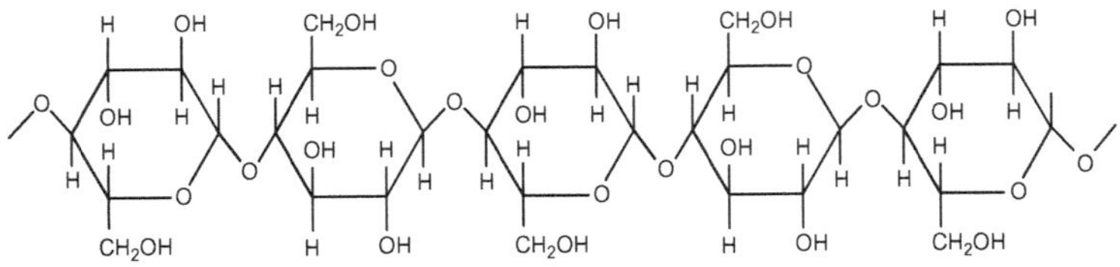

Fig. 5 Structure of cellulose.
From http://www.easychem.com.au/production-of-materials/biomass-research/cellulose.

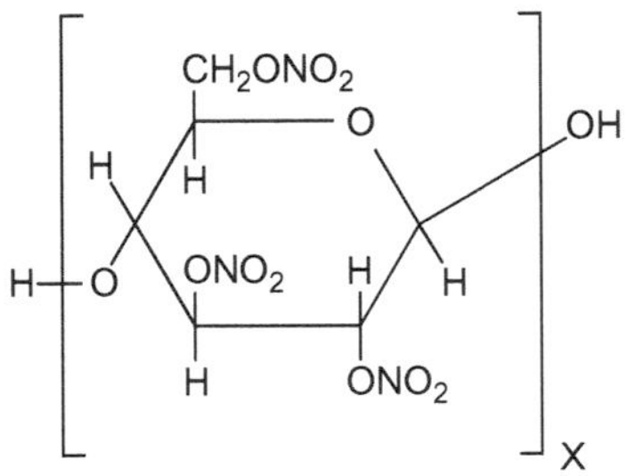

Fig. 6 Structure of cellulose nitrate.
From https://www.nfsa.gov.au/preservation/guide/research/faded-silver.

When a cellulose molecule is chemically modified by strong nitric acid, sulfuric acid, and water, the resultant product is cellulose nitrate. The structure of cellulose nitrate is shown in Fig. 6 [11,12]. Cellulose nitrate has been found useful in a variety of materials. Some of them include adhesives, explosives, film bases, lacquers and coatings, and solid plastics. Depending on the degree of nitration of the cellulose nitrate molecule used in manufacture can result in different products. In addition, by varying the ratio of cellulose nitrate to plasticizer and the varying amounts of solvent in the formulation can also lead to a different product. The level of nitration of a cellulose nitrate molecule is usually given as the percentage of nitrogen atoms per glucose residue in the molecule [11,12,14].

Once cellulose nitrate is synthesized, it is then mixed with camphor; a resin from the evergreen *Cinnamomum camphora*. The structure of camphor is shown in Fig. 7 [11,15]. Camphor is bicyclic material belonging to terpenoid family. It is white and crystalline substance with a specific gravity of 0.986–0.996. Camphor sublimes at room temperature, melts at 176°C, and boils at 209°C [12,15]. Synthetic camphor is produced by converting α-pinene from turpentine to camphene to isobornyl acetate to isoborneol to camphor. Synthetic camphor is chemically identical to natural camphor but is optically inactive (racemic) [16].

Some of the characteristics that make camphor the best solvent for cellulose nitrate are [11,12,15] as follows:

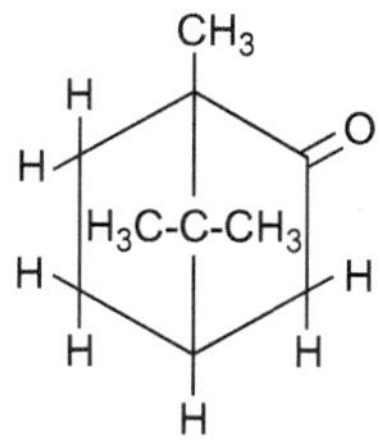

Fig. 7 Structure of camphor.
From J.A. Reilly, Celluloid objects: their chemistry and preservation, J. Am. Inst. Conserv. 30 (2) (1991) pp. 145–162. http://cool.conservation-us.org/jaic/articles/jaic30-02-003.html; http://www.ottawavalleybuttonclub.ca/Articles/tabid/544/articleType/ArticleView/articleId/409/The-History-and-Properties-of-Celluloid.aspx; Encyclopedia Americana, 1983, p. 303.

- Excellent gelatinizing agent of quick and deep penetration.
- Readily soluble in ethyl alcohol, methyl alcohol, and acetone.
- It has a high melting point (176°C) and boiling point (209°C).
- Unpleasant odor but no cumulative or deleterious effects on the makers and handlers of the celluloid.
- Volatility is extremely low.
- Solvent resistant to harsh chemical such as nitric acid, acids, and alkalis.
- Reduces the explosiveness of cellulose nitrate when mixed with it.
- Does not separate out of celluloid upon heating and cooling or upon long storage.
- Light has no effect upon it.

Colorants such as pigments and dyes can be effectively added to celluloid without much change in its properties. Some other fillers and additives include alcohol-soluble antacids for translucent celluloid. Zinc oxide and zinc carbonate are added to opaque celluloid to make it stabilized and do not require added antacid [16]. Stabilizers for translucent celluloid included aryl phenyl ureas, lactates (sodium, potassium, calcium, strontium, magnesium), and many other substances [17].

Fig. 8 shows thermal degradation process of celluloid [11]. Typical degradation process involves recrystallization of cellulose nitrate molecules and the expulsion and sublimation of camphor. This results in product getting brittle and noticeable

$$RCH_2CH_2ONO_2 + H_2O \rightarrow RCH_2CH_2OH + HNO_3$$

$$+OH \rightarrow RCHCH_2 + H_2O + NO_3$$

$$\rightarrow RCH_2CHO + HNO_3$$

Fig. 8 Thermal degradation of celluloid.
From J.A. Reilly, Celluloid objects: their chemistry and preservation, J. Am. Inst. Conserv. 30 (2) (1991) pp. 145–162. http://cool.conservation-us.org/jaic/articles/jaic30-02-003.html.

crazing in transparent or translucent products. In extreme cases such as degradation of nitrocellulose, it gives off oxides of nitrogen which in contact with water forms nitric acid. Nitric acid is very corrosive to metal, fabric, paper, etc., and often results in substantial destruction of surrounding materials. The effects are worse when the celluloid products are stored in closed containers. The fumes are more concentrated, and the by-product catalyzes further reaction, leading to further damage [11].

To prevent degradation of celluloid, proper storage of celluloid in ventilated containers without excess humidity or high temperatures is required. However, once the degradation process has started, it is impossible to reverse. Coloring agents that prevent light transmission through celluloid objects appear to give some protection.

Some of the factors that can contribute to degradation of celluloid include [11]:

- The purity of ingredient materials
- The rinsing and drying processes, which may leave agents catalyzing degradation in the finished materials
- The kneading process

4.3 Lignin

Lignin is a complex polymer comprised of aromatic alcohols known as monolignols. It is the most abundant natural polymer typically found in plant cell walls. Unlike the rest of the polymers, lignin is not composed of carbohydrate monomers. In a typical biomass like wood, lignin is approximately 17%–32% of the weight of the moisture. However, the lignin content depends on different factors such as species of wood, age of wood, etc. In soft wood, average lignin content is 28%, while in hardwood; average lignin content is 24% [18,19].

Lignin plays an important role in a series of biological events in plants. They are as follows:

1. Provides construction of plants such as grasses, bushes, and rigid trees.
2. Lignin is less hydrophilic than other biomass.
3. Lignin also acts as a barrier against insects and fungi.
4. It helps transport liquids all over the plant tissues.

Lignin has a highly complex chemical nature and its structure depends on its derivatives. The lignin structure is amorphous and the color is dark brown. Nearly all of the lignin in plants is insoluble in inert solvents. Lignin structure is primarily composed of up to three different phenyl propane monomers as shown in Fig. 9 [20]:

1. Coniferyl alcohol
2. Syringyl alcohol
3. Coumaryl alcohol

Each monomer is dominant in a typical lignin biomass. For example, coniferyl alcohol is mostly found in softwoods, syringyl in hardwoods, and coumaryl alcohol in grasses and agricultural crops. These individual units can be composed of various

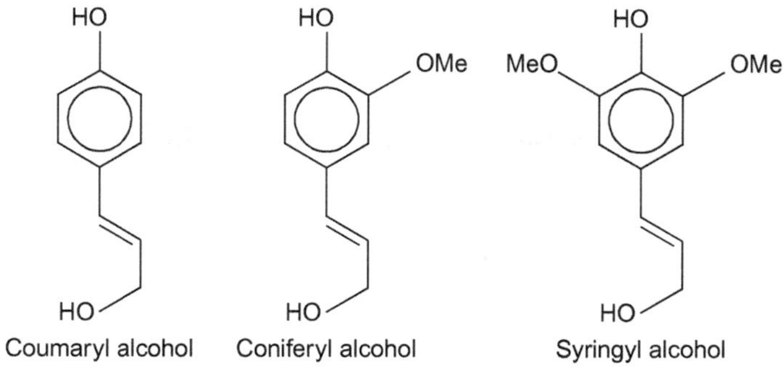

Fig. 9 Three different phenyl propane monomers of lignin.
From http://www.pulppapermill.com/lignin-structure-properties-function-and-uses/; R.H. Kaul, V. Ibrahim, Lignin-degrading enzymes: an overview, in: S.-T. Yang, H.E. Enshasy, N. Thongchul (Eds.), Bioprocessing Technologies in Biorefinery for Sustainable Production of Fuels, Chemicals, and Polymers, John Wiley & Sons, 2013 (Chapter 10).

probable bonding patterns. Unlike the cellulose which is linear in shape, lignin is highly branched or cross-linked polymer. It contains many reactive functional groups as like phenol propane unit, hydroxyl groups, methoxyl groups, and carbonyl groups. Under suitable conditions, each of these groups exhibits a characteristic reaction [20].

4.4 Hemicellulose

Hemicelluloses are a group of heterogeneous polysaccharides which are formed through biosynthetic routes different from that of cellulose. In contrast to cellulose which is a homopolysaccharide, hemicelluloses are heteropolysaccharides. Like cellulose most hemicelluloses function as supporting material in the cell walls. Hemicelluloses are relatively easily hydrolyzed by acids to their monomeric components consisting of D-glucose, D-mannose, D-galactose, D-xylose, L-arabinose, and small amounts of L-rhamnose in addition to D-glucuronic acid, 4-*O*-methyl-D-glucuronic acid, and D-galacturonic acid [21,22].

Due to its amorphous structure, hemicellulose has higher reactivity than that of cellulose. It is easily decomposed into monomer components, such as manose, xylose, and some glucuronic acid. When hemicellulose is exposed to acids, the glycosidic bonds get cleaved, resulting in hemicellulose degradation.

Hemicellulose has a complex structure and is composed of five sugar units: (1) xylose, (2) mannose, (3) galactose, (4) rhamnose, and (5) arabinose. The structure of these units is shown in Fig. 10 [23,24].

The monomeric sugar units together form five basic structural units of hemicellulose.

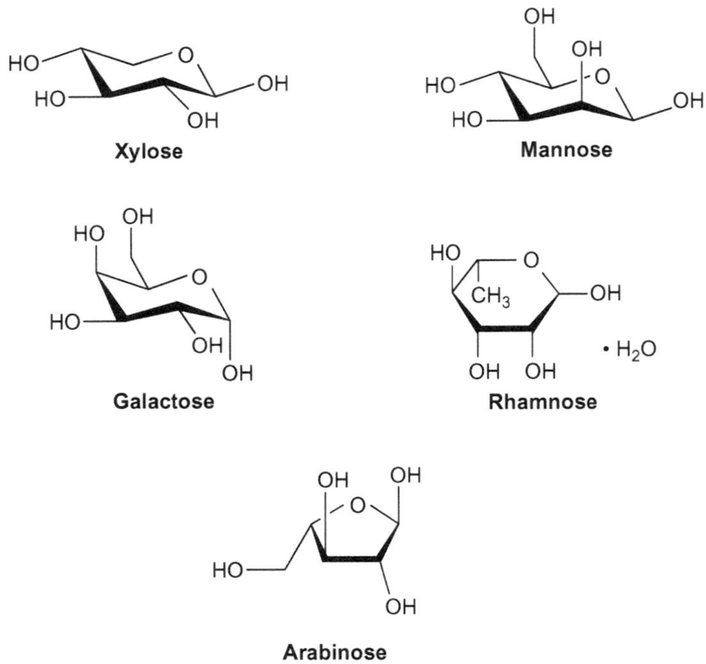

Fig. 10 Monomer sugars found in hemicellulose.
From http://pubs.rsc.org/en/content/articlehtml/2013/gc/c2gc36364j; http://www.sigmaaldrich.com/catalog/product/sigma/r3875?lang=en&region=U.

1. Glucuronoxylan
2. Galactoglucomannan
3. Arabinoglucuronoxylan
4. Xyloglucan
5. Arabinoxylan

1. *Glucuronoxylan*: Also known as *O*-acetyl-4-*O*-methylglucuronoxylan, it is the most common chemical structure in softwood hemicellulose (Fig. 11) [25]. It constitutes to approximately 15%–30% of hemicellulose. Glucuronoxylan consists of β-D-xylose backbones linked through β-1,4-glycosidic bonds. Uronic acid occurs in every one-tenth of xylose since C_2 and C_3 atoms in xylose are susceptible to acetylation. Side chains are relatively strong and hard to dissolve in acid conditions [26].
2. *Galactoglucomannan*: It is the typical xylose structure found in hardwood hemicellulose (Fig. 12) [23]. It constitutes to approximately 20%–25% of hardwood hemicellulose; however, its content is less than 5% in softwood hemicellulose. It is mainly a chain skeleton structure consisting of β-D-xylose and β-D-rhamnose units connected to β-1,4-glycosidic bonds. Acetyl content in galactoglucomannan is around 6% and three to four hexose units correspond to one acetyl on average [26].

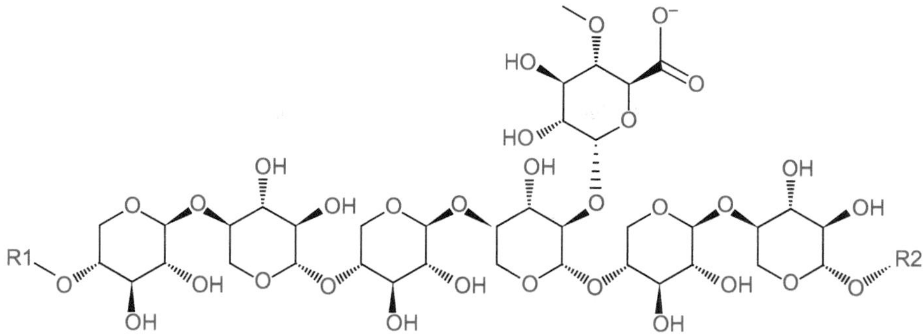

Fig. 11 Structure of glucuronoxylan.
From http://vm-trypanocyc.toulouse.inra.fr/LEISH/NEW-IMAGE?object=Glucuronoxylans.

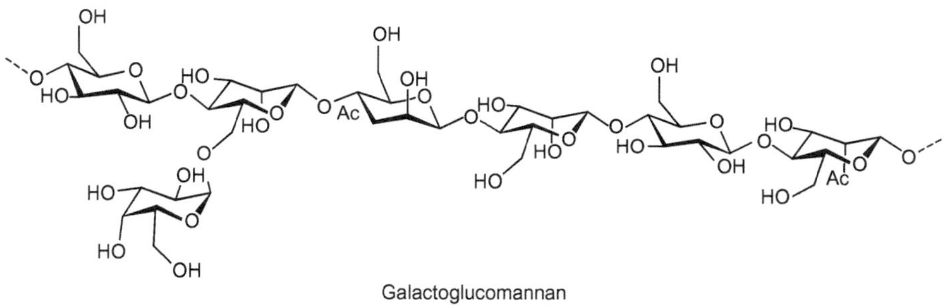

Fig. 12 Structure of galactoglucomannan.
From http://pubs.rsc.org/en/content/articlehtml/2013/gc/c2gc36364j.

3. *Arabinoglucuronoxylan*: It is the typical structure found in nonwood hemicellulose such as agricultural biomass (Fig. 13) [27]. It also exists in small amounts in soft wood. It mainly consists of a linear β-1,4-D-xylose skeleton structure, including 4-*O*-methyl-α-D-pyran glucuronic acid linked by α-1,2-glycosidic bonds and α-L-arabinofuranose linked by α-1,3-glycosidic bonds [26].
4. *Xyloglucan*: These structures are largely found in hardwood plants such as dicotyledons and monocotyledons. As shown in Fig. 14 [28], the structure of xyloglucan primarily consists of a D-glucose skeleton linked by β-1,4 glycosidic bonds, including D-xylose, L-arabinose, and D-galactose units [26].

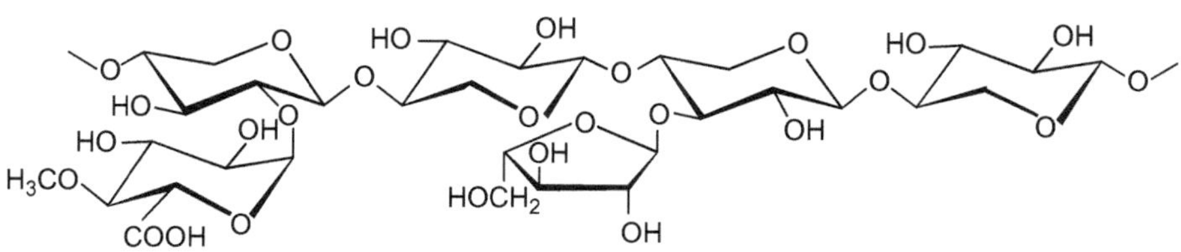

Fig. 13 Structure of arabinoglucuronoxylan.
From http://www.mdpi.com/1999-4907/2/4/929/htm.

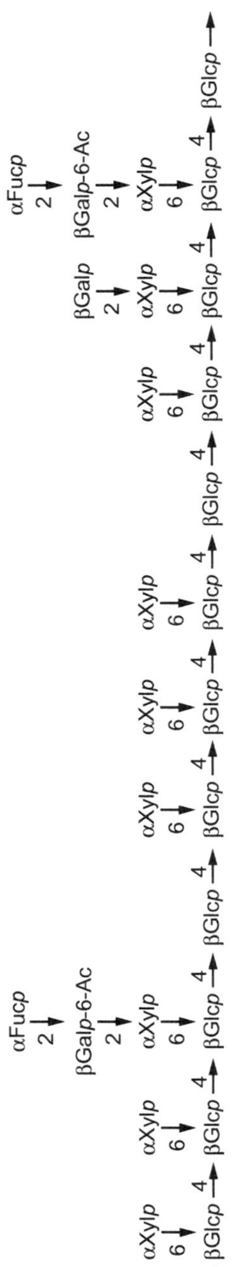

Fig. 14 Structure of xyloglucan.
From https://www.researchgate.net/figure/227969668_fig5_Figure-4-Xyloglucan-structure-depicting-a-b-1-4-d-glucosyl-backbone-bGlcp-with.

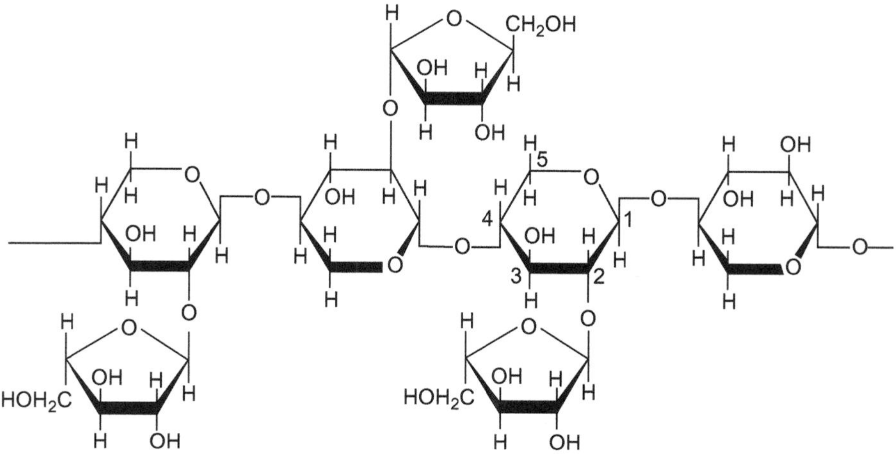

Fig. 15 Structure of arabinoxylan.
From https:/www.researchgate.net/figure/257561875_fig7_Figure-12-Molecular-structure-of-arabinoxylan.

5. *Arabinoxylan*: These structures are found in cereal hemicellulose, and in hardwood with higher L-arabinose content. In typical arabinoxylan structure (Fig. 15) [29], the linear β-1,4-D-xylopyranose is replaced with an α-L-arabinofuranose unit at 2-*O* or and α-D-pyran glucuronic acid unit at 3-*O* position. During the lignification process, there is a tendency for xylan to connect to the lignin structure through esterification of aldehyde acid side chains [26].

4.5 Lignocellulosic

Lignocellulosic is a biomass which is made up of three main components: (1) cellulose, (2) lignin, and (3) hemicellulose. With all the three components, lignocellulosic is a complex structure. As shown in Fig. 16 [30,31], the cellulose is surrounded by a monolayer of hemicellulose and embedded in a matrix of hemicellulose and lignin. Furthermore, lignin specifically creates a barrier to enzymatic attack while the highly crystalline structure of cellulose is insoluble in water while the hemicellulose and lignin create a protective sheath around the cellulose [32–35].

Table 4 summarizes different types of lignocellulosic biomass and their chemical composition [36].

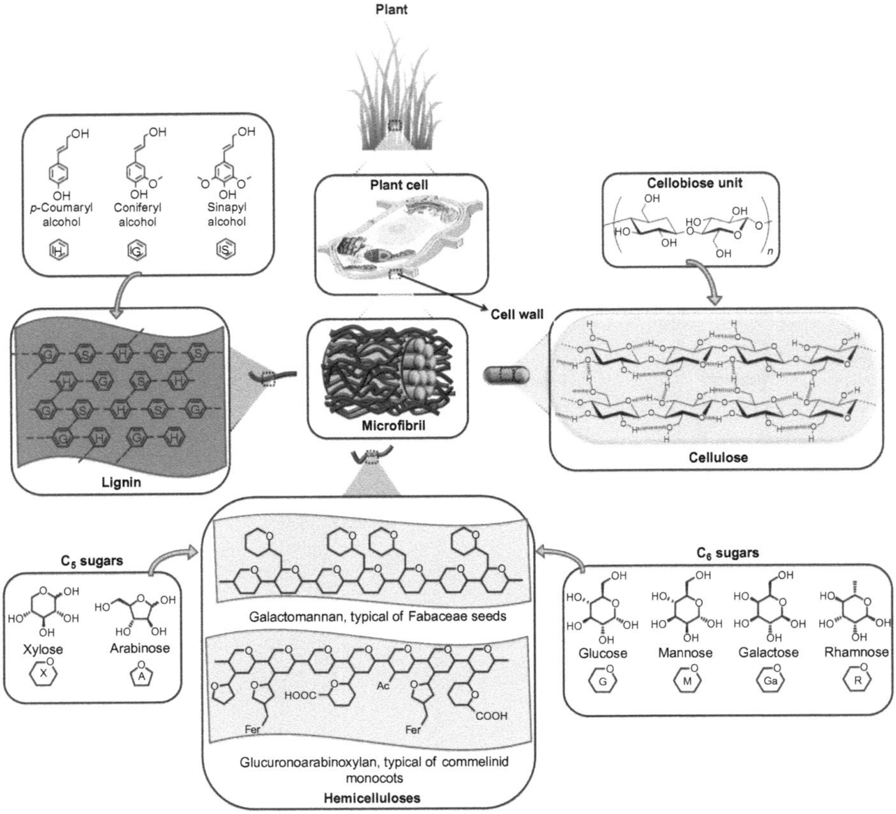

Fig. 16 Formation of lignocellulosic.
From http:/pubs.rsc.org/en/content/articlehtml/2015/py/c5py00263j; H.V. Scheller, P. Ulvskov, Annu. Rev. Plant Biol. 61 (2010) 263–289; https:/www.customgrocerybags.com/blog/cook-county-recently-passed-a-construction-and-demolition-recycling-ordinance/.

Table 4 Composition of lignocellulose

Lignocellulosic material	Cellulose (%)	Hemicellulose (%)	Lignin (%)
Hardwood stems	40–55	24–40	18–25
Softwood stems	45–50	25–35	25–35
Nut shells	25–30	25–30	30–40
Corn cobs	45	35	15
Grasses	25–40	35–50	10–30
Paper	85–99	0	0–15
Wheat straw	30	50	15
Sorted refuse	60	20	20
Leaves	15–20	80–85	0
Cotton seed hairs	80–95	5–20	0
Newspaper	40–55	25–40	18–30
Waste paper from chemical pulps	60–70	10–20	5–10
Solid cattle manure	1.6–4.7	1.4–3.3	2.7–5.7
Switch grass	45	31.4	12

From http:/www.katzen.com/ethanol101/Lignocellulosic%20Biomass.pdf.

References

[1] Eastman cellulose esters, technical literature. http://www.eastman.com/Literature_Center/P/PCI105.pdf.
[2] D.L. Kaplan, Biopolymers from Renewable Resources, Springer, New York, NY, USA, 1998. p. 80.
[3] O. Olabisi, K. Adewale, Handbook of Thermoplastics, Marcel Dekker, New York, NY, USA, 1997. p. 340.
[4] E. Lokensgard, Industrial Plastics: Theory and Applications, fifth ed., Delmar Cengage Learning, Boston, MA, USA, 2010. p. 439.
[5] Cellulose acetate. http://wpage.unina.it/avitabil/testi/AcetilCellulosa.pdf.
[6] Manufacturing process of acetate rayon. http://textilelibrary.weebly.com/acetate-rayon-mfg-process.html.
[7] M.M.L. Loo, R. Hashim, C.P. Leh, Recycling of valueless paper dust to a low grade cellulose acetate: effect of pretreatments on acetylation. https://www.ncsu.edu/bioresources/BioRes_07/BioRes_07_1_1068_Loo_HP_Recycling_Paper_Dust_Cellulose_Acetate_Pretreat_2309.pdf.
[8] Cellulose acetate butyrate, Sigma-Aldrich. http://www.sigmaaldrich.com/catalog/product/aldrich/419044?lang=en&region=US.
[9] Cellulose acetate propionate, Sigma-Aldrich. http://www.sigmaaldrich.com/catalog/product/aldrich/330183?lang=en&region=US.
[10] J.E. Boyd, Celluloid: the eternal substitute. http://www.chemheritage.org/discover/media/magazine/articles/29-3-celluloid-the-eternal-substitute.aspx.
[11] J.A. Reilly, Celluloid objects: their chemistry and preservation, J. Am. Inst. Conserv. 30 (2) (1991) 145–162. http://cool.conservation-us.org/jaic/articles/jaic30-02-003.html.
[12] S. Smith, The history and properties of celluloid. http://www.ottawavalleybuttonclub.ca/Articles/tabid/544/articleType/ArticleView/articleId/409/The-History-and-Properties-of-Celluloid.aspx.
[13] http://www.easychem.com.au/production-of-materials/biomass-research/cellulose.
[14] https://www.nfsa.gov.au/preservation/guide/research/faded-silver.
[15] Encyclopedia Americana, 1983, p. 303.
[16] G.S. Brady, H.R. Clauser, Materials Handbook, 11th ed., McGraw-Hill, New York, 1977.
[17] Worden, Nitrocellulose Industry, vols. 1 & 2, Van Nostrand, New York, 1911.
[18] http://www.pulppapermill.com/lignin-structure-properties-function-and-uses/.
[19] A. Sakakibara, Y. Sano, Chemistry of lignin, in: D.N.-S. Hon, N. Shiraishi (Eds.), Wood and Cellulosic Chemistry, second ed., CRC Press, New York, NY, USA, 2001, p. 109 (Chapter 4).
[20] R.H. Kaul, V. Ibrahim, Lignin-degrading enzymes: an overview, in: S.-T. Yang, H. E. Enshasy, N. Thongchul (Eds.), Bioprocessing Technologies in Biorefinery for Sustainable Production of Fuels, Chemicals, and Polymers, John Wiley & Sons, Hoboken, NJ, USA, 2013 (Chapter 10).
[21] E. Sjostrom, Wood polysaccharides, in: Wood Chemistry: Fundamentals and Applications, second ed., Academic Press, London, UK, 1993, p. 63 (Chapter 3).
[22] S. Sungsoo-Cho, L. Prosky, M. Dreher, Sources and uses of fiber, in: Complex Carbohydrates in Foods, Marcel Dekker, New York, NY, USA, 1999, p. 355 (Chapter 24).
[23] http://pubs.rsc.org/en/content/articlehtml/2013/gc/c2gc36364j.
[24] http://www.sigmaaldrich.com/catalog/product/sigma/r3875?lang=en&region=U.
[25] http://vm-trypanocyc.toulouse.inra.fr/LEISH/NEW-IMAGE?object=Glucuronoxylans.

[26] S. Wang, Z. Luo, Biomass components and characteristics, in: Green Alternative Energy Resources, Pyrolysis of Biomass, vol. 1, Walter de Gruyter GmbH, Berlin, Germany, 2017, p. 11 (Chapter 1).
[27] http://www.mdpi.com/1999-4907/2/4/929/htm.
[28] https://www.researchgate.net/figure/227969668_fig5_Figure-4-Xyloglucan-structure-depicting-a-b-1-4-d-glucosyl-backbone-bGlcp-with.
[29] https://www.researchgate.net/figure/257561875_fig7_Figure-12-Molecular-structure-of-arabinoxylan.
[30] http://pubs.rsc.org/en/content/articlehtml/2015/py/c5py00263j.
[31] H.V. Scheller, P. Ulvskov, Hemicelluloses, Annu. Rev. Plant Biol. 61 (2010) 263–289.
[32] https://lignofuel.wordpress.com/category/lignocellulosic-structure/.
[33] A. Barakat, H. de Vries, X. Rouau, Dry fractionation as an important step in current and future lignocellulose biorefineries: a review, Bioresour. Technol. 134 (2013) 362–373.
[34] V.B. Agbor, N. Cicek, R. Sparling, A. Berlin, D.B. Levin, Biomass pretreatment: fundamentals toward application, Biotechnol. Adv. 29 (2011) 675–685.
[35] S. Morais, E. Morag, Y. Barak, D. Goldman, Y. Hadar, R. Lamed, Y. Shoham, D. B. Wilson, E.A. Bayer, Deconstruction of lignocellulose into soluble sugars by native and designer cellulosomes, mBio 3 (6) (2012) 508–512.
[36] http://www.katzen.com/ethanol101/Lignocellulosic%20Biomass.pdf.

Biomass conversion approaches

5

5.1 Introduction

Biomass resources are derived from three sources—primary, secondary, and tertiary. Biomass is termed as primary when it is directly produced by photosynthesis. Some examples are included in Fig. 1. Perennial short rotation woody crops, herbaceous crops, and residues resulting from the harvesting of agricultural crops and forest trees are some examples of primary biomass. The biomass that is generated from the processing of primary biomass is termed as secondary. The processing can be physical, chemical, or biological. They include sawdust generated from the milling operation, black liquor from pulping processes, and manure production by animals. All the postconsumer residue streams such as used vegetable oils, packaging wastes, and construction and demolition debris are considered tertiary biomass [1].

Biofuels can be divided into three types:

1. Solid
2. Liquid
3. Gaseous fuels

Solid and gaseous fuels are typically used in producing electrical power that directly or indirectly support turbine-equipped power plants. Liquid fuels can be used directly in the existing transportation such as road, railroad, and aviation as well as in engine and turbine electrical power generators.

5.2 Classes of raw biomass

The raw materials for bio-based industrial production are supplied by plants as plant parts, separated components, and fermentable sugars [2]. Raw biomass is broadly classified into five types:

1. Wood
2. Energy crops
3. Agricultural residues
4. Food wastes
5. Industrial wastes

5.2.1 Wood

Wood biomass is a renewable source of energy and is the most important source of energy for large portion of households. Typically, four forms of wood are used shown in Fig. 2 [3–5]:

1. Firewood
2. Logwood

Technology and Applications of Polymers Derived from Biomass. https://doi.org/10.1016/B978-0-323-51115-5.00005-0

Fig. 1 (A) Primary biomass resource, (B) secondary biomass resource, and (C) tertiary biomass resource.
From (A) https://www.nrs.fs.fed.us/disturbance/market_induced_change/biofuels_bioenergy_bioproducts; (B) http://articles.extension.org/pages/70355/wood-processing-residues; (C) https://en.wikipedia.org/wiki/Biomass#/media/File:Rotor_stump_remover_recover_a_tree_roots_for_a_biomass_central..JPG.

3. Wood chips
4. Wood pellets

5.2.2 Energy crops

Nonfood crops that provide an additional potential source of feedstock for the production of second-generation biofuels are termed as energy crops. Examples of first-generation energy crops include corn and soybeans while grassy and woody energy crops are considered as second-generation energy crops.

Two main grassy energy crops include switchgrass and Miscanthus. Switchgrass is the most common as it requires relatively low water and nutrients. Miscanthus is a grass typically found in Asia and is gaining popularity in Europe as a potential feedstock for second-generation biofuel production. Poplar, willow, and eucalyptus fall into woody energy crops category. These energy crops are becoming popular because of the low level of input required when compared with annual crops [6].

Fig. 2 Examples of wood biomass: (A) firewood, (B) logwood, (C) wood chips, and (D) wood pellets.
From (A) https://en.wikipedia.org/wiki/Firewood#/media/File:%C5%A0palek_na_%C5%A1t%C3%ADp%C3%A1n%C3%AD.jpg; (B) https://en.wikipedia.org/wiki/Wood#/media/File:Gelugu_(coconut_wood)_in_Klaten,_Java.jpg (C) https://commons.wikimedia.org/wiki/File:Wood_chips.JPG; (D) https://commons.wikimedia.org/wiki/File:Wood_pellets-small_huddle_PNr%C2%B00108.jpg.

Some of the energy crops that are grown in different parts of the world as shown in Table 1 are [7] as follows:

- Short rotation coppice (SRC)
- Sorghum
- Miscanthus
- Szarvasi grass
- *Silphium perfoliatum*
- Igniscum
- Vetch-Rye

Short rotation coppice (SRC) is a densely planted, high-yielding variety of either willow or poplar as shown in Fig. 3. Although, it is common practice to harvest SRC every 3 years however, harvest period can vary between 2 and 5 years. SRC is a woody, perennial crop, the rootstock or stools remaining in the ground after harvest with new shoots emerging the following spring. A plantation could last up to 30 years before replanting becomes necessary, although this depends on the productivity of the stools [8].

Table 1 **List of energy crops [7]**

Crop type	Sorghum	Miscanthus	Szarvasi grass	*Silphium perfoliatum*	Igniscum	Vetch-Rye
Origin	Africa	East Asia	Hungary	North America	Japan	Europe
Other names	Kaffir corn, Sudan grass, sugar sorghum	Chinese silver grass, silk grass	Tall wheat grass	Compass plant or cup plant		
Annual or multiyear crop	Annual	Multiyear. Can remain on one field for up to 20 years	Multiyear	Multiyear. Land use possible over a period of at least 10 years	Multiyear	Annual

From http://www.claas.co.uk/fascination-claas/special-topics/biomass/energy-crops.

Fig. 3 Short rotation coppice.
From https://en.wikipedia.org/wiki/Short_rotation_coppice#/media/File:Coppice.jpg.

Fig. 4 Miscanthus.
From http://www.newenergyfarms.com/crops/miscanthus/.

Miscanthus also referred to as elephant grass is a high yielding energy crop that resembles bamboo. It grows over 3 m tall and produces a crop every year without the need for replanting (Fig. 4). Miscanthus crops are woody, perennial, rhizomatous grasses, which can be harvested annually for at least 15 years [9]. Miscanthus is becoming popular for the use as biofuel because of its rapid growth, low mineral content, and high biomass yield [10].

S. perfoliatum, commonly referred to as cup plant, is a coarse, sunflower-like-plant and typically occurs in low woods and thickets, meadows, and prairie stream/pond peripheries (Fig. 5) [7]. It grows on tough, erect stems to 4–8′ tall and is distinguished from the other Silphiums by its square/quadrangular stems and pairs of cup-forming leaves. Flower heads (to 3″ diameter) featuring light yellow rays (20–40) and darker yellow center disks bloom in summer on the upper part of the plant [11,12].

Fig. 5 *Silphium perfoliatum.*
From https://en.wikipedia.org/wiki/Silphium_perfoliatum#/media/File:Cupplants.jpg.

Fig. 6 Sorghum.
From https://en.wikipedia.org/wiki/Sorghum#/media/File:Sorghum.jpg.

Sorghum is a C4 plant shown in Fig. 6 with the following interesting characteristics [13]:

(i) Growth cycle is short (about 4 months) facilitating double cropping
(ii) Can be easily grown from seeds
(iii) Production can be completely mechanized
(iv) Produce sugar in the stalk and starch in the grain
(v) High water and nutrient use efficiency
(vi) Bagasse produced from sweet sorghum has high biological value when used as forage
(vii) Wide adaptability to different environments

Szarvasi grass known by tall wheatgrass is part of Pontic-Mediterranean grass species. There are two subspecies of Szarvasi:

1. *Elymus elongates* Runemark subsp. *elongates*
2. *E. elongates* Runemark subsp. *ponticus*

(A) (B)

Fig. 7 (A) Szarvasi grass, and (B) the roots of *E. elongates* subsp. *ponticus* and Szarvasi grass. From S. Csete, Tall wheatgrass cultivar Szarvasi-1 (*Elymus elongates* subsp. *ponticus* cv. Szarvasi-1) as a potential energy crop for semi-arid lands of eastern Europe, Sustainable Growth and Applications in Renewable Energy Sources, InTech, 2011, pp. 270–273; https://www.dreamstime.com/stock-photo-energy-plant-szarvasi-grass-renewable-resource-multiple-purpose-image31223530.

E. elongates Runemark subsp. *elongates* is shorter and more fragile while *E. elongates* Runemark subsp. *ponticus* is taller and more robust. Originally, Szarvasi grass was bred with *E. elongates* subsp. *ponticus* populations from Hungary to be drought tolerant. The breeding process was conducted in Szarvas, East Hungary [14]. Fig. 7A and B shows Szarvasi grass and root system of *E. elongates* subsp. *ponticus* and Szarvasi grass [14,15].

Igniscum is part of Sakhalin Knotweed (*Fallopia sachalinensis*, Fam. Polygonaceae) family [16]. As shown in Fig. 8, Igniscum is an herbaceous perennial plant that can grow up to 4 m tall. The hollow stems are reddish-brown at the base fading to

Fig. 8 Igniscum.
From http://www.desertconsult.de/projects/igniscum.html.

light green higher up, with nodes along the length and some branching. The leaves are between 15 and 40 cm long, heart shaped at the stem end, with a blunt tip and wavy edges, arranged opposite to one another.

5.2.3 Agricultural residues

Agricultural wastes are the residues typically generated from cereals, cotton, groundnut, jute, legumes, coffee, tea, fruits, and palm oil. Some other residues include straw, stem, leaves, husk, shell, peel, and pulp. Some of the examples are shown in Fig. 9 [17–28].

An example for this kind of residue includes rice straw and rice husks produced at the processing plant that is easily converted into energy. In the case of maize, biomass remain in the fields while sugarcane harvesting leads to harvest residues in the fields

Rice straws

Rice husks

Maize cobs

Maize husk

Fig. 9 Examples of agricultural residues.
From https://commons.wikimedia.org/wiki/File:Straw_of_the_rice.08Oct9.jpg; http://irri.org/rice-today/a-second-life-for-rice-husk; https://commons.wikimedia.org/wiki/File:YellowCorn.jpg; https://commons.wikimedia.org/wiki/File:E8025-Milyanfan-corn-huskers.jpg; https://commons.wikimedia.org/wiki/File:Embrapa_news_sp_of_cassava_(Jos%C3%A9_Cruz_ABr)_25jan2008.jpg; http://en.agroghana.com; http://www.wikihow.com/Roast-Peanuts; http://nyp-corp.com/blog/tag/jute; Texas A&M AgriLife Extension Service economist says large cotton supplies, weather will weigh heavily in 2013, http://agrilifecdn.tamu.edu/wp-content/uploads/8241854144_65bd4780a6_b.jpeg; http://www.kingsagriseeds.com/minimize-waste-with-stover-and-straw; http://www.agrimarketing.com/show_story.php?id=57169; https://en.wikipedia.org/wiki/Bagasse#/media/File:Bagasse_in_Hainan_-_02.jpg.

(Continued)

Cassava stalks

Cassava peels

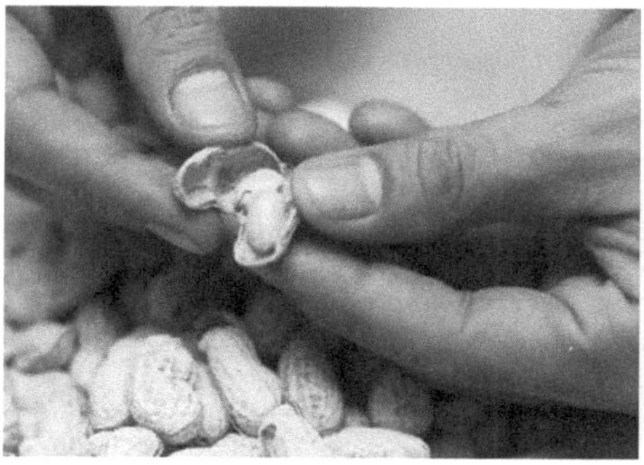

Groundnut shells

Jute stalk

Cotton stalks

Soybean stubble

Soybean pods

Sugarcane bagasse

Fig. 9—continued

while processing produces fibrous bagasse, both of which are good sources of energy. All these lignocellulosic materials can be converted into biofuels by a wide range of technologies [6].

5.2.4 Food wastes

A new emerging area of untapped energy is food waste that ends up in landfills. The food waste starts to rot and thereby release greenhouse gases into the atmosphere. Another area is the municipal waste stream which has seen gradual increase in food waste. This is pushing for proper food waste management strategy that is environment friendly and leads to sustainable disposal. Food waste is difficult to treat or recycle since it contains high levels of sodium salt and moisture and is mixed with other waste during collection. Some of the leaders involved in developing new advanced biogas technologies include Sweden, Austria, Denmark, Germany, and England [29].

Some of the major generators of food wastes are (Fig. 10) [30–34] as follows:

1. Hotels
2. Restaurants
3. Supermarkets
4. Residential blocks
5. Cafeterias
6. Airline caterers

Fig. 10 Examples of food waste generated by hotels, restaurants, supermarkets, residential areas, cafeteria, and airline caterers.
From https://commons.wikimedia.org/wiki/File:GI_Market_food_waste.jpg; https://en.wikipedia.org/wiki/Food_waste; https://upload.wikimedia.org/wikipedia/commons/5/5b/Recovering_wasted_food.JPG; https://en.wikipedia.org/wiki/Food_waste#/media/File:Not_avoidable_food_waste.jpg.

Fig. 11 Black liquor from pulp and paper processing.
From https://en.wikipedia.org/wiki/Environmental_impact_of_paper#/media/File:PulpAndPaperMill.jpg; https://en.wikipedia.org/wiki/Black_liquor#/media/File:Svartlut_76.jpg.

5.2.5 Industrial waste

Industrial wastes are residues and by-products generated from processing industries and can be used as biomass energy sources. Some of the examples include waste that is generated from:

1. Meat production
2. Confectionery production
3. Pulp and fiber from sugar and starch extraction
4. Filter sludges
5. Coffee grounding process
6. Pulp and paper processing

The industrial wastewater discharged is highly heterogeneous containing compounds from wood or other raw materials, processed chemicals, as well as compound formed during processing. Black liquor is a by-product that is generated from pulp and paper processing (Fig. 11) [35–37].

5.3 Conversion of biomass into energy

There are four fundamental forms of biomass energy conversion technologies:

1. Photosynthesis
2. Direct combustion process
   a. Cofiring
   b. Cogeneration
   c. Direct-fired gas turbine technology
3. Thermochemical process
   a. Pyrolysis
   b. Carbonization
   c. Gasification
   d. Catalytic liquefaction

4. Biochemical process
   a. Anaerobic fermentation
   b. Methane production in landfills
   c. Ethanol fermentation
   d. Biodiesel

5.3.1 Photosynthesis

A typical biomass generally stores energy absorbed from the sun. Energy from the sun is absorbed by plants which converts carbon dioxide (CO_2) and water into chemical energy such as glucose as shown in Eq. (1). This conversion of radiant energy into chemical energy happens through a process known as photosynthesis [38].

$$6CO_2 + 6H_2O \xrightarrow[\text{(Sunlight)}]{\text{Radiant energy}} C_6H_{12}O_6 + 6O_2 \tag{1}$$

Fig. 12 shows a typical process of converting biomass into energy using photosynthesis [39]. The chemical energy generated is in the form of heat. The biomass can either be burned directly or converted into useable forms of biogas such as methane gas and transportation fuels like ethanol and biodiesel [38]. Methane gas is a component of landfill gas or biogas that forms when garbage, agricultural waste, and human waste decompose in landfills or in special containers called digesters [38]. Crops such as corn and sugarcane are fermented to produce fuel ethanol for use in vehicles. Biodiesel, another transportation fuel, is produced from vegetable oils and animal fats [38].

5.3.2 Direct combustion process

Direct combustion is a process in which a biomass is converted into heat. The biomass fuel is burned in excess air to produce heat. During the first stage of combustion, combustible gases are generated and burns as flames [40].

Furnaces for direct combustion are classified into two broad categories:

1. Two-stage combustion process—Dutch ovens, fuel cell furnaces
2. Suspension and fluidized bed furnace

The first category includes a two-stage combustion process. Examples include some ovens and furnaces such as Dutch ovens, spreader stoker, and fuel cell furnaces that utilize two-stage combustion process. During the first stage, drying and partial gasification happens during the first stage followed by complete combustion during the second stage. Some of the more advanced versions use rotating or vibrating grates to facilitate ash removal, with some requiring water cooling [41].

The second category includes suspension and fluidized bed furnaces which are generally used with fine particle biomass feedstocks and liquids. In suspension furnaces, the particles are burnt while being kept in suspension by the injection of turbulent preheated air which may already have the biomass particles mixed in it. In fluidized bed furnaces, bed of sand is preheated to temperatures of 500–900°C providing the combustion medium, into which the biomass fuel is either dropped or injected if

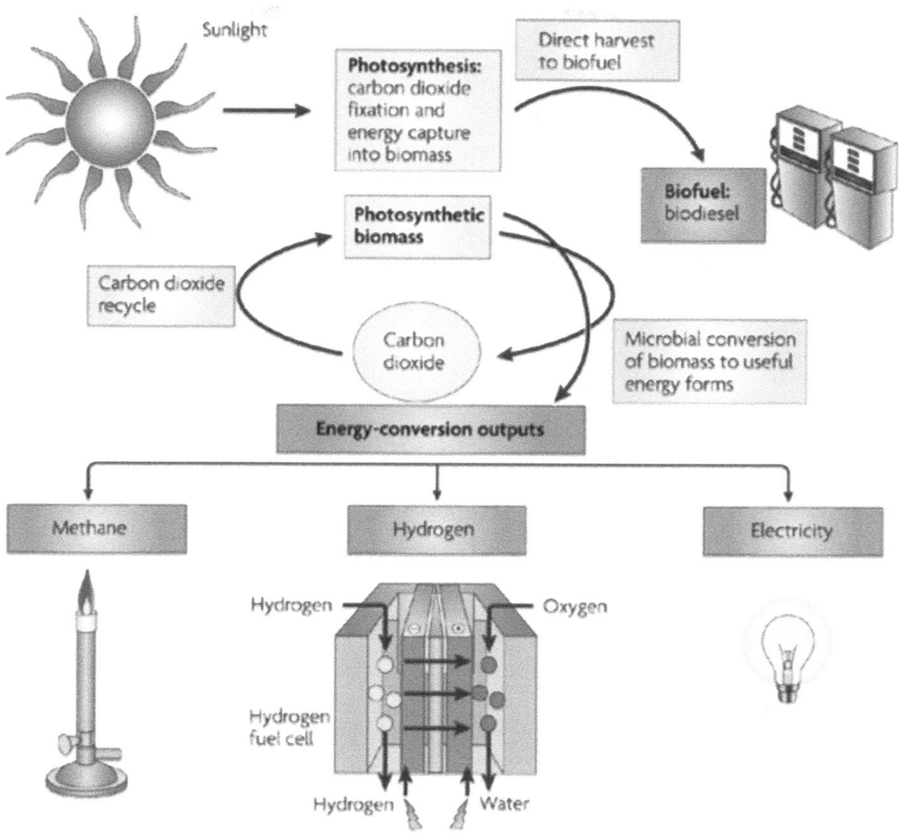

Fig. 12 An example showing a typical process of converting biomass into energy using photosynthesis.
From http://www.eia.gov/energyexplained/?page=biomass_home.

particulate or fluid. These furnaces require methods of preheating the air or sand and may need water cooling for less bulky biomass feedstocks and liquid [41].

5.3.2.1 *Cofiring*

Cofiring is a very common approach that is used for converting organic matter into energy. In a typical cofiring process, wood, waste, and other plant matter are burned. The chemical energy that is released is converted in the form of heat.

Cofiring processes can be classified into three types:

1. Direct cofiring,
2. Indirect cofiring,
3. Parallel cofiring.

Fig. 13 shows schematics for a typical cofiring process [42].

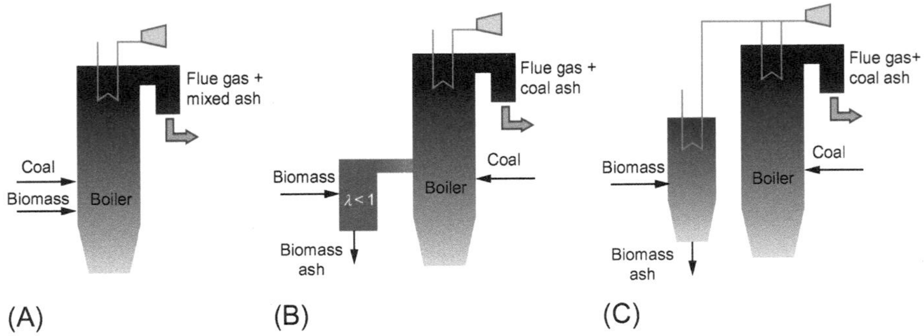

Fig. 13 A typical schematics for (A) direct cofiring, (B) indirect cofiring, and (C) parallel cofiring processes.
From E. Karampinis, P. Grammelis, M. Agraniotis, I. Violidakis, E. Kakaras, Co-firing of biomass with coal in thermal power plants: technology schemes, impacts, and future perspectives, in: P. Lund, J.A. Byrne, G. Berndes, Advances in Bioenergy: The Sustainability Challenge, Wiley Publishers, 2016, p. 235, ISBN: 978-1-11895-787-5.

1. Direct cofiring
   This type of cofiring process is the simplest, cheapest, and most common. It can be implemented in three different ways [43]:
   a. Mix biofuel with coal and then the mixed fuel is processed through integrated coal milling and firing operation.
   b. The second way involves separate handling and injection of the biofuel into the pulverized fuel pipe system before the burners.
   c. The third option involves the separate handling of biofuel with combustion through dedicated burners.
2. Indirect cofiring
   This type of cofiring process is less commonly used approach. Here, the fuel generated by gasification of solid biomass is burned with coal in the same furnace. Some of its other characteristics are [43,44] as follows:
   a. Requires the use of additional equipment such as gasifier making it more expensive than the direct cofiring process.
   b. Process allows for greater variety and higher percentages of biomass to be used.
3. Parallel cofiring
   This type of cofiring process is the most expensive in comparison with direct and indirect cofiring processes. A significantly higher capital investment is needed. Installation of separate combustion and boiler units to supply steam to the same steam cycle is required. Parallel cofiring process allows for a greater variety and higher percentages of biomass to be used [44].

5.3.2.2 Cogeneration

Cogeneration also termed as combined heat and power (CHP) is a process where heat and power are obtained from the same fuel at the same time. A generalized CHP system is shown in Fig. 14 [45].

Variety of fuels can be used for cogeneration such as bagasse, natural gas, coal, and biomass. Power and heat generated through CHP are considered most efficient and

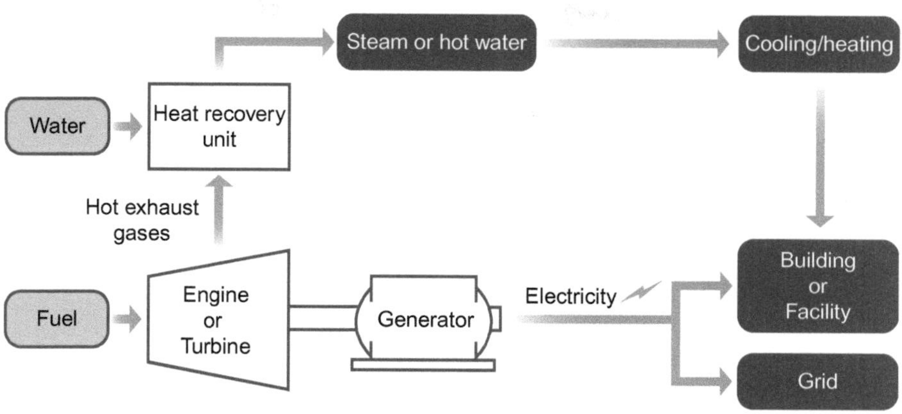

Fig. 14 A generalized CHP system.
From The basics of combined heat and power, The News Magazine, 19 Oct 2015. http://www.achrnews.com/articles/130800-the-basics-of-combined-heat-and-power and from the U.S. EPA, via the EPA Combined Heat and Power Partnership (www.epa.gov/chp).

beneficial especially when biomass fuels are used. Installation of CHP that runs biomass fuel consists of three major components [46]:

1. Preparation of biomass receiving and feedstock.
2. Energy conversion of the biomass into steam for direct combustion systems or into biogas for the gasification systems.
3. Conversion of the steam or syngas or biogas into electric power and process steam or hot water

Combined heat and power system has advantages and limitations. Both are discussed in Table 2 [47,48]:

5.3.2.3 Direct-fired gas turbine technology

Direct-fired gas turbine technology is used for converting biomass into energy in applications that only require up to 5 MW of energy. The preferred fuel for this process is low-ash, debarked wood particles that are less than 0.3 cm long and contains moisture less than 15%. Other types of processed biomass can also be used [49]. An efficiency greater than 70% can be achieved by using this process [49,50].

A typical direct-fired gas turbine setup is shown in Fig. 15 [50].

Following are the key process steps [49,50]:

- The compressor provides pressurized combustion air to the pressurized combustion chamber where biomass is burned. Pressurized combustion chamber can operate at turbine required pressures.
- Cyclone separator transfers hot combustion gases from the combustion chamber to the hot section of the turbine.
- Flue gas that is generated from the turbine can be utilized in two ways: (1) directly as a source of thermal energy or (2) fed into a heat recovery steam generator to produce process steam.

Table 2 **Advantages and limitations of combined heat and power system (CHP)**

Advantages	Limitations
• Efficiency o Requires less fuel to produce an energy output o Recovers wasted heat, achieve 60%–80% total system efficiencies o Reduces transmission and distribution losses • Environmental o Greenhouse gas emission and other air pollutants are significantly reduced • Economic o Reduces energy costs o Reduces capital costs o Guaranteed revenue • Reliability o Reduced dependency on electric grid	• CHP is not an actual energy source, only a means of extending energy • Only suitable when both electricity and hot water are available • Capital intensive

From R. P. Siegel, "Combined Heat and Power: Pros and Cons", Apr 30, 2012. http://www.triplepundit.com/special/energy-options-pros-and-cons/combined-heat-power-pros-cons; Catalog of CHP Technologies, US Environmental Protection Agency. https://www.epa.gov/chp/catalog-chp-technologies.

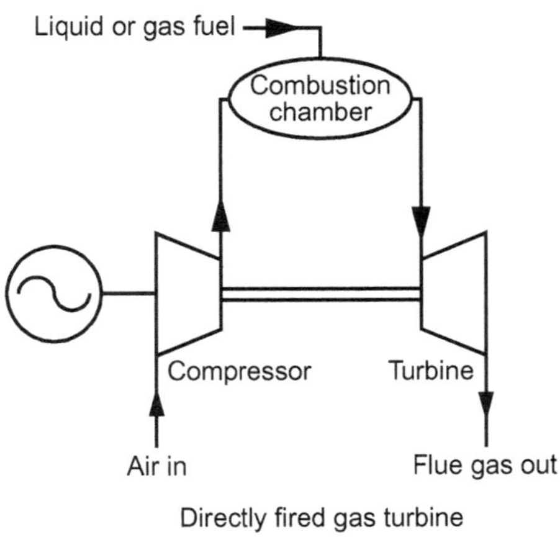

Fig. 15 Direct-fired gas turbine process.
From K.A. Al-attab, Z.A. Zainal, Externally fired gas turbine technology: a review, Appl. Energy 138 (2015) 475.

5.3.3 Thermochemical process

A process is referred to as thermochemical when the heat is used to chemically convert biomass into energy and other chemical by-products. Thermochemical process is comprised of five main processes:

1. Combustion
2. Pyrolysis
3. Carbonization
4. Gasification
5. Liquefaction

Each process is exposed to different reaction conditions producing specific products. Typical reaction conditions include temperature, pressure, heating rate, residence time, and gas flow rate [51].

5.3.3.1 Combustion

Combustion is a term that is used when there is as an exothermic chemical reaction followed by large heat generation and light. Carbon, hydrogen, oxygen, combustible sulfur, and nitrogen constituents are some of the biomass fuels. During this process, the reaction is spontaneous which is followed by the heat generated by the oxidation reaction. These oxidation reactions are initiated when biomass reacts with air or oxygen. Once the oxidation reaction is initiated, it proceeds by gas phase reaction, surface reaction, or both. This is then followed by fusion, evaporation, and pyrolysis.

Evaporation, mixing, diffusion, convection, heat conduction, radiation, and luminescence happen rapidly during an actual combustion reaction. Burning happens directly in their respective phases depending on the fuel type. For a liquid fuel such as heavy oil, burning happens as an inflammable gas in gas phase and is called as evaporation combustion. The heavy oils undergo decomposition combustion prior to burning in evaporation combustion. Heat is produced when fuel is decomposed [52].

Direct combustion of biomass results into solid forms such as [52]:

- Evaporation combustion—the fuel with low fusing point fuses and evaporates when heated and then reacts with oxygen in gas phase and burns.
- Decomposition combustion—gas produced from thermal decomposition by heating (H_2, CO, H_2O, and CO_2) reacts with oxygen in gas phase, forms flame, and burns. Usually, char remains after these forms of combustion and burns by surface combustion.
- Surface combustion—happens when the fuel is comprised of carbon with little volatile portions such as charcoal is thermally decomposed. Oxygen, carbon dioxide (CO_2), or steam diffuses through the pores that exist inside or on solid surface of the component, and burns by surface reaction.
- Smoldering combustion—thermal decomposition reaction happens at temperature lower than the ignition temperature of volatile component of the reactive fuels such as wood. Flammable combustion happens if ignition is forced to smoke or temperature exceeds ignition point.

5.3.3.2 *Pyrolysis*

Pyrolysis happens when a biomass is thermally decomposed without oxygen. The reaction is preceded by combustion and gasification processes. Some of the by-products of biomass pyrolysis include bio-char; bio-oil; and gases including methane, hydrogen, carbon monoxide, and carbon dioxide.

Depending on the thermal environment and the final temperature, different by-products are produced. Heat transfer is a critical to pyrolysis operation as the reactions are endothermic and require sufficient heat transfer surface in order to achieve required process heat needs. Pyrolysis will mainly yield bio-char when temperatures are less than 600°C. At these temperatures, the heating rate is low. When the temperatures are greater than 800°C, the by-product of pyrolysis is mainly gases. Bio-oil is produced at an intermediate temperature and under relatively high heating rates [53,54]. Pyrolysis is an operation that can be performed at smaller scale and often at remote locations. This is important keeping in mind the density of the biomass resource and thereby, enhances the ability to reduce transport and handling costs [54].

Types of pyrolysis

Pyrolysis operation is primarily of two types:

1. Slow pyrolysis: It takes several hours to complete and the main by-product is bio-char.
2. Fast pyrolysis: It is a quick process and yields approximately, 60% bio-oil, 20% bio-char, and 20% syngas. Some of the fast pyrolysis processes include open-core fixed bed pyrolysis, ablative fast pyrolysis, cyclonic fast pyrolysis, and rotating core fast pyrolysis systems [55,56].

Basic components of pyrolysis system are shown in Fig. 16 [57]

Some of the essential features of a fast pyrolysis process are as follows:

- Very high heating and heat transfer rates, which require a finely ground feed.
- Carefully controlled reaction temperature of around 500°C in the vapor phase.
- Rapid cooling of the pyrolysis vapors to give the bio-oil product.

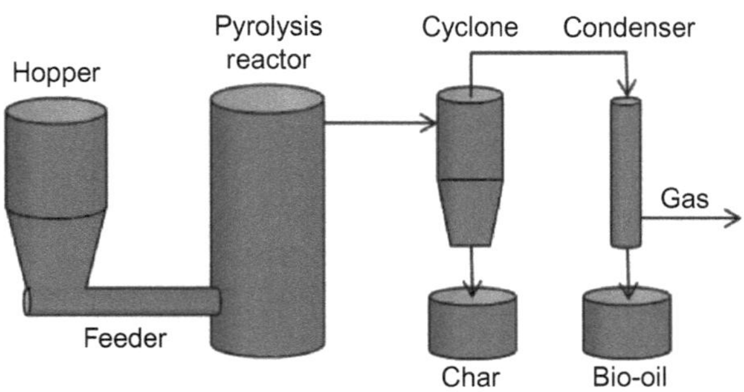

Fig. 16 Basic components of a pyrolysis system.
From F.L.P. Resende, Reactor configurations and design parameters for thermochemical conversion of biomass into fuels, energy, and chemicals, Reactor and Process Design in Sustainable Energy Technology, Elsevier, 2014, p. 7.

Pyrolysis reactors

There are four main types of reactors used in biomass pyrolysis:

1. Bubbling fluidized bed reactor
2. Circulating fluidized bed reactor
3. Auger reactor
4. Ablative reactors

Bubbling fluidized bed reactor Some of the important features of bubbling fluidized bed are as follows:

- Provide good mixing and uniform conditions throughout the bed
- Efficient with high degree of temperature control
- The system is scalable
- Features short residence

A typical bubbling fluidized bed reactor is a vertical cylindrical tube with a perforated distributor plate at the bottom as shown in Fig. 17. Material in particulate form such as sand is placed inside the tube, and gas is allowed to flow from the distributor plate upward. There are two ways the gas is introduced into the reactor:

1. The gas is injected at a low velocity. In this case, the gas will simply flow around the particles and leave at the top. This situation is termed as fixed bed.
2. The gas is injected at a high velocity. In this case, the force exerted by gas upward on each particle equals the particle weight suspending the particles. This situation is termed as minimum fluidization velocity. Any additional increase in velocity creates gas bubbles that rapidly flow upward in the system [57].

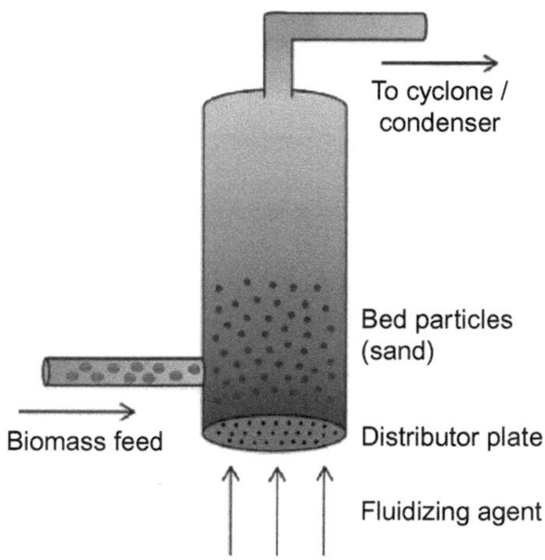

Fig. 17 Bubbling fluidized bed reactor.
From F.L.P. Resende, Reactor configurations and design parameters for thermochemical conversion of biomass into fuels, energy, and chemicals, Reactor and Process Design in Sustainable Energy Technology, Elsevier, 2014, p. 7.

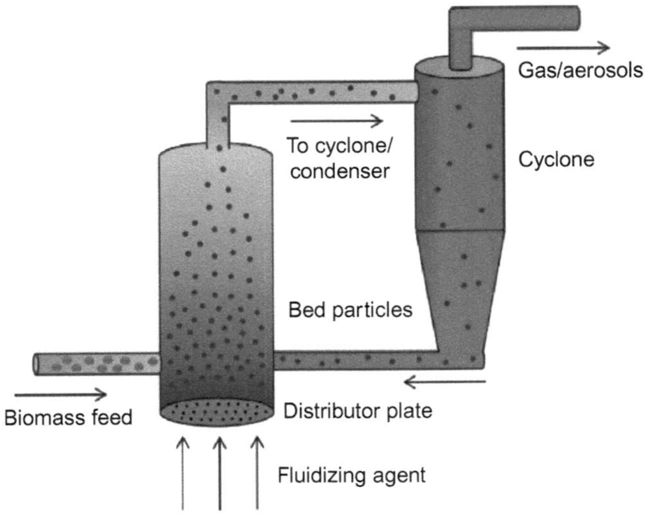

Fig. 18 Circulating fluidized bed reactor.
From F.L.P. Resende, Reactor configurations and design parameters for thermochemical conversion of biomass into fuels, energy, and chemicals, Reactor and Process Design in Sustainable Energy Technology, Elsevier, 2014, p. 7.

Circulating fluidized bed reactor The construction of circulating fluidized bed reactor is similar to a bubbling fluidized bed reactor as shown in Fig. 18. The elutriation process carries both sand and char particles outside of the bed. The reactor is connected to the cyclone which separates char and sand from the main gas stream. The separated char is combusted to generate energy which heats the sand. The hot sand is then recirculated back into the reactor [57].

Some of the features of circulating fluidized bed reactors are as follows:

- Maintains good temperature control and uniform conditions
- Easily scalable
- Only used for commercial applicable because of the control requirements

Auger reactor Auger reactor is the most popular pyrolysis reactor because of its basic construction as shown in Fig. 19. In this reactor, the biomass is fed continuously to rotating augers filled with preheated sand. When the biomass is mixed with the preheated sand, the auger rotation moves the product along the auger axis. When it reaches the end of the reactor, the gases and volatiles exit from the top and char and sand are collected at the bottom. Because of the auger design, vapors take much longer time to exit in comparison to fluidized bed type. This increases the possibility of secondary reactions, and thereby increases the yield of char [57].

Ablative reactors Ablative reactors are the ones in which the heat transfer takes place mainly by solid-solid contact. Solid biomass such as entire wood rods is heated up by direct contact with a hot metallic surface. The main significance of solid-to-

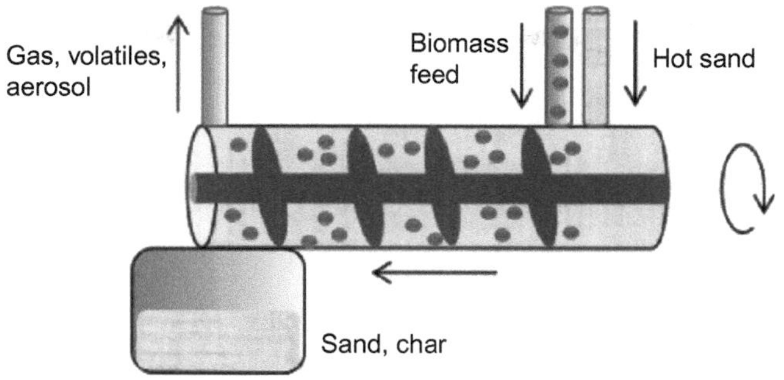

Fig. 19 Auger reactor.
From F.L.P. Resende, Reactor configurations and design parameters for thermochemical conversion of biomass into fuels, energy, and chemicals, Reactor and Process Design in Sustainable Energy Technology, Elsevier, 2014, p. 7.

solid contact is high heat transfer rates; however, limited surface area of contact makes the scale-up very difficult. Ablative reactors are only recommended for small-scale applications [57].

There are two types of ablative reactors:

(1) Vortex reactor
(2) Cone reactor

(1) Vortex reactor

Fig. 20 shows a design of a vortex reactor. In this type of reactor, a stream of gas carrying biomass enters the tube at the top with a high velocity. The mixture of gas and biomass is then fed into a reactor in such a way that the biomass particles are forced against its internal walls. Heaters are located on the outside which keeps the reactor wall hot internally. By doing this, it ensures that the biomass is heated and undergoes pyrolysis by solid–solid contact. When the gas/biomass mixture makes an impact on the walls, char on the walls gets abraded by the high velocity of the incoming biomass. This exposes the wall to further incoming particles [57].

(2) Cone reactor

The cone reactor as shown in Fig. 21 has a cone upside down with its base open and hence, termed as cone reactor. This reactor works on the principles of centrifugation. The biomass is fed along with sand into the reactor from the bottom, rotating in a spiral trajectory until it leaves the reactor at the top. The heaters are located on the outside that heats the biomass/sand mixture producing char [57].

5.3.3.3 Carbonization

Carbonization is the process when a solid biomass such as wood, bark, bamboo, rice husks, etc., is heated at 400–600°C in the absence of air or oxygen producing charcoal. The by-products from this process are tar, pyroligneous acid, and combustible gases [52].

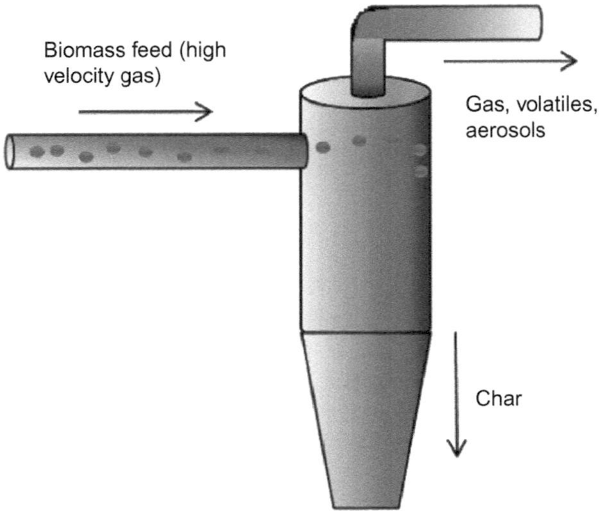

Fig. 20 Vortex reactor.
From F.L.P. Resende, Reactor configurations and design parameters for thermochemical conversion of biomass into fuels, energy, and chemicals, Reactor and Process Design in Sustainable Energy Technology, Elsevier, 2014, p. 7.

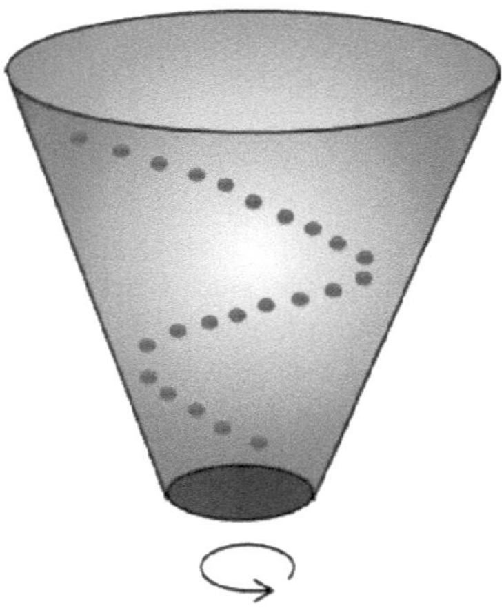

Fig. 21 Cone reactor.
From F.L.P. Resende, Reactor configurations and design parameters for thermochemical conversion of biomass into fuels, energy, and chemicals, Reactor and Process Design in Sustainable Energy Technology, Elsevier, 2014, p. 7.

Similar to pyrolysis, reactions happen in the presence of inert gas like nitrogen. In the case of wood, cellulose, hemicellulose, and lignin are decomposed after most water is evaporated at below 200°C. This produces liquid fraction and gaseous fraction, consisting mainly of carbon monoxide and carbon dioxide, at 200–500°C [58–62].

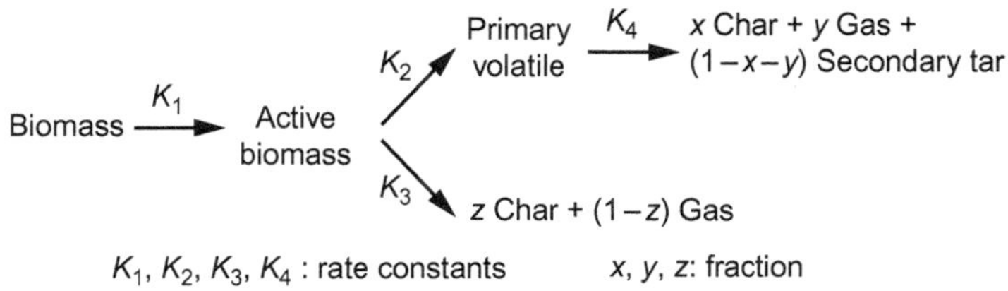

Fig. 22 Modified Broide-Shafizadeh model.
From E. Kantarelis, W. Yang, W. Blasiak, Biomass pyrolysis for energy and fuels production, in: E. Dahlquist (Ed.), Technologies for Converting Biomass to Useful Energy, CRC Press, Taylor and Francis Group, LLC, 2013, p. 267 (Chapter 8).

The overall scheme of carbonization is represented by modified Broide-Shfizadeh model in Fig. 22. The model shows that the product distribution depends on (a) steps of the decomposition resulting from wood constituents into gas, liquid, and solid fractions at the first stage as well as (b) subsequent decomposition of the liquid fraction at second stage, and (c) the ratio of rate constants for first stage to that for second stage.

Other key factors impacting the distribution are moisture, size of feed, heating rate, and operating temperature. Size of feed, heating rate, and operating temperature are important as the yield of tar is increased with decrease in size and increase in the rate. Higher temperature decreases charcoal yields and increase in the tar yields at temperature below 500°C [63].

5.3.3.4 Gasification

Gasification is a two-stage process of converting solid biomass into fuel gas. Air and fuel are two components that are added in at air-to-fuel ratio of 1.5:1 to 1.8:1 to allow combustion of biomass. Gasification consists of two stages, namely, oxidation and reduction. Optimum conditions of air with biomass are used during each step. Volatiles are removed from the biomass at temperatures between 1100°C and 1200°C. This produces gases such as carbon monoxide, hydrogen, carbon dioxide, and water vapors. Charcoal generated during the process converts the carbon dioxide and water vapor into carbon monoxide, hydrogen, and methane as shown in the following reactions [63]:

$$C + CO_2 \rightarrow 2CO$$

$$C + H_2O \rightarrow CO + H_2$$

$$C + 2H_2 \rightarrow CH_4$$

Gasification process is classified based on conditional factors listed in Table 3 [52].

Fixed bed gasifier as shown in Fig. 23 features a simple structure and low equipment cost that functions based on the combustion or gasification of solid fuel. Wood chips of about 2.5–5 cm are generally used as the raw material. They are fed through

Table 3 **Classification of gasification method**

Classification	Conditional factors
Gasification pressure	Normal pressure (0.1–0.12 MPa), high pressure (0.5–2.5 MPa)
Gasification temperature	Low temperature (700°C and below), high temperature (700°C and above) High temperature decomposition (ash fusion point and above)
Gasification agent	Air, oxygen, steam and combination of them, carbon dioxide for a specific time
Heating (temperature zone formation)	• Direct gasification (heat generation from reaction of heating raw material and oxygen) • Indirect gasification (heating raw material and gasification agent by external heat)
Gasifier types	Fixed bed, flow bed, circulating flow bed, entrained bed, mixing bed, rotary kiln, twin tower, molten furnace

From S. Yokoyama, Part 4: Thermochemical conversion of biomass, Section 4.1, Asian Biomass Handbook, The Japan Institute of Energy, Ministry of Agriculture, Forestry, and Fishery, 2008, p. 94. http://www.jie.or.jp/biomass/AsiaBiomassHandbook/English/Part-4_E.pdf.

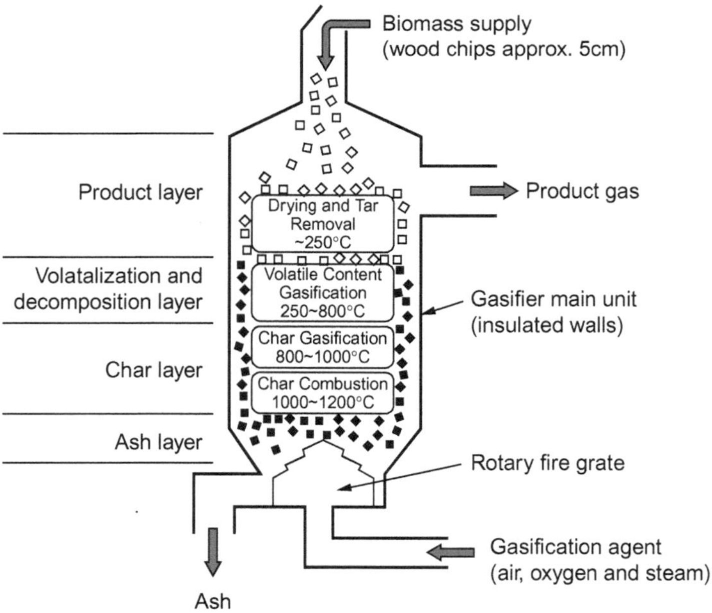

Fig. 23 Fixed bed gasifier.
From S. Yokoyama, Part 4: Thermochemical conversion of biomass, Section 4.1, Asian Biomass Handbook, The Japan Institute of Energy, Ministry of Agriculture, Forestry, and Fishery, 2008, p. 94. http://www.jie.or.jp/biomass/AsiaBiomassHandbook/English/Part-4_E.pdf.

the upper feed port and layered in the furnace. The gasifying agent such as air, oxygen, steam, or a mixture is supplied from the bottom. The gasification reaction starts from the bottom and proceeds toward the top. Individual layers are formed at the bottom because of gasification of raw material. Ash, char, volatilized and decomposed materials are the by-products. The product gas exits the top [52].

5.3.3.5 Catalytic liquefaction

Catalytic liquefaction is another way to produce bio-oils and is summarized in Fig. 24 [64]. This process consists of the catalytic thermal decomposition of large molecules to produce unstable shorter molecules. These smaller species are used to polymerize again into a bio-oil. The biomass is mixed with water and basic catalysts like sodium carbonate, and the process is carried out at temperatures lower than pyrolysis but higher pressures and longer residence times [65,66].

5.3.4 Biochemical process

Biochemical conversion uses enzymes to break down structural carbohydrates into sugars, which are then transformed into alcohols, organic acids, or hydrocarbons by microorganisms in fermentation. The conversions typically take place at atmospheric pressure and temperatures ranging from ambient to 70°C.

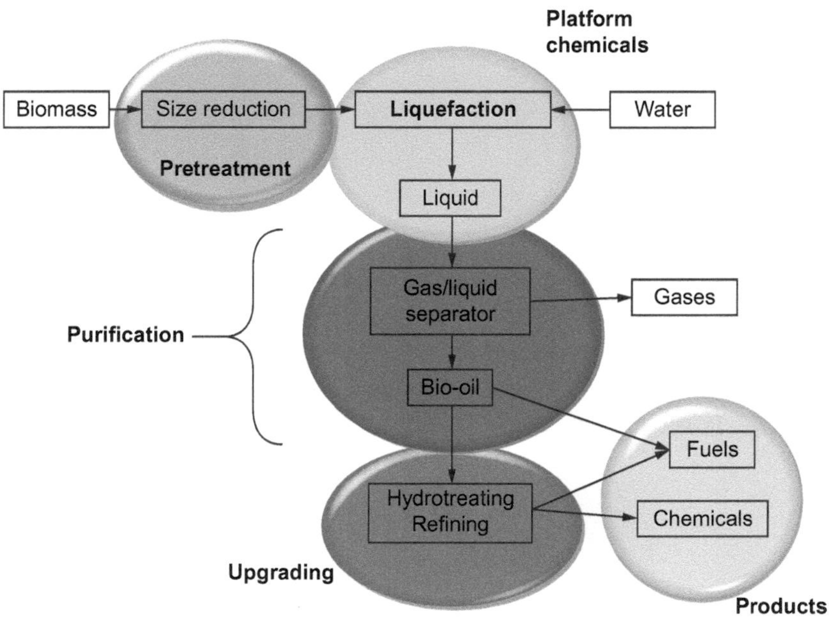

Fig. 24 Liquefaction of biomass.
From D.M. Alonso, J.Q. Bond, J.A. Dumesic, Catalytic conversion of biomass to biofuels. Green Chem. R. Soc. Chem. 12 (2010) 1499–1500.

5.3.4.1 Anaerobic digestion

The natural biological process where an organic waste is transformed into biofertilizer and biogas in the absence of air is termed as anaerobic digestion. Almost any type of organic material including biodegradable waste such as municipal solid waste, animal manure, poultry litter, food wastes, sewage, and industrial wastes can be used with anaerobic digestion process. A typical anaerobic digester is shown in Fig. 25. An anaerobic digestion plant produces two outputs, biogas and digestate. Digestate can be further processed to produce liquor and a fibrous material. The fiber, which can be processed into compost, is a bulky material with low levels of nutrients and can be used as a soil conditioner or a low-level fertilizer. A high proportion of the nutrients remain in the liquor, which can be used as a liquid fertilizer [67]. Similarly, biogas can be used for producing electricity and heat, as a natural gas substitute and as a transportation fuel [67].

5.3.4.2 Ethanol fermentation

Anaerobic fermentation is the process of breaking down the glucose within organic materials in the absence of air or oxygen converting sugar into alcohol through a series of chemical reactions. This process requires enzymes such as yeast or bacteria that are added to the biomass material. These enzymes feed on the sugars to produce ethanol and carbon dioxide. The ethanol is distilled and dehydrated to obtain a higher concentration of alcohol to achieve the required purity. The solid residue from the fermentation process can be used as cattle feed, and in the case of sugarcane, the bagasse can be used as a fuel for boilers or for subsequent gasification [68,69].

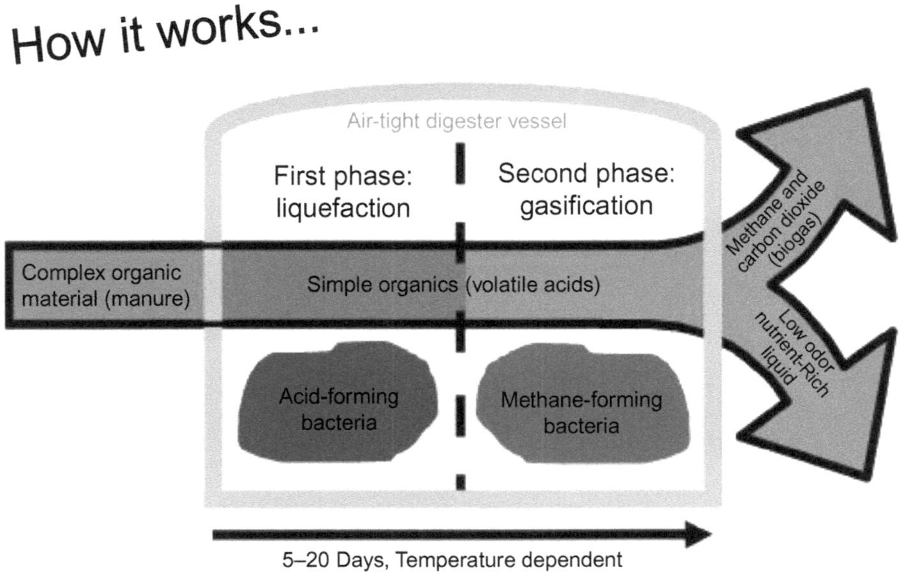

Fig. 25 Anaerobic digestion.
From http://www.bioenergyconsult.com/biochemical-conversion-technologies/.

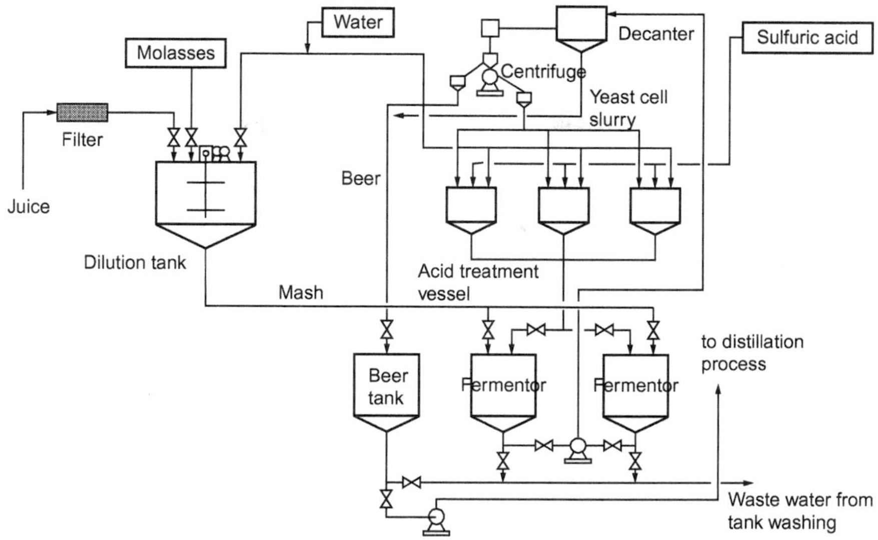

Fig. 26 Ethanol fermentation by Melle-Boinot process.
From S. Yokoyama, Part 4: Thermochemical conversion of biomass, Section 4.1, Asian Biomass Handbook, The Japan Institute of Energy, Ministry of Agriculture, Forestry, and Fishery, 2008, p. 94. http://www.jie.or.jp/biomass/AsiaBiomassHandbook/English/Part-4_E.pdf.

Fig. 26 shows Melle-Boinot ethanol fermentation process. This is a continuous or semicontinuous fermentation process popular in Brazil. In this process, the yeast cells are recovered from beer through centrifuge and recycled to fermentation tank after sterilization of contaminated bacteria with dilute sulfuric acid at pH 3 [52,67]. Ethanol fermentation at high concentration of yeast cells can produce product containing 6%–8% of ethanol from cane juice in about 15 h of fermentation period [67].

5.3.4.3 Methane production in landfills

Landfill gas is generated when an organic material gets decomposed in the landfills. It is primarily composed of methane, carbon dioxide, and a trace amount of nonmethane organic compounds [70]. Fig. 27 shows the process of converting methane into electricity. Methane gas is generated when trash is decomposed in the landfills. The gas rises to the top of the landfills and is collected in the gas collection wells. The gas is then burned to produce heat and electricity before it is released to the atmosphere. This entire process helps to prevent an impact on the environment [71].

5.4 Bio-oil

Bio-oil is the by-product during pyrolysis of biomass. It is dark brown liquid with composition like biomass. Because of its higher density than woody materials, it is difficult to store and transport. Bio-oil is not suitable for direct use in standard internal

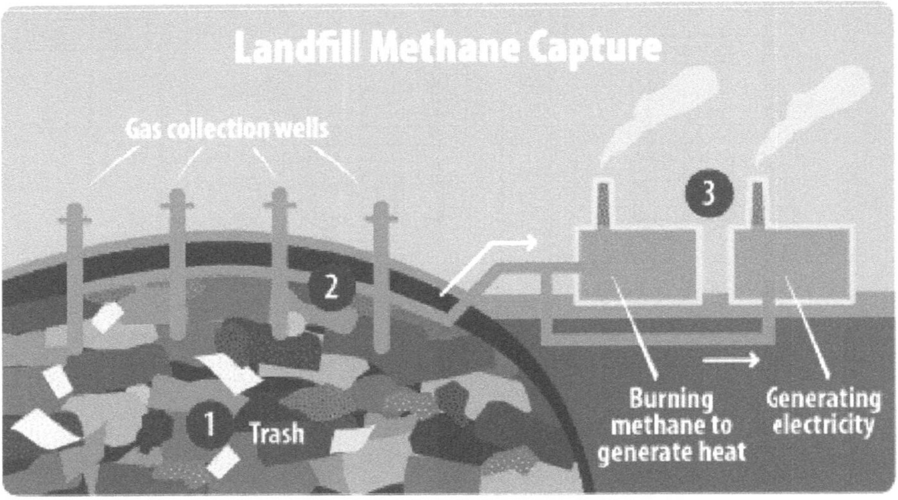

Fig. 27 Principle of converting methane from landfill.
From https://www3.epa.gov/climatechange/kids/solutions/technologies/methane.html.

combustion engines however; it can be upgraded to syngas and biodiesel by undergoing gasification process. It is commonly used as a feedstock for cofiring process because it is easy to handle and is cheaper than solid fuel [54].

For example, cofiring of bio-oil has been demonstrated in 350-MW gas-fired power station in Netherlands, when 1% of the boiler output was successfully replaced. It is in such applications that bio-oil can offer major advantages over solid biomass and gasification due to the ease of handling, storage, and combustion in an existing power station when special start-up procedures are not necessary [54].

5.5 Bio-char

Charring happens when a biomass is heated in the absence of or with limited air to temperatures greater than 250°C. This results in a by-product known as bio-char. This material distinguishes itself from charcoal or other carbon products in that it is intended for use as a soil application [72].

Fig. 28 shows bio-char bioenergy life cycle. Typical feedstock for bio-char production consists of cellulosic biomass such as wood chips, corn stover, rice and peanut hulls, tree bark, paper mill sludge, animal manure and most urban, agricultural and forestry biomass residues. In addition to bio-char, bioenergy is also produced in the form of either syngas, or bio-oils, which can be used to produce heat, power, or combined heat and power [73].

Bio-char is enriched in carbon, phosphorus, and other metals such as calcium and magnesium. Combustion of wood biomass and agricultural residues can lead to emission of large amount of carbon dioxide. Bio-char can absorb the carbon dioxide in the soil leading to reduction in greenhouse gases [74].

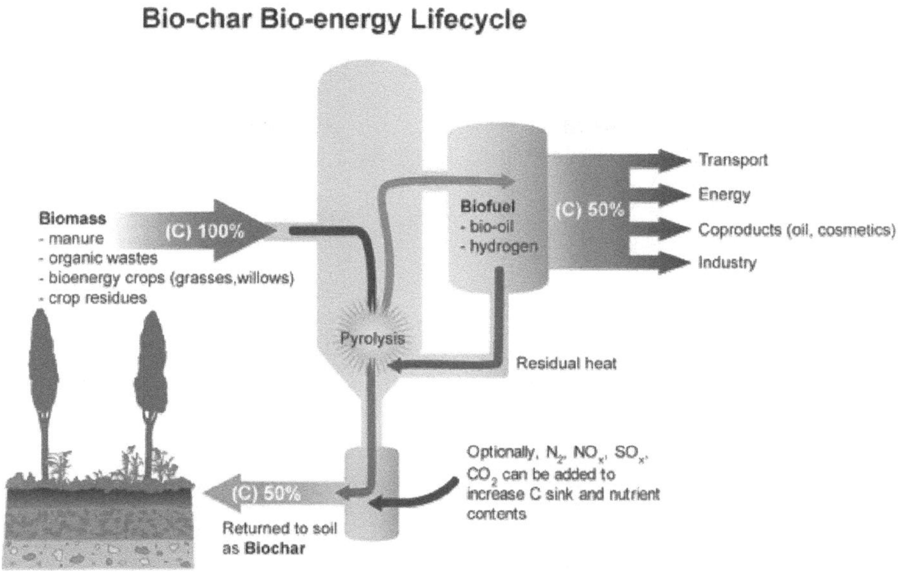

Fig. 28 Bio-char bioenergy life cycle.
From http://www.biochar-international.org/sites/default/files/Production_web.pdf.

There are several advantages of bio-char [75]:

- Soil fertility—increase the available nutrients for plant growth, water retention, and reduce the amount of fertilizer by preventing the leaching of nutrients out of the soil.
- Reduction in greenhouse gas emission—reduces methane and nitrous oxide emissions from soil.
- Replacement for other biomass energy systems.
- Soil amendment to increase plant growth yield.

5.6 Biofuels

Fuel that is produced from a biomass is termed as biofuel and exists as solid, liquid, or gas. Some examples of biofuel include ethanol, methanol, and biodiesel, and gaseous fuel, such as hydrogen and methane [76].

1. First approach—grow crops that are rich in sugar or starch followed by fermentation producing ethanol.
2. Second approach—growing plants that contain high amount of vegetable oil such as oil palm, soybean, and algae and convert them to biodiesel.

5.6.1 Bioethanol

Different forms of sugar including sugarcane, corn, and sweet potatoes are used in the production of bioethanol. Other forms of biomass that are used in fermentation

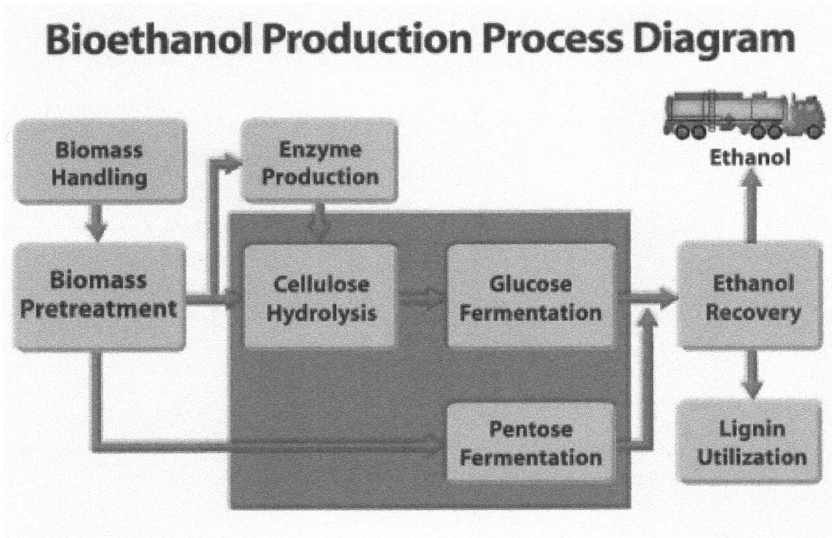

Fig. 29 Bioethanol production process [67].
From G.W. Huber, S. Iborra, A. Corma, Chem. Rev. 106 (2006) 4044.

processes are starchy materials such as wheat, barley, oat, and rice along with ligno-cellulosic materials such as agricultural wastes and woody materials [67].

Lignocellulosic biomass is slightly different because it contains lignin, a non-fermentable fraction in addition to cellulose and hemicellulose that can be converted to sugars. Lignins are high in energy content and therefore can be utilized to produce electricity and/or heat. Although the decomposition of lignin into fermentable sugars is challenging, the fermentation, distillation, and dehydration process steps are basically identical for bioethanol from either biomass [67].

Fig. 29 shows bioethanol production process. There are three major steps involved in the production of bioethanol:

- Pretreatment
- Enzymatic hydrolysis
- Fermentation

At the start of the process, the biomass is pretreated to improve the accessibility of enzymes. After pretreatment, biomass undergoes enzymatic hydrolysis for conversion of polysaccharides into monomer sugars, such as glucose and xylose. Subsequently, sugars are fermented to ethanol using different microorganisms [67].

5.6.2 Biodiesel

It is an alternative to fossil fuel and can be produced from a variety of natural crops including rapeseed, soybean, mustard, flax, sunflower, canola, palm oil, hemp, and waste vegetable oils. A typical biodiesel cycle is shown in Fig. 30 [77].

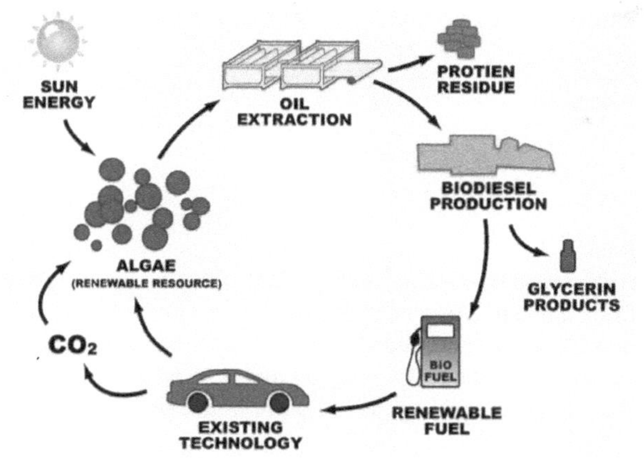

Fig. 30 The biodiesel cycle. From https://www.intechopen.com/books/biomass-now-sustainable-growth-and-use/lignocelluloses-feedstock-biorefinery-as-petrorefinery-substitutes.

Biodiesel can be produced from waste vegetable oils from restaurants and food processing industries. The biomass is treated for impurities before it is converted to biodiesel. Biodiesel from oils and fats can be produced in the following three ways:

1. Base catalyzed transesterification of the oil.
2. Direct acid catalyzed transesterification of the oil.
3. Conversion of the oil to its fatty acids and then to biodiesel.

Most of the biodiesel that is produced is through base-catalyzed transesterification process because it is most economical and requires only low temperatures and pressures. Catalyzed transesterification process yields 98% conversion. Fig. 31 shows the biodiesel production process using transesterification [78].

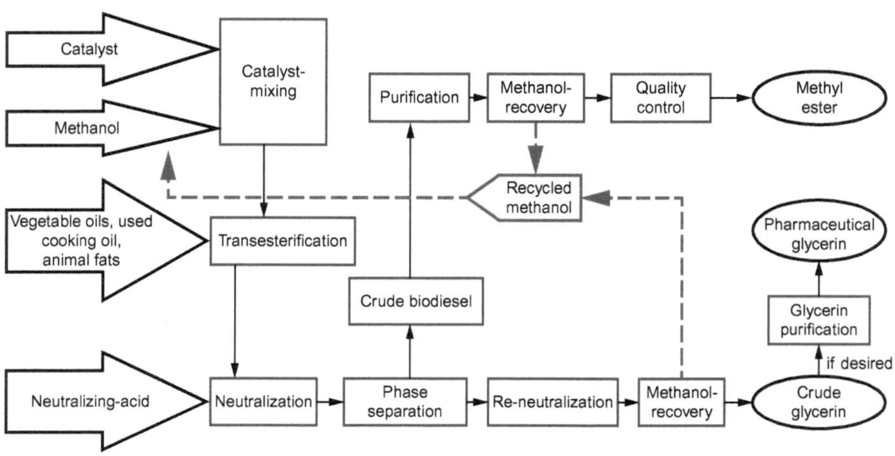

Fig. 31 The biodiesel production process.
From http://www.alternative-energy-news.info/technology/biofuels/biodiesel-fuel/.

The typical biodiesel production process involves five primary steps [78]:

1. Mixing of alcohol and catalyst
2. Separation of biodiesel and glycerin
3. Alcohol removal
4. Glycerin neutralization
5. Methyl ester wash

Step 1: Mixing of alcohol and catalyst

The first step involves mixing of alcohol and catalyst. The catalyst is then dissolved in the alcohol using a mixer. The alcohol/catalyst mix is then charged into a closed reaction vessel and the oil or fat is added. The system is then sealed to the atmosphere to prevent the loss of alcohol. The reaction mix is kept just above the boiling point of the alcohol to speed up the reaction and the reaction takes place from 1 to 8 h [78].

Step 2: Separation

The second step involves separation of two products, glycerin and biodiesel. Each has a substantial amount of the excess methanol that was used in the reaction. The glycerin phase is much denser than biodiesel phase and the two are separated using gravity. The glycerin settles at the bottom of the settling vessel due to its heavy density.

Step 3: Alcohol removal

The third step is to remove excess alcohol in each phase. This is done by flash evaporation process or by distillation. The alcohol is recovered using distillation equipment and is reused.

Step 4: Glycerin neutralization

The fourth step is to neutralize the glycerin by-product containing unused catalyst and soaps with an acid. Water and alcohol are removed to produce 80%–88% pure glycerin that is ready to be sold as crude glycerin.

Step 5: Methyl ester wash

The last step is to purify biodiesel by washing gently with warm water to remove residual catalyst or soaps and dried. This is normally the end of the production process resulting in a clear amber-yellow liquid with a viscosity like petro diesel.

5.7 Biorefinery

The term biorefinery is defined as a system that encompasses complete biomass conversion process and equipment to produce fuels, power, and value-added chemicals from biomass. A typical biorefinery is shown in Fig. 32. It works on similar principles of petroleum refinery which produces multiple fuels and products from petroleum [79]. A biorefinery is capable of producing multiple fuels and products such as liquid transportation fuel such as biodiesel or bioethanol [79]. Biorefineries help utilize the optimum potential of wastes and solve the problems of waste management and greenhouse gas emissions by converting them through enzymatic or chemical treatments into either hydrogen or liquid fuels [79].

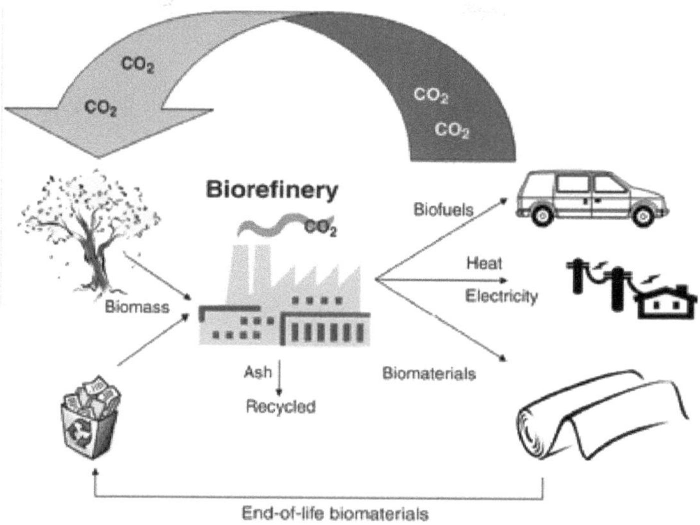

Fig. 32 Principles of biorefinery.
From http://www.ecomena.org/biorefinery.

References

[1] http://www.globalproblems-globalsolutions-files.org/gpgs_files/pdf/UNF_Bioenergy/UNF_Bioenergy_5.pdf.

[2] Committee on Biobased Industrial Products, National Research Council, Biobased Industrial Products: Research and Commercialization Priorities, National Academic Press, Washington, DC, 2000. pp. 5–6.

[3] http://www.motherearthnews.com/homesteading-and-livestock/home-business-selling-firewood-zmaz76soztak.

[4] https://www.recipeshubs.com/food?t=cherry-logs.

[5] http://caldwelltreecare.com/contact-us/free-tree-mulch/, https://www.ruralking.com/hardwood-wood-pellet-fuel-40lb.html.

[6] http://www.bioenergyconsult.com/tag/agricultural-residues/.

[7] http://www.claas.co.uk/fascination-claas/special-topics/biomass/energy-crops.

[8] http://www.crops4energy.co.uk/short-rotation-coppice-src/.

[9] http://www.esru.strath.ac.uk/EandE/Web_sites/03-04/biomass/background%20info4.html.

[10] http://www.recrops.com/Miscanthus.

[11] http://www.missouribotanicalgarden.org/PlantFinder/PlantFinderDetails.aspx?kempercode=g650.

[12] http://www.sweetfuel-project.eu/.

[13] https://www.pioneer.com/home/site/us/products/sorghum/.

[14] S. Csete, Tall wheatgrass cultivar Szarvasi-1 (*Elymus elongates* subsp. ponticus cv. Szarvasi-1) as a potential energy crop for semi-arid lands of eastern Europe, in: Sustainable Growth and Applications in Renewable Energy Sources, InTech, Rijeka, Croatia, 2011, pp. 270–273.

[15] https://www.dreamstime.com/stock-photo-energy-plant-szarvasi-grass-renewable-resource-multiple-purpose-image31223530.
[16] http://www.desertconsult.de/projects/igniscum.html.
[17] http://nextgenbioenergy.com/two-platform-technologies/ifes/rice-straw.
[18] http://irri.org/rice-today/a-second-life-for-rice-husk.
[19] http://www.chronicle.co.zw/maize-price-reduction-welcome/.
[20] https://www.dreamstime.com/stock-photo-corn-husks-dry-corn-crumb-crust-corn-field-image64568493.
[21] http://www.agrifarming.in/cassava-cultivation-information-guide/.
[22] http://en.agroghana.com/.
[23] http://www.royalbiocoal.com/groundnut-shell-1160980.html.
[24] http://nyp-corp.com/blog/tag/jute/.
[25] http://kitairu.net/cn/agriculture-and-food/farm-machinery-and-equipment/farm-machinery/188312.html.
[26] http://www.kingsagriseeds.com/minimize-waste-with-stover-and-straw/.
[27] http://www.agrimarketing.com/show_story.php?id=57169.
[28] https://dir.indiamart.com/impcat/sugarcane-bagasse.html.
[29] http://www.energycentral.com/c/gn/renewable-energy-food-wastes.
[30] http://www.bioenergyconsult.com/tag/hotel-waste/.
[31] http://livingvalues.fourseasons.com/2011/03/15/knocking-out-waste-in-philadelphia/.
[32] http://www.tidyplanet.co.uk/who-we-work-with/case-studies/exclusive-hotels-south-lodge/.
[33] https://www.biorefinerycentre.org/.
[34] http://www.rotochopper.com/feedstocks-products/food-waste.html.
[35] http://www.bioenergyconsult.com/tag/industrial-wastes/.
[36] http://www.lovelyasatree.com/paper_issues.htm.
[37] http://www.nationalobserver.com/2016/06/20/news/playing-hardball-jd-irving-way.
[38] http://www.nature.com/scitable/content/photosynthesis-creates-biomass-that-can-be-converted-14464254.
[39] http://www.eia.gov/energyexplained/?page=biomass_home.
[40] http://www.bioenergyconsult.com/tag/biomass-combustion-process.
[41] http://www.fao.org/docrep/t1804e/t1804e06.htm—ethanol fermentation.
[42] E. Karampinis, P. Grammelis, M. Agraniotis, I. Violidakis, E. Kakaras, Co-firing of biomass with coal in thermal power plants: technology schemes, impacts, and future perspectives, in: P. Lund, J.A. Byrne, G. Berndes (Eds.), Advances in Bioenergy: The Sustainability Challenge, Wiley Publishers, West Sussex, UK, ISBN 978-1-11895-787-5, 2016, p. 235.
[43] J. Koppejan, S. Van Loo, The Handbook of Biomass Combustion and Co-Firing, Earthscan Publishers, London, UK, ISBN 978-1-84407-249-1, 2008. p. 207.
[44] Biomass co-firing technology brief, IEA-ETSAP and IRENA© technology brief E21, January 2013. www.irena.org/Publications.
[45] The basics of combined heat and power, The News Magazine, 19 Oct 2015. http://www.achrnews.com/articles/130800-the-basics-of-combined-heat-and-power.
[46] S. Zafar, What is Biomass Cogeneration, BioEnergy Consult, Sep 9, 2015. http://www.bioenergyconsult.com/tag/what-is-biomass-cogeneration.
[47] R.P. Siegel, Combined heat and power: pros and cons, Apr 30, 2012. http://www.triplepundit.com/special/energy-options-pros-and-cons/combined-heat-power-pros-cons.
[48] Catalog of CHP technologies, US Environmental Protection Agency. https://www.epa.gov/chp/catalog-chp-technologies.

[49] D.L. Klass, Biomass for Renewable Energy, Fuels, and Chemicals, Academic Press, San Diego, CA, USA, 1998. p. 210.
[50] K.A. Al-attab, Z.A. Zainal, Externally fired gas turbine technology: a review, Appl. Energy 138 (2015) 475.
[51] http://www.engineering.iastate.edu/brl/files/2011/10/brl_thermochemical.pdf.
[52] S. Yokoyama, Part 4: Thermochemical conversion of biomass, Section 4.1, in: Asian Biomass Handbook, The Japan Institute of Energy, Ministry of Agriculture, Forestry, and Fishery, 2008, p. 94. http://www.jie.or.jp/biomass/AsiaBiomassHandbook/English/Part-4_E.pdf.
[53] http://www.bioenergyconsult.com/biomass-pyrolysis-process.
[54] http://doctor-biochar.blogspot.com/2013/10/bamboo-for-biochar-production.html.
[55] M.I. Jahirul, M.G. Rasul, A.A. Chowdhury, N. Ashwath, Biofuels production through biomass pyrolysis—a technological review, Energies 5 (2012) 4952–5001.
[56] J.T. Oladeji, E.A. Itabiyi, P.O. Okekunle, A comprehensive review of biomass pyrolysis as a process of renewable energy generation, J. Nat. Sci. Res. 5 (5) (2015) 99–105.
[57] F.L.P. Resende, Reactor configurations and design parameters for thermochemical conversion of biomass into fuels, energy, and chemicals, in: Reactor and Process Design in Sustainable Energy Technology, Elsevier, Oxford, UK, 2014, p. 7.
[58] A.V. Bridgwater, S.A. Bridge, in: A.V. Bridgwater, G. Grassi (Eds.), Biomass Pyrolysis Liquids Upgrading and Utilization, Elsevier Applied Science, Netherlands, 1991, p. 22.
[59] J. Lede, Reaction temperature of solid particles undergoing an endothermal volatilization. Application to the fast pyrolysis of biomass, Biomass Bioenergy 7 (1994) 49–60.
[60] C.F. Pomeroy, S.S. Sofer, O.R. Zaborsky (Eds.), Biomass Conversion Processes for Energy and Fuels, Plenum, New York, NY, 1981, pp. 201–211.
[61] T. Suzuki, M. Miyamoto, W.-M. Luo, T. Yamada, T. Yoshida, A.V. Bridgwater, D.G. B. Boocock (Eds.), Science in Thermal and Chemical Biomass Conversion, vol. 2, CPL Press, Birmingham, UK, 2006, pp. 1580–1591.
[62] T. Suzuki, K. Suzuki, Y. Takahashi, M. Okimoto, T. Yamada, N. Okazakik, Y. Shimizu, M. Fujiwara, Nickel-catalyzed carbonization of wood for coproduction of functional carbon and fluid fuels I, J. Wood Sci. 53 (2007) 54–60.
[63] E. Kantarelis, W. Yang, W. Blasiak, Biomass pyrolysis for energy and fuels production, in: E. Dahlquist (Ed.), Technologies for Converting Biomass to Useful Energy, CRC Press, Taylor and Francis Group, LLC, Boca Raton, FL, USA, 2013, p. 267 (Chapter 8).
[64] D.M. Alonso, J.Q. Bond, J.A. Dumesic, Catalytic conversion of biomass to biofuels, Green Chem. R. Soc. Chem. 12 (2010) 1499–1500.
[65] D.C. Elliott, Historical development in hydroprocessing bio-oils, Energy Fuels 21 (3) (2007) 1792–1815.
[66] G.W. Huber, S. Iborra, A. Corma, Synthesis of transportation fuels from biomass: chemistry, catalysts, and engineering, Chem. Rev. 106 (9) (2006) 4044–4098.
[67] http://www.bioenergyconsult.com/biochemical-conversion-technologies/.
[68] http://www.biomassinnovation.ca/fermentation.html.
[69] http://www.fao.org/docrep/t1804e/t1804e06.htm.
[70] https://www.epa.gov/lmop/basic-information-about-landfill-gas.
[71] https://www3.epa.gov/climatechange/kids/solutions/technologies/methane.html.
[72] http://csanr.wsu.edu/biochar/.
[73] http://www.biochar-international.org/sites/default/files/Production_web.pdf.
[74] J. Lehmann, S. Joseph, Biochar for environmental management: an introduction, in: Biochar for Environmental Management: Science, Technology and Implementation, Routledge, Oxon, UK, 2015, p. 2 (Chapter 1).

[75] http://biomassmagazine.com/articles/3091/a-new-climate-change-mitigation-tool.
[76] http://www.aepc.gov.np/old/index.php?option=com_content&view=article&id=245:introduction-of-biofuel&catid=72:biofuel&Itemid=99.
[77] http://www.esru.strath.ac.uk/EandE/Web_sites/02-03/biofuels/what_biodiesel.htm.
[78] http://www.alternative-energy-news.info/technology/biofuels/biodiesel-fuel/.
[79] http://www.ecomena.org/biorefinery/.

Further reading

[1] https://www.nrs.fs.fed.us/disturbance/market_induced_change/biofuels_bioenergy_bioproducts/.
[2] http://articles.extension.org/pages/70355/wood-processing-residues.
[3] https://www.customgrocerybags.com/blog/cook-county-recently-passed-a-construction-and-demolition-recycling-ordinance/.
[4] http://www.newenergyfarms.com/crops/miscanthus/.

Manufacturing Industrial Chemicals from Biomass

6

6.1 Introduction

Man-made materials can be obtained from one of these three resources: minerals, fossil resources, and plant-based resources (wood, cotton, etc.). The fossil resources and energy generate undesirable carbon dioxide emissions to the atmosphere resulting in global warming. In addition, fossil resources are difficult to dispose thereby, causing environmental pollution.

The choice of material and its source has therefore become a challenge. Use of plant-based materials also known as biomass is on the rise. Plants naturally convert the carbon dioxide from atmosphere into biopolymers and other compounds such as sugars and lipids, and easily convert them into materials.

Materials produced from the biomass can be classified based on the original content of the biomass or the process used:

1. Sugar derived
2. Plant derived
3. Pyrolysis derived
4. Gasification derived
5. Catalyst derived

6.2 Sugar-derived industrial chemicals

6.2.1 2,5-Furandicarboxylic acid

2,5-Furandicarboxylic acid (FDCA) is an important monomer that can potentially substitute terephthalic acid in the production of polyesters and other polymers that contain an aromatic component [1]. It is derived from sugar such as fructose, glucose, and cellulose [2–4]. Fig. 1 shows two routes to synthesize 2,5-FDCA:

(1) Oxidation of 5-hydroxymethylfurfural (HMF)
(2) Direct conversion of fructose

Route 1: This route involves two-step reactions. First, fructose is converted into HMF by an acid-catalyzed dehydration reaction in supercritical acetone, water with phase modifiers, or high boiling solvents [5–7]. This HMF is then converted into 2,5-FDCA through an oxidation reaction in the presence of gold, cobalt, chromium, or platinum catalysts [8–12].

Route 2: This route involves the direct conversion of fructose into 2,5-FDCA in one pot [13,14].

Technology and Applications of Polymers Derived from Biomass. https://doi.org/10.1016/B978-0-323-51115-5.00006-2

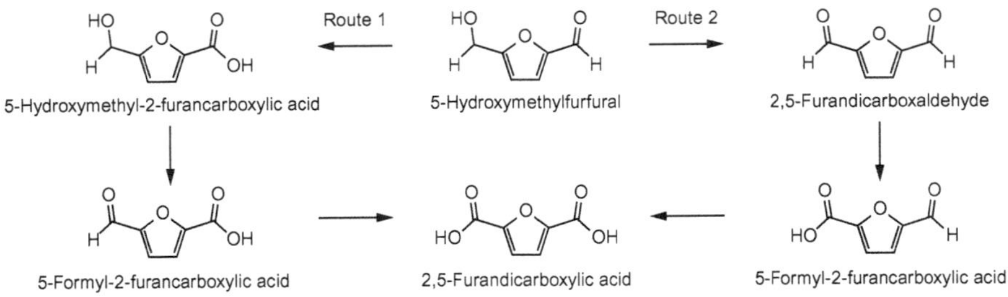

Fig. 1 Different routes to synthesize 2,5-FDCA.
From Y.Y. Gorbanev, S.K. Klitgaard, J.M. Woodley, C.H. Christensen, A. Riisager, ChemSusChem 2(7) (2009) 672–675; C.H.R.M. Wilsens, Exploring the Application of 2,5-Furandicarboxylic Acid as a Monomer in High Performance Polymers: Synthesis, Characterization, and Properties, Technische Universiteit, Eindhoven University of Technology, 2015.

2,5-FDCA has two carboxylic acid groups, which makes it a suitable monomer for polycondensation reactions with diols or diamines. An attached aromatic ring to the carboxylic group gives 2,5-FDCA high glass transition temperatures and good thermal stability. These properties can be useful in the design of renewable rigid materials.

Fig. 2 illustrates rigid monomers derived from HMF that are suitable for polycondensation reactions and for the development of renewable rigid materials [2].

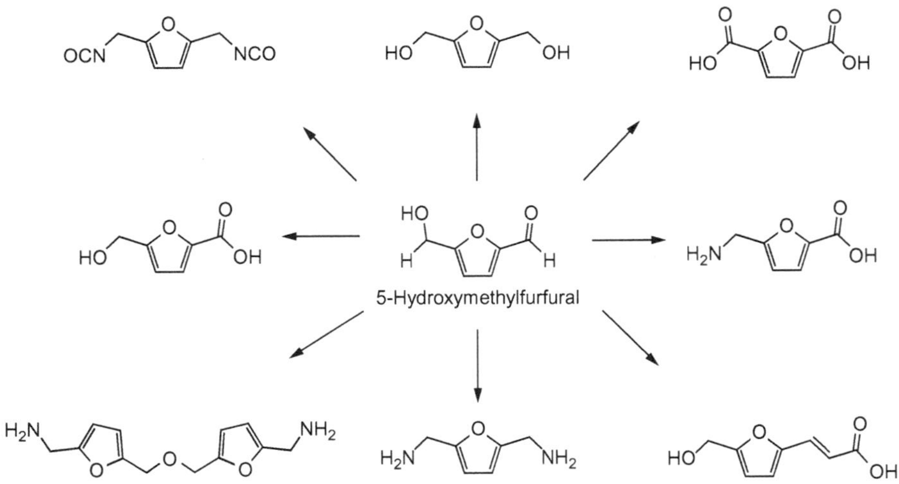

Fig. 2 Rigid monomers derived from HMF for polycondensation reactions.
From A. Gandini, Polym. Chem. 1 (2010) 245–251.

6.2.2 3-Hydroxypropionic acid

3-Hydroxypropionic acid (HPA) is an important element in the production of acrylic monomers, 1,3-propanediol, propene, malonic acid, β-propiolactone, and 3-hydroxypropionic esters. It is a beta hydroxyl acid that is water soluble and dehydrates to acrylic acid when distilled.

3-HPA can be produced by the following chemical routes:

1. Oxidation of PDO in the presence of Pd-containing supported catalyst
2. Oxidation of 3-hydroxypropionaldehyde in the presence of Pd-containing supported catalyst
3. Hydration of acrylic acid in the presence of acid catalysts

These chemical routes are expensive and an alternate pathway for large-scale production of 3-HPA is shown in Fig. 3. This high molecular weight HPA and its derivatives are developed by University of Minnesota using a ring-opening polymerization. It is used in the industrial production of various chemicals such as acrylates and can be produced by engineered microbes. This high molecular weight HPA is more structurally sound and has low toxicity than other approaches.

Another approach for large-scale production of 3-HPA can be done through fermentation of sugars by employing genetically modified microorganisms. The process requires an improved microbial biocatalyst as well as reduction in the production cost.

6.2.3 L-Aspartic acid

Production of L-aspartic acid was first started by the Tanaba Seiyaku Company, Japan in 1973. This process uses aspartase contained in whole microorganisms and is involved in the immobilization of *Escherichia coli* on polyacrylamide gel. The

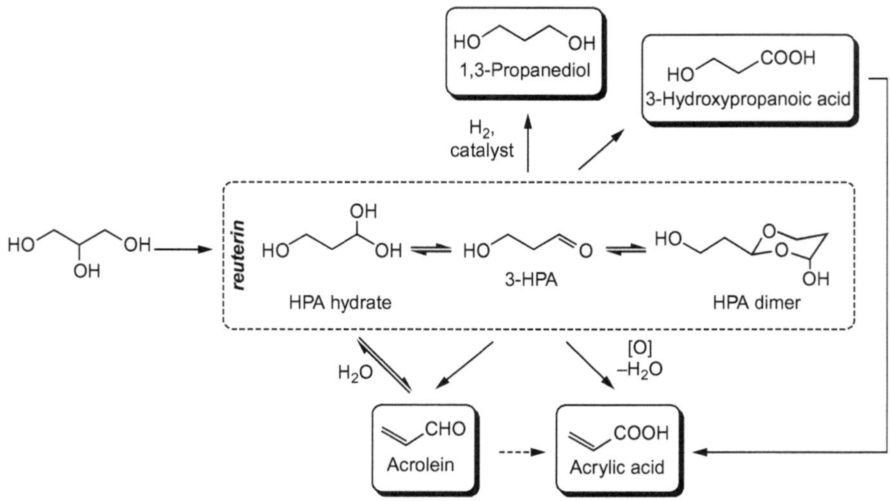

Fig. 3 Production of HPA and relative derivatives.
From S.W. Snyder, Commercializing Biobased Products: Opportunities, Challenges, Benefits, and Risks, Catalysis's Role in Bioproducts Update, 2015, p. 81 (Chapter 5).

immobilized cells are then treated to increase their cell permeability. Ammonium fumarate formed by dissolving fumaric acid substrate into 25% ammonia solution interacts with the immobilized *E. coli* in the reactor. This reaction is exothermic and it is important that the reactor is designed to remove the heat that is produced [15].

Direct fermentation of sugars using bacterial strain is another way to produce L-aspartic acid. First, fumaric acid is dissolved in ammonia solution to form ammonium fumarate and then interacts with immobilized biocatalysts that have high aspartase activity in a fixed bed reactor. Some of the microbes that are employed for industrial bioconversion of fumaric acid to L-aspartic acid include strains of *Brevibacterium*, *Corynebacterium*, *E. coli*, and *Pseudomonas*. Conversion of fumaric acid to aspartic acid is shown in the following reaction [15]:

$$\text{Fumaric} + NH_3 \rightarrow L - \text{aspartic acid}$$

The conversion of fumaric acid to aspartic acid is more economical than the direct fermentation of sugars [15].

6.2.4 Glucaric acid

Glucaric acid is derived from glucose using the following methods:

- Oxidation of glucose with nitric acid or facile nitroxide mediated
- Photooxidation with titanium photocatalysts

In each of these methods, glucose is used as a feedstock; however, in some cases, oxidation of biomass under acid environment using catalyst is employed. Use of sugar beet molasses in a packed bed with vanadium pentoxide as a catalyst with nitric acid, sulfuric acid, and sodium nitrite is one such method [16].

Fig. 4 shows a biosynthetic route to D-glucaric acid from D-glucose. It consists of three heterologous genes that had been constructed in recombinant *E. coli*. D-glucose was imported onto *E. coli* through the native phosphotransferase system, generating glucose-6-phosphate. Glucose-6-phosphate was then isomerized to *myo*-inositol-1-phosphate by *myo*-inositol-1-phosphate synthase from *Saccharomyces cerevisiae*. An endogenous phosphatase dephosphorylated *myo*-inositol-1-phosphate to obtain *myo*-inositol, which was oxidized to D-glucuronic acid by *myo*-inositol oxygenase from *Mus musculus*. D-Glucuronic acid was further oxidized by urinate dehydrogenase from *Pseudomonas syringae* to produce D-glucaric acid [17].

6.2.5 Glutamic acid

Amino acids are chemicals that form proteins which make up tissue, cells, and transmit chemical information within our bodies to our brains. Amino acids can be either essential or nonessential.

Essential amino acids are the ones in which the organisms are incapable of building themselves from other chemicals within their bodies and must ingest food. On the

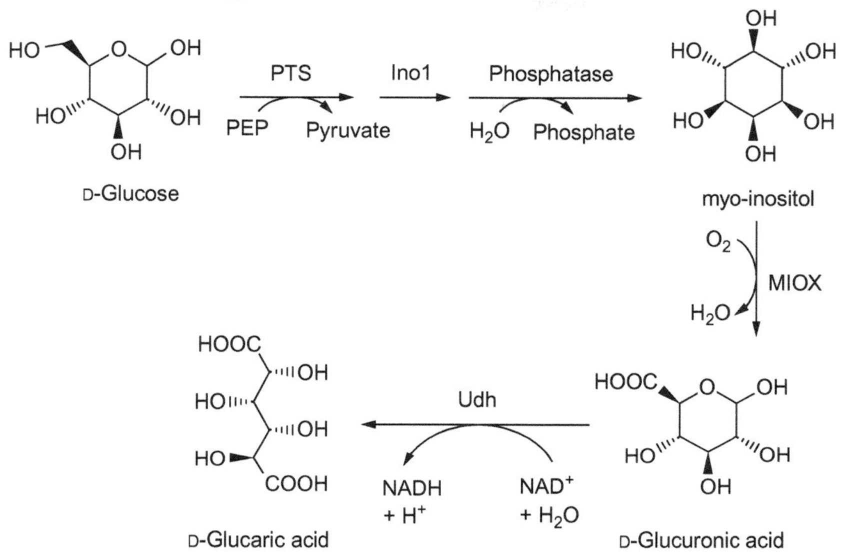

Fig. 4 Biosynthetic route to prepare D-glucaric acid from D-glucose.
From T.S. Moon, S.-H. Yoon, A.M. Lanza, J.D. R-Mayhew, K.L.J. Prather, Production of glucaric acid from a synthetic pathway in recombinant *Escherichia coli*, Appl. Environ. Microbiol. (2009) 589–595.

other hand, nonessential amino acids are the ones in which organisms can form in their body from other chemicals. Glutamic acid falls in nonessential amino acid category.

The chemical formula for glutamic acid is represented by:

$C_5H_9NO_4$

Fig. 5 shows the structure of glutamic acid [18,19]:

There are three possible configurations of glutamic acid as shown in Fig. 6 [20]:

L-Glutamic acid was first produced by fermentation process using strains of *Corynebacterium* and *Brevibacterium*. Glutamic acid concentration yields reached greater than 150 g/L during the fermentation process that lasted 40–60 h [20]. Glutamic acid is an important intermediate in Krebs cycle with αα-ketoglutarate as both are interconvertible by *trans* amination process.

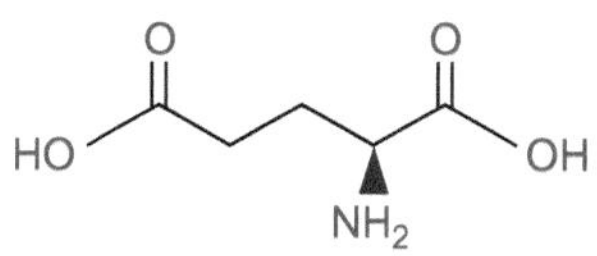

Fig. 5 Structure of glutamic acid.
From Glutamic Acid, Tutorvista. http://chemistry.tutorvista.com/biochemistry/glutamic-acid.html.

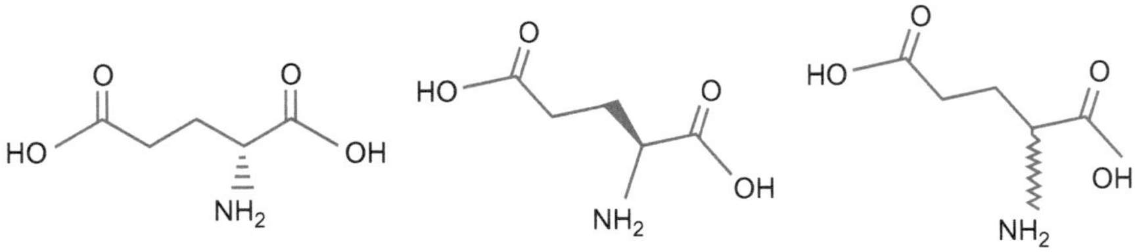

Fig. 6 Different configurations of glutamic acid.
From J.-J. Zhong, Developing a sustainable bioprocessing strategy based on a generic feedstock, Biomanufacturing, Springer, 2004, p. 224.

$$\text{Glutamate} + H_2O + NADP^+ \rightarrow \alpha\alpha - \text{ketoglutarate} + NADPH + NH_3 + H^+$$

L-Glutamic acid can also be produced by fermentation process using strains of *biotin* which is derived from corn steep liquor. The concentration of biotin has a significant influence on the yield of glutamic acid. In this fermentation process, α-ketoglutaric acid is an intermediate in the Krebs cycle and is the precursor of glutamic acid.

Fig. 7 shows Krebs cycle to produce L-glutamic acid. In this cycle, the initial conversion of α-ketoglutaric acid to glutamic acid is accomplished in the presence of different dehydrogenase such as glutamic acid dehydrogenase, ammonia, and nicotinamide adenine dinucleotide dehydrogenase. The living cells assimilate nitrogen by incorporating it into ketoglutaric acid and then to glutamic acid and glutamine. Reaction between ammonia and α-ketoglutaric acid in one of the tricarboxylic acid leads to the formation of glutamic acid [21].

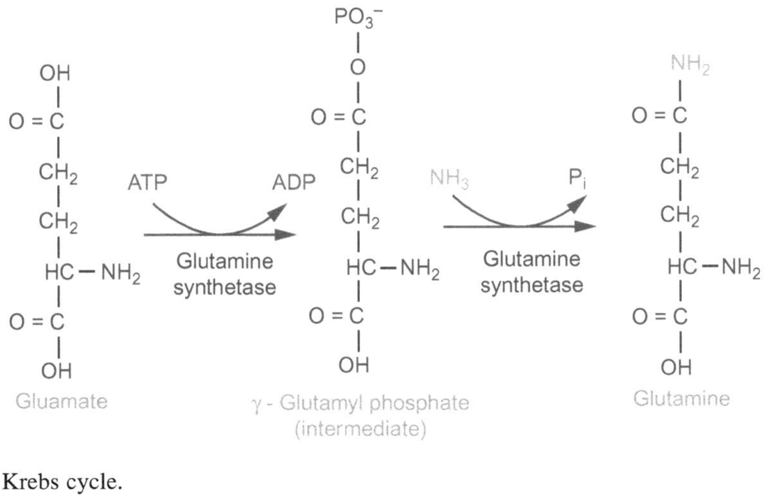

Fig. 7 Krebs cycle.
From G. Najafpour, Industrial Microbiology, Biochemical Engineering and Biotechnology, Elsevier, 2007, p. 9.

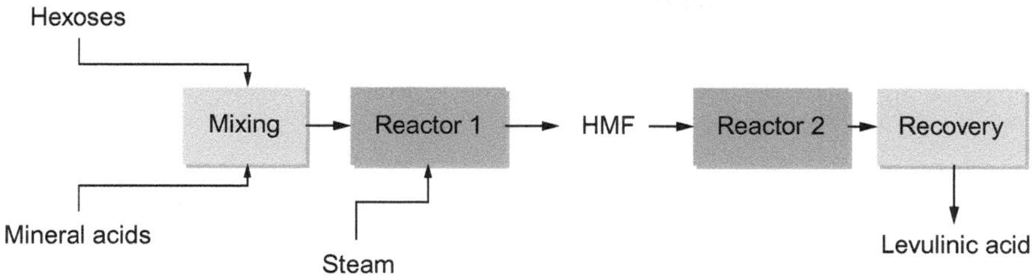

Fig. 8 Production of levulinic acid by Biofine process.
From K. Karimi, Introduction to lignocellulose-based bioproducts, Lignocellulose-Based Bioproducts, Springer, 2015, pp. 24–25.

6.2.6 Levulinic acid

Levulinic acid belongs to C5 chain fatty acid containing a ketone carbonyl and an acidic carboxyl group. Levulinic acid has been found useful in synthesis of alternative fuels, fuel additives, solvents, dyes, flavoring agents, and different resins.

Levulinic acid is produced by degradation of hexoses in the presence of mineral acids. During this process, 5-hydroxymethylfurfural (HMF) is obtained as an intermediate. Biofine Corporation developed an improved process for production of levulinic acid with significantly higher yield compared to traditional methods [22].

Fig. 8 shows a schematic of production of levulinic acid by Biofine process. Hexoses are fed to the first reactor where they are converted mainly to HMF in the presence of 1.5% mineral acid. HMF is continuously removed using steam and fed to the second reactor. The residence time in the first reactor is 15–25 s, while the reaction occurs at 210–230°C. Longer residence times of 15–30 min as well as lower temperatures of 145–230°C are applied in the second step to convert HMF to levulinic acid [22].

6.2.7 3-Hydroxybutyrolactone

3-Hydroxybutyrolactone is produced from starch oxidation through catalysts. Fig. 9 shows oxidative cleavage of 4-linked hexopyranose sugar, such as maltose, into (S)-3-Hydroxy-γ-butyrolactone. The reaction sequence involves:

- Base-catalyzed isomerization
- Elimination at C4
- Oxidative cleavage

The configuration at C3 in (S)-3-Hydroxy-γ-butyrolactone is derived from C5 in the hexose system, which is D in all readily available oligosaccharides and polysaccharides.

Starting material for (R)-3-Hydroxy-γ-butyrolactone is restricted to L-monosaccharides such as L-arabinose, which is derived from sugar beet pulp. Fig. 10 shows process of producing (R)-3-Hydroxy-γ-butyrolactone [23].

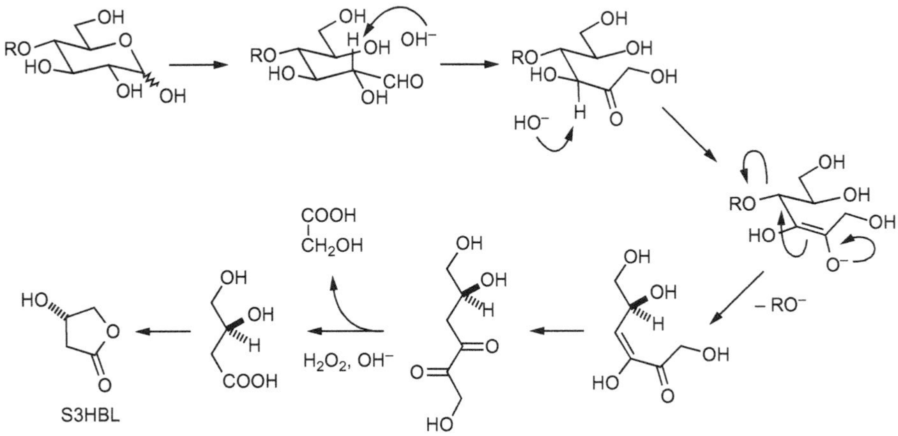

Fig. 9 Production of (S)-3-γ-Hydroxy-butyrolactone.
From R.A. Sheldon, I. Arends, U. Hanefeld, Chemicals from renewable raw materials, Green Chemistry and Catalysis, Wiley, 2007, pp. 368–370.

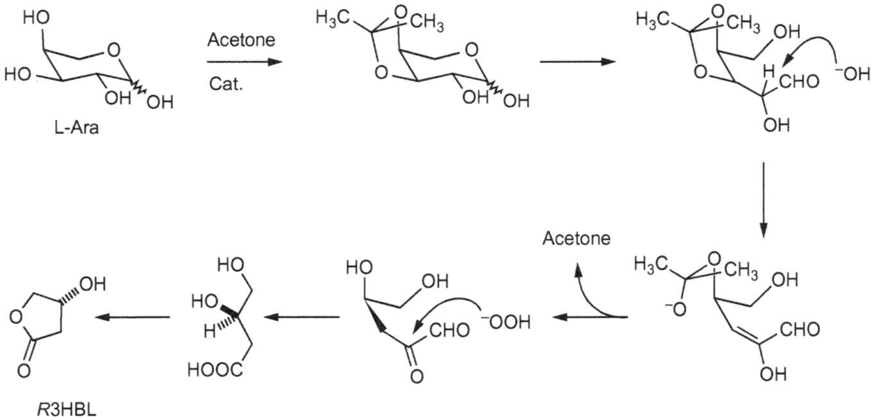

Fig. 10 Production of (R)-3-γ-Hydroxy-butyrolactone.
From R.A. Sheldon, I. Arends, U. Hanefeld, Chemicals from renewable raw materials, Green Chemistry and Catalysis, Wiley, 2007, pp. 368–370.

6.2.8 *1,2,3-Propanetriol*

1,2,3-Propanetriol also known as glycerol is an alcohol produced either by microbial fermentation, by chemical synthesis from fossil feedstocks, or recovered as a by-product from a soap manufacturing process. Glycerol has found its use in cosmetic, paint, automotive, food, tobacco, pharmaceutical, pulp and paper, leather and textile industries, or it can be used as a feedstock in the production of various chemicals.

Glycerol is a by-product of the fermentation of sugar to ethanol in a redox-neutral process. The role of NADH-consuming glycerol formation is to maintain the cytosolic redox balance especially under anaerobic conditions, compensating for cellular reactions that produce nicotinamide-adenine dinucleotide (NADH) [24].

Substantial overproduction of glycerol from monosaccharides can be obtained by yeast [24]:

(1) Forming a complex of acetaldehyde with the bisulfite ion
(2) Growing at pH values around 7 or above
(3) Osmotolerant yeasts

Substantial overproduction of glycerol by *S. cerevisiae* from monosaccharides can be achieved by [24]:

- Combining acetaldehyde with the bisulfite ion
- Growing the cell at pH values around 7 or above.

Following are two methods to yield glycerol:

(1) Method 1: Based on the trapping of acetaldehyde by bisulfite ions, yields the following reaction:

$$\text{Hexose} + \text{Bisulfate} \rightarrow \text{acetaldehyde} - \text{bisulfite} + CO_2 + H_2O + \text{glycerol}$$

(2) Method 2: Based on the following reaction operated at pH values of 7 or above:

$$2\text{Hexose} \rightarrow 2\text{glycerol} + \text{ethanol} + \text{acetic acid} + 2CO_2 + 2H_2O$$

6.2.9 *Sorbitol*

Sorbitol exists as a white crystal or crystalline powder with a molecular formula $C_6H_{14}O_6$ and chemical structure as shown in Fig. 11.

Some of the physical properties of sorbitol include:

- Melting point is greater than 95°C
- Soluble in water, but slightly soluble in organic solvents
- Good resistance to acid and heat

Sorbitol is produced commercially by the hydrogenation of glucose. D-Sorbitol is oxidized by nicotinamide-adenine dinucleotide (NAD+) to D-fructose in the presence

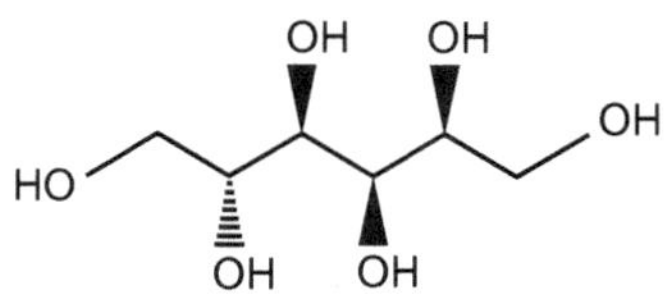

Fig. 11 Chemical structure of sorbitol.
From Sorbitol, Huakang Pharma. http://www.huakangpharma.com/en/index.php/product/1.

of sorbitol dehydrogenase (SDH) with the formation of reduced nicotinamide-adenine dinucleotide (NADH) [25].

$$\text{D}-\text{Sorbitol}+\text{NAD}^{+}\xrightleftharpoons{(\text{SDH})}\text{D}-\text{fructose}+\text{NADH}+\text{H}^{+}$$

An additional reaction is required to utilize the NADH product. In this third reaction, in the presence of diaphorase, NADH reduces iodonitrotetrazolium chloride (INT) to an INT-formazan compound [25].

$$\text{NADH}+\text{INT}+\text{H}^{+}\xrightarrow{(\text{diaphorase})}\text{NAD}^{+}+\text{INT}-\text{formazan}$$

6.2.10 Xylitol

Xylitol is a naturally occurring C5 sugar alcohol that is used commercially as a natural sweetener in various food products. Xylitol is industrially manufactured by reducing pure xylose achieved from hardwood hemicellulosic hydrolyzate in the presence of a Raney nickel catalyst [26].

The chemical process to manufacture xylitol is shown in Fig. 12. It consists of four major steps [26]:

- Hydrolysis of lignocellulosic biomass by mineral acid
- Purification and separation of the hydrolyzate to obtain pure xylose as solution or in crystalline form
- Catalytic reduction of the xylose to xylitol
- Crystallization and separation of the xylitol.

6.3 Plant-derived industrial chemicals

6.3.1 Oleochemicals

There are four classes of basic oleochemicals:

1. Acids
2. methyl (or alkyl) esters
3. Alcohols
4. Amines

Fig. 13 shows the relationship between the oils and fats and the basic oleochemicals [27].

6.3.1.1 Acids

Fats undergo hydrolysis to form free acid in a homogeneous medium. This is a continuous reaction carried under pressure (2–6 MPa) and at a temperature of 250°C. At high temperatures, the products become discolored and both fatty acids and the glycerol may be subsequently distilled [27].

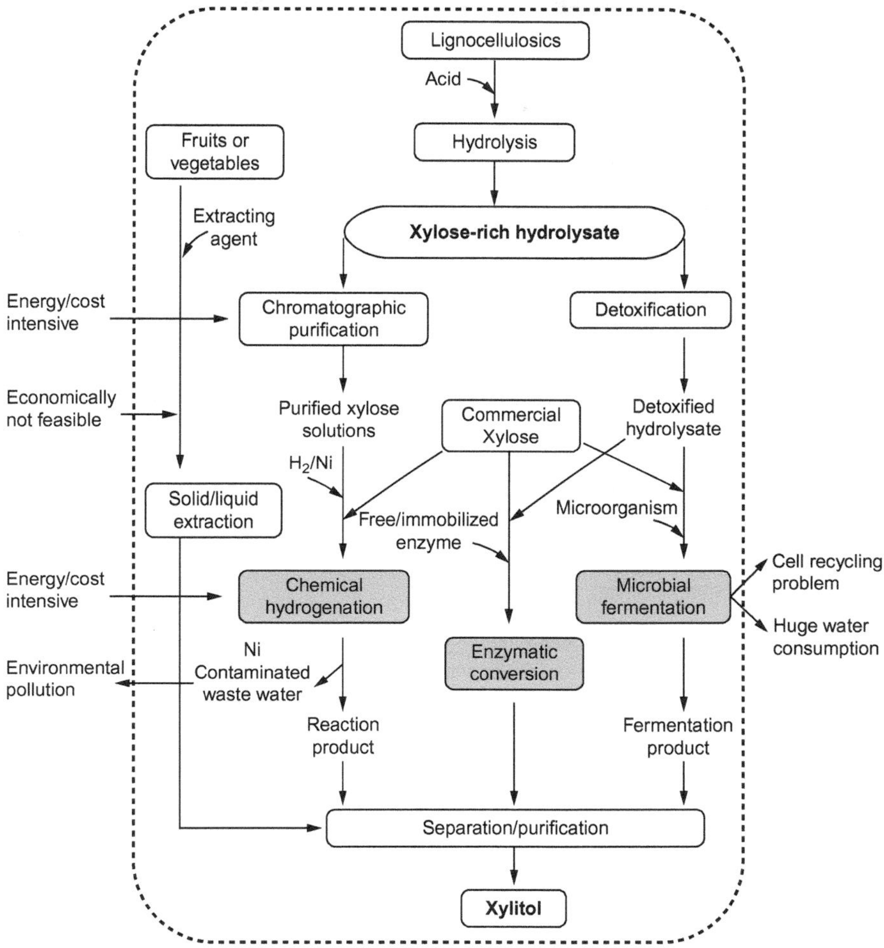

Fig. 12 Chemical process to manufacture xylitol.
From I.S.M. Rafiqul, A.M. Mimi Sakinah, Processes for the production of Xylitol-A—a review, Food Rev. Int. 29 (2013) 127–156, Taylor & Francis Group.

6.3.1.2 Methyl or alkyl esters

Esters can be produced by several routes. The most common way is when triacylglycerol reacts with methanol in the presence of catalyst. Methanolysis is carried out on a large scale to make methyl esters for use in biodiesel. Oil low in free to acid can be converted directly to methyl esters with an alkaline catalyst. Under the optimum conditions the reaction requires 6–8 min and takes place during passage through the reaction equipment [27].

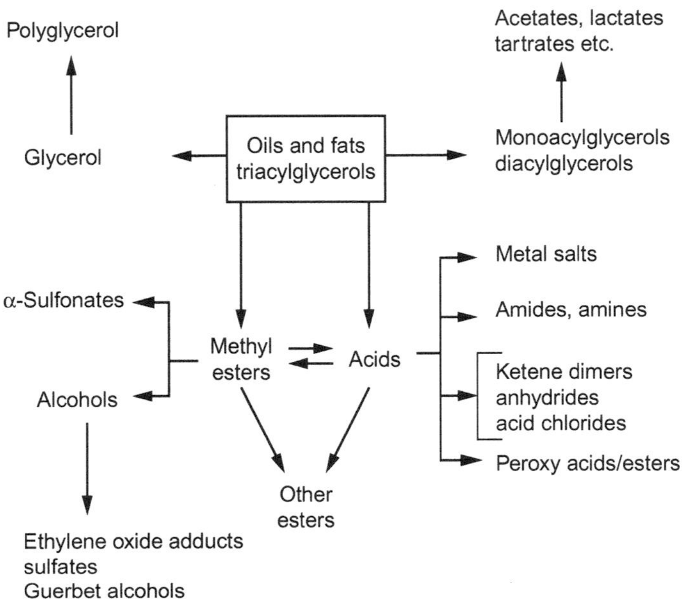

Fig. 13 Schematics of basic oleochemicals.
From F.D. Gunstone, R.J. Hamilton, Basic oleochemicals, Oleochemical Manufacture and Applications, CRC Press, 2001, pp. 4–8.

6.3.1.3 Alcohols

There are several ways to produce alcohol:

- The oleochemical route to alcohols from lauric oils, tallow, or palm stearin involves hydrogenolysis of glycerol ester, methyl ester, free acid, or wax ester.
- Catalytic reduction of methyl esters at high temperature and pressure using copper chromite as catalyst.

6.3.1.4 Fatty amines

There are three steps to produce fatty amines:

Step 1: Acids are converted to nitriles, probably via amides, by reaction with ammonia at 280–360°C in the presence of zinc oxide, manganese acetate, bauxite, or cobalt salts as catalyst.
Step 2: The nitriles are then reduced to primary or secondary amines by hydrogenation in the presence of ammonia and a nickel or cobalt catalyst.
Step 3: These are converted to tertiary amines by catalytic reaction with formaldehyde and to quaternary ammonium compounds by further reaction with alkylating agents such as methyl chloride or sulfate or benzyl chloride.

6.3.2 Polyhydroxyalkanoates (PHA)

Polyhydroxyalkanoates (PHAs) are a family of intracellular biopolymers produced by bacterial fermentation of sugar or lipids. They are produced by many bacteria to store carbon and energy. The structure of polyhydroxyalkanoates (PHAs) is shown in Fig. 14 [28]. Molecular weight of these polyesters can range from tens into the hundreds of thousands depending on growth conditions, bacterial strain, and carbon source [29]. Bacterially synthesized PHAs have found importance because they can be produced from a variety of renewable resources and are considered biodegradable.

PHAs are formed mainly from saturated and unsaturated hydroxyalkanoic acids. The monomer of PHA can be branched or unbranched 3-hydroxyalkanoic acids or those with substituted side chains as 4- and 5-hydroxyalkanoic acids. PHAs are homo-, co-, and terpolymers depending on the kind of monomer. Wide variety of PHAs with different properties can be created from variety of monomers and by varying constitutional isomerism.

Fig. 15A and B shows process of PHA synthesis from bacteria. Accumulation of PHA is an inherent response to the stress conditions faced by bacterial cells. Stress conditions faced by bacterial cells are generated in vitro by exposing bacteria to nutrient limitations, due to which they switch their metabolic pathways and cause PHA production as their carbon and energy reserves to name a few of these substrates are bagasse, molasses, corn cob, and other agricultural wastes [30].

Different microorganisms can be used to produce PHAs. There are more than 300 different microorganisms that generate PHAs as natural energy reserves. PHAs tend to degrade if they lack carbon or energy. Microorganism's stability and biological safety, its PHA production rates, PHA extractability, the molecular weight of the agglomerated PHA, as well as the spectrum of useable carbon sources will dictate the type of microorganism for an application is needed [31].

The first step of the bacterial fermentation process is inoculation. In this process, the bacteria required for the subsequent metabolization process multiply and grow in an aqueous medium enriched with a balanced nutrition supply and air under optimum physical conditions. The next step involves actual PHA synthesis under conditions not conducive to growth and multiplication and a relative oversupply of Carbon C. The PHAs are usually stored in intracellular inclusion bodies and can account for up to 90% of dry cellular weight. Their molecular weight generally ranges from 100,000 to 500,000 g/mol. However, molecular weights of considerably more than 1,000,000 g/mol are obtained under special conditions. The complete fermentation process typically takes two days [31].

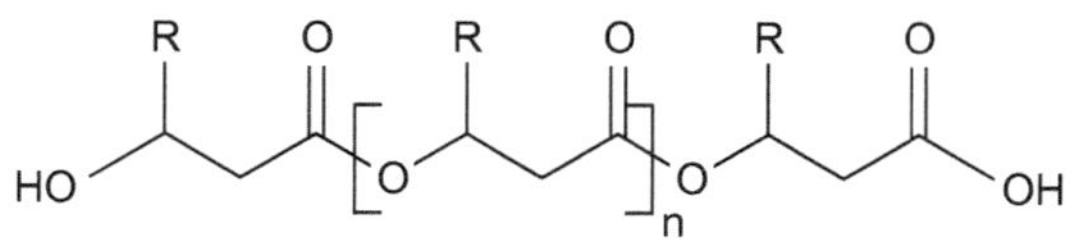

Fig. 14 Structure of polyhydroxyalkanoates (PHA).
From What are PHAs, Renewable and Biodegradable Polymer for Engineered Solution, PolyFerm Canada. http://www.polyfermcanada.com/pha.html.

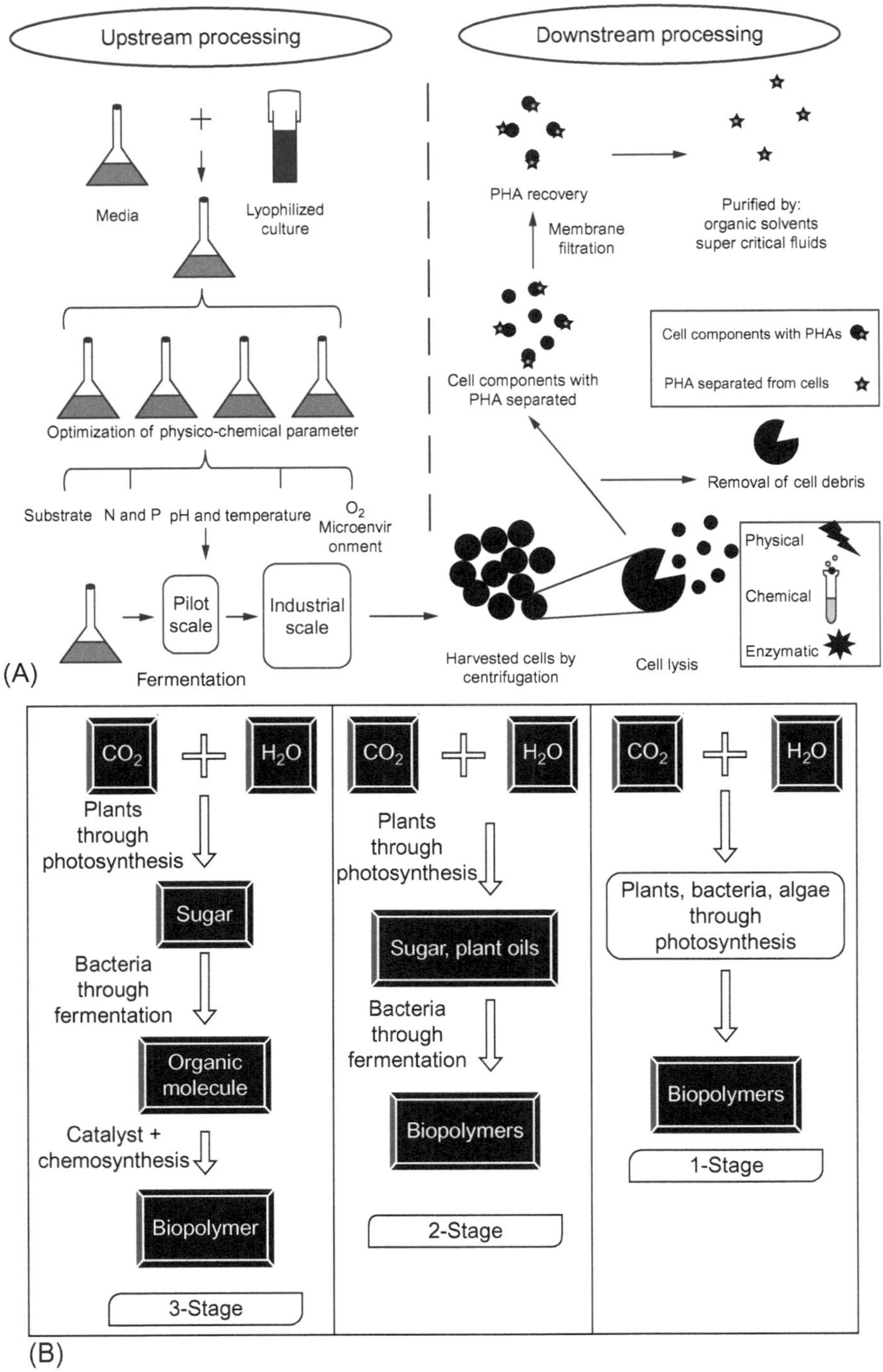

Fig. 15 (A) Synthesis of PHA and (B) systems for PHA synthesis.
From A. Girdhar, M. Bhatia, S. Nagpal, A. Kanampalliwar, A. Tiwari, Process parameters for influencing polyhydroxyalkanoate producing bacterial factories: an overview, J. Pet. Environ. Biotechnol. 4(5) (2013).

Some of the general characteristics of PHAs are as follows:

- Water insoluble and relatively resistant to hydrolytic degradation
- Good ultraviolet resistance but poor resistance to acids and bases
- Soluble in chloroform and other chlorinated hydrocarbons
- Biocompatible and hence suitable for medical applications
- Sinks in water, facilitating its anaerobic biodegradation in sediments
- Nontoxic
- Less sticky than traditional polymers when melted

6.4 Catalytic-derived industrial chemicals

6.4.1 Succinic acid

Succinic acid is also produced by anaerobic fermentation of several wild-type strains [32–34]. Succinic acid is considered as one of the major bio-based platform chemicals and lead to different products as shown in Fig. 16 [35].

Succinic acid is commonly produced as part of two-step reaction:

1. Maleic anhydride is hydrogenated to succinic anhydride as shown in Fig. 17. Choice of the catalyst and process conditions for the hydrogenation of maleic anhydride are important. The catalyst typically used for this step is an alloy of nickel, zirconium, aluminum, and silicon [36].

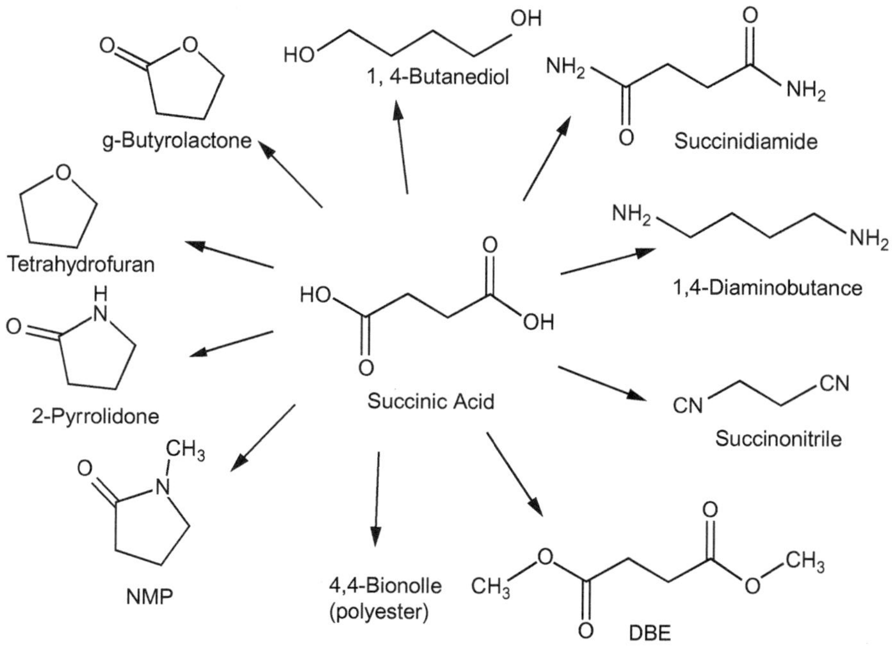

Fig. 16 Different products of succinic acid.
From L. Stols, M.I. Donnelly, Appl. Environ. Microbiol. 63 (1997) 2695–2701.

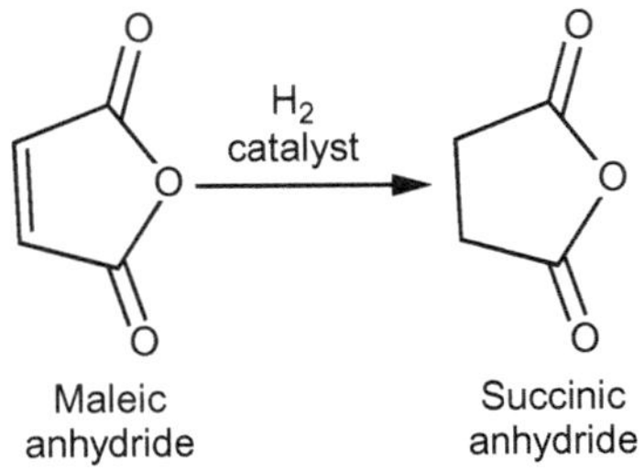

Fig. 17 Hydrogenation of maleic anhydride.
From S. Vaswani, Bio-Based Succinic Acid, PEP Review 2010–2014, U. S. Department of Energy. https://www.ihs.com/pdf/RW2010-14_220240110917062932.pdf.

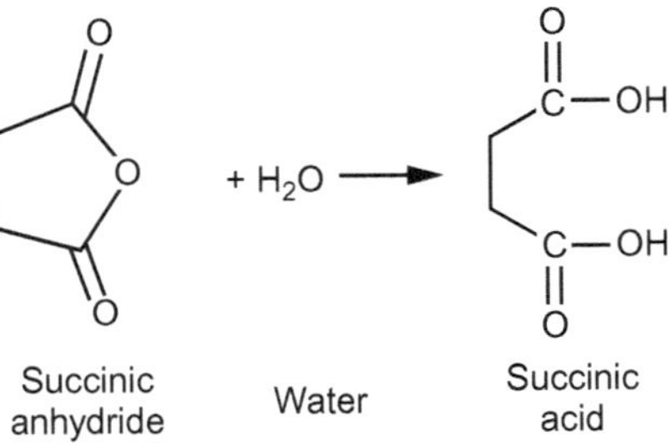

Fig. 18 Hydration of succinic anhydride to succinic acid.
From S. Vaswani, Bio-Based Succinic Acid, PEP Review 2010–2014, U. S. Department of Energy. https://www.ihs.com/pdf/RW2010-14_220240110917062932.pdf.

2. Hydration of succinic anhydride to succinic acid (Fig. 18) [36].

Succinic acid is used in a variety of applications in (a) food, (b) pharmaceuticals, (c) agricultural, and (d) industrial. Table 1 shows different uses for each application [37].

6.4.2 *Fumaric acid*

Fumaric acid is found naturally in many plants and is named after the genus *Fumaria*. It is derived as a by-product in the manufacture of phthalic acid and maleic anhydride or by the catalytic isomerization of maleic acid [38]. Mineral acids, peroxy compounds with bromides and bromates, and sulfur-containing compounds can be used as catalysts.

A high purity fumaric acid can be prepared by cooling the reaction mixture and then separating, washing, and drying the crystals. Another way to produce fumaric acid is by the fermentation of glucose and molasses with certain strains of *Rhizopus nigricans* and *R. japonicas*.

Chemical structure of fumaric acid is shown in Fig. 19 [38]:

6.4.3 *Malic acid*

Chemical structure of malic acid is shown in Fig. 20:

Table 1 Different applications of succinic acid

Area of application	Uses
Food	• As a flavoring agent in food and beverages • For dry gelatin desserts, for preparing cake flavorings • As bread softening agents • As sodium succinate as flavor enhancer • As catalyst for food seasoning preparation • In the synthesis of modified starch • As micro encapsulation of flavoring oils • For protein gelatination • In the preservation of chicken against Salmonella • In the synthesis of modified starch • Dog food preservative
Pharmaceuticals	• In the preparation of active calcium succinate, which acts as anticarcinogenic agent and as deodorant for removal of fish odor • Starting material for Active Pharmaceutical Ingredients (APIs) including adipic acid, N-methyl pyrrolidinone, 2-pyrrolidinone, succinate salts, 1,4-butanediol, tetrahydrofuran and gamma-butyrolactone • As an additive in formulation, as an insulinotropic agent, and in drug control release polymers • Medicines of sedative, antispasmer, antiplegm, antiphogistic, anrhoter, contraception, and cancer curing • In the preparation of Vitamin A • In the preparation of antiinflammatory erthyrodiol derivates • Antidote for toxic substance
Agriculture	• Seed treatment to enhance germination, growth, and yield • Plant growth regulator and accelerator • Root stimulator for grapes • Soil chelating agent • Crop dust antistat • Cut flower preservative • Component of barley seed antismut treatment
Industrial	• For curing epoxy resins, polyester synthesis, and other applications • As an intermediate in the manufacture of dyes, perfumes, lacquers, photographic chemicals, alkyd resins, plasticizer, metal treatment chemical, vehicle water cooling systems, alkyd resins and coatings, and succinates • Plasticizer for polymers • Biodegradable solvents and lubricants • Engineering plastics • Adhesive and powder coating • Corrosion inhibitor

From J.W. Lee, H.U. Kim, S. Choi, J. Yi, S.Y. Lee, Curr. Opin. Biotechnol. 22 (2011) 758–767; S.A. Ashter, Types of biodegradable polymers, Introduction to Bioplastics Engineering, Elsevier Publishers, 2016, pp. 104–112 (Chapter 5).

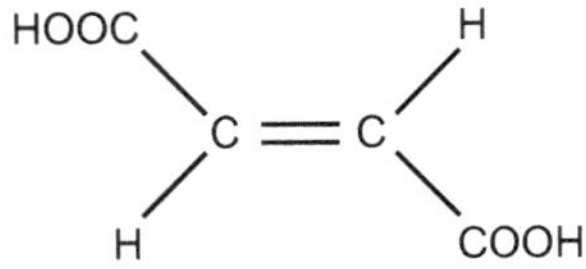

Fig. 19 Chemical structure of fumaric acid.
From J.A. Maga, A.T. Tu, Food acidulants, Food Additive Toxicology, Marcel Dekker, 1994, p. 39.

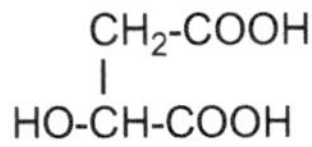

Fig. 20 Chemical structure of malic acid.
From Y.H. Hui, G.G. Khachatourians, Microorganisms for Organic Acid Production, Food Biotechnology: Microorganisms, Wiley, 2005, p. 263.

There are several routes to produce malic acid:

1. Hydration of maleic or fumaric acid at elevated temperature and pressure.
2. Immobilized cells of a species of *Brevibacterium* are treated with bile extract to suppress the production of succinic acid, an unwanted by-product.

The continuous production of L-malic acid from fumaric acid is efficient because the immobilization process maintains high enzyme activity at high enzyme yield, and it exhibits high operational stability.

6.4.4 Itaconic acid

Itaconic acid also known as 2-methylidenebutanedioic acid is an unsaturated dicarbonic acid. The chemical structure is shown in Fig. 21 [39].

Itaconic acid is produced from pyruvate via three different routes:

1. Incorporating pyruvate into the tricarboxylic acid cycle
2. Condensation of pyruvate and acetyl coenzyme A (acetyl-CoA) to citramalate
3. Transferring acetyl-CoA on succinate to form 1, 2, 3-tricarboxypropanoate followed by dehydrogenation reaction

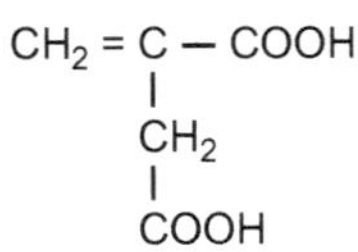

Fig. 21 Chemical structure of 2-methylidenebutanedioic acid.
From M. Okabe, D. Lies, S. Kanamasa, E.Y. Park, Biotechnological production of itaconic acid and its biosynthesis in *Aspergillus terreus*. Appl. Microbiol. Biotechnol. 84 (2009) 597–606.

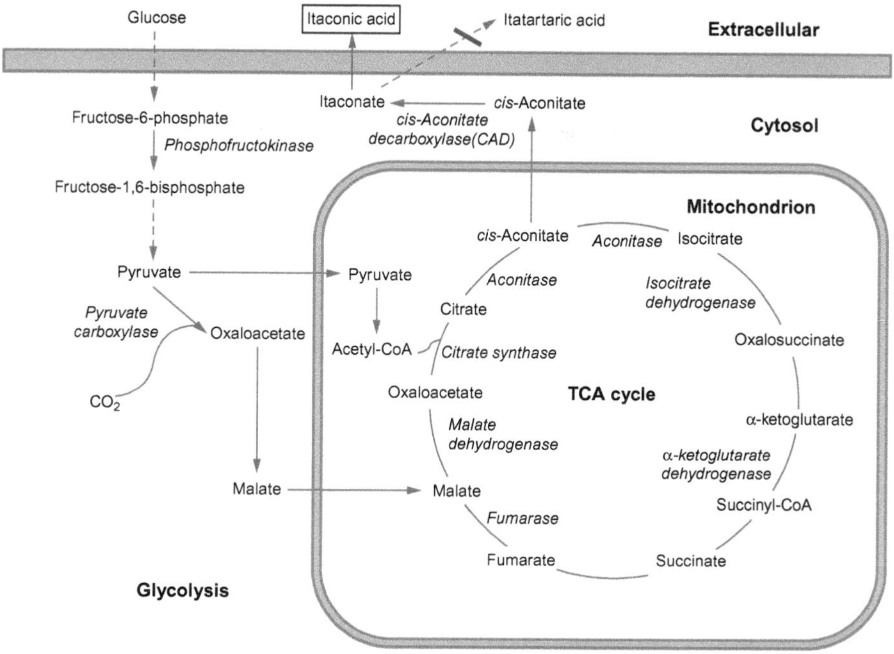

Fig. 22 Biosynthesis pathway of itaconic acid via aconitase and *cis*-aconitate decarboxylase. From T. Klement, J. Buchs, Itaconic acid—a biotechnological process in change, Bioresour. Technol. 135 (2013) 422–431.

Fig. 22 shows biosynthesis pathway of itaconic acid via aconitase and *cis*-aconitate decarboxylase. *A. terreus* is cultivated at high substrate concentrations and a low pH value under phosphate-limited conditions. Glucose is converted into two molecules of pyruvate via glycolysis. One pyruvate molecule is decarboxylated and further converted to acetyl-CoA while the other pyruvate molecule is used in an anaplerotic reaction to replenish the oxaloacetate pool. Afterward, both oxaloacetate and acetyl-CoA are combined by the citrate synthase (CS) in the first step of the TCA cycle to yield citrate [40].

During the glycolysis, phosphofructokinase (PFK) and the citrate synthase both catalyze irreversible reactions as shown in Fig. 22. For enzyme *Aspergillus niger*, inhibitory effect of PFK on CS is overcome by a high oxaloacetate concentration. Enzyme pyruvate carboxylase (PC) activity generates new oxaloacetate and replenishes the loss of TCA intermediates. For enzyme *Aspergillus terreus*, PC is located in the cytosol where the cleaved carbon dioxide from pyruvate or *cis*-aconitate is utilized [40].

6.5 Pyrolysis-derived industrial chemicals

Pyrolysis is an efficient method to convert biomass into energy. For example, char can be produced from conventional slow pyrolysis, and liquids or gases from fast pyrolysis. As the main component of biomass, cellulose pyrolysis plays an important role in the investigation of biomass pyrolysis.

There are two important chemicals derived from cellulose pyrolysis:

1. Levoglucosan
2. Levoglucosenone

6.5.1 Levoglucosan

Pyrolysis of cellulose results in levoglucosan, either as an intermediate or as a product. Researchers have studied different mechanisms for levoglucosan formation. They are as follows:

- Glucose intermediate mechanism
- Free radical mechanism
- Ionic mechanism
- Levoglucosan chain-end mechanism

Fig. 23 summarizes the formation mechanism of levoglucosan from cellulose via glucose intermediate, via free-radical mechanism, via ionic mechanism, and via levoglucosan chain-end mechanism [41].

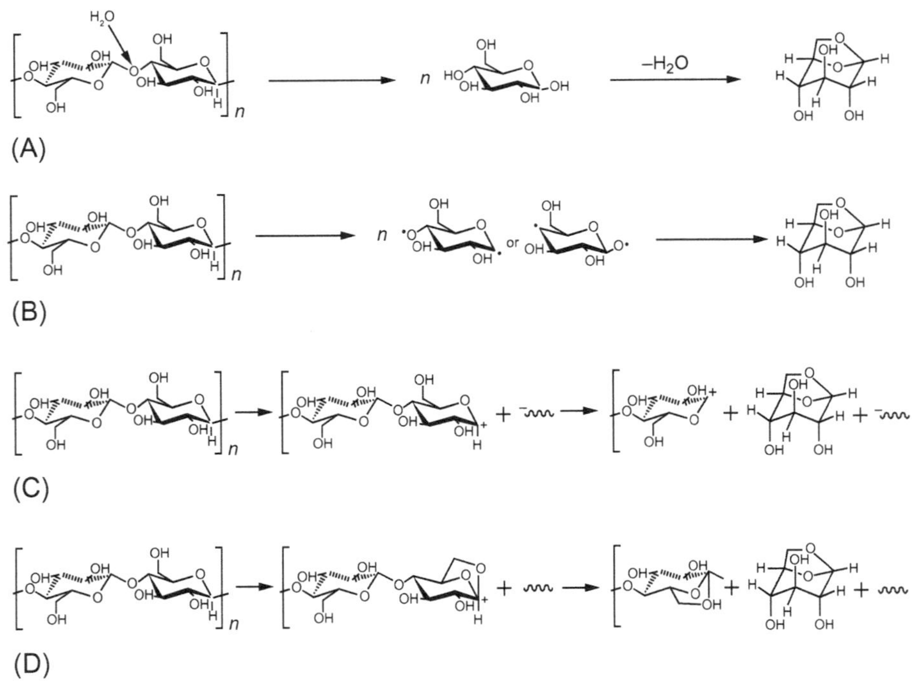

Fig. 23 Formation mechanism of levoglucosan (A) via glucose intermediate, (B) via free radical, (C) via ionic, and (D) via levoglucosan chain end.
From X. Zhang, W. Yang, C. Dong, Levoglucosan formation mechanisms during cellulose pyrolysis, J. Anal. Appl. Pyrolysis 104 (2013), 19–27, Elsevier.

Mayes et al. developed levoglucosan formation mechanism from glucose intermediate which describe that the cellulose chain will break into dehydroglucose diradicals and then levoglucosan can be formed from these dehydroglucose diradicals. Oldham's work has shown that levoglucosan can be formed from glucose which was produced by hydration of cellulose chain. Free radical mechanism has been proposed by Shen et al. This mechanism proposes that levoglucosan radical with one unpaired electron will be transferred into levoglucosan by reacting with hydroxyl radical. This mechanism is not representative for levoglucosan formation because the limitation of hydroxyl groups existed in biomass pyrolysis process [41].

Ionic mechanism proposes that the cellulose can produce levoglucosan by an ionic intermediate. Regarding radical cleavage and ionic cleavage, it is proved that under the gas atmosphere, which is the atmosphere for real biomass gasification, ionic cleavage needs quite more energy than radical cleavage. Two transglycosylation steps are part of levoglucosan chain-end mechanism. In the first transglycosylation step, a cellulose chain is depolymerized into a levoglucosan-end intermediate and a short cellulose chain. During the second transglycosylation step, the levoglucosan-end intermediate which is formed from the first step will be unzipped into a levoglucosan molecule, with another levoglucosan-end intermediate formed by the remaining part [41].

6.5.2 *Levoglucosenone*

Chemical structure of levoglucosenone is shown in Fig. 24. It is formed by a combination of transglycosylation and dehydration reactions of the D-glucosyl residues in cellulose. The transglycosylation reactions at 300–500°C in the presence of an acid catalyst promote the dehydration and charring reactions producing levoglucosenone [42]. Higher yields of levoglucosenone can be obtained if pyrolysis happens at atmospheric pressure.

6.6 Gasification-derived industrial chemicals

In a gasification process, the biomass undergoes thermal treatment to produce high yield of gaseous products and small quantities of char and ash. Eqs. (1)–(4) show reaction steps of thermal decomposition of the lignocellulosic resulting in the production of volatiles and char [43].

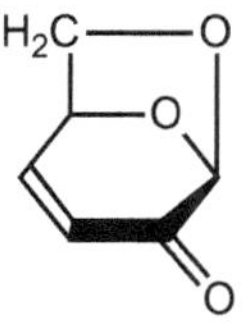

Fig. 24 Chemical structure of levoglucosenone.
From F. Shafizadeh, R.H. Furneaux, T.T. Stevenson, Some reactions of levoglucosenone, Carbohydr. Res. 71(1) (1979) 169–191, Elsevier.

$$C + H_2O \leftrightarrow CO + H_2 \quad (1)$$

$$C + CO_2 \leftrightarrow 2CO \quad (2)$$

$$CO + H_2O \leftrightarrow H_2 + CO_2 \quad (3)$$

$$CH_4 + H_2O \leftrightarrow CO + 3H_2 \quad (4)$$

6.6.1 Synthesis gas

In syngas, although the total hydrogen content of natural gas is high, the amount of free hydrogen is low. Because of this characteristic, natural gas is not as easy to burn as some manufactured gases with their high free hydrogen content. The high hydrogen content of natural gas results in more water vapor produced in the combustion gases with a correspondingly lower efficiency [44]. Table 2 shows chemical composition of syngas [44].

Syngas can be produced by two approaches:

- Primary method
- Secondary method

The primary method involves gasification process and tar elimination is carried out simultaneously in gasifier. Although primary methods eliminate the need for downstream cleanup, they cannot effectively solve the purpose of tar reduction without affecting the useful gas composition and heating value. As a result, the syngas yields in primary methods will be relatively low compared with those in secondary methods.

The secondary method involves gas cleanup in a separate reformer in downstream gasifier. Extensive studies on secondary methods of biomass gasification have also been conducted. By steam gasification of pine sawdust using an updraft gasifier combined with a porous ceramic reformer, Gao et al. obtained syngas with a maximum yield of 1.72 N m^3/kg biomass and lower heating value (LHV) of 11.73 MJ/N m^3 at 950°C. Secondary methods are effective in reducing tar content and improving syngas yield, but additional equipment required will increase the investment [45].

Table 2 Chemical composition of syngas

Composition	Coal gas (%)	Biogas (%)	Natural gas (%)
H	14.0	18.0	–
CO	27.0	24.0	–
CO_2	4.5	6.0	–
O_2	0.6	0.4	–
CH_4	3.0	3.0	90.0
N_2	50.9	48.6	5.0
C_2H_6	–	–	5.0

From M.A. Yaqoob, Production of Syngas From Biomass and Its Purification, University of Engineering and Technology, Lahore, 2011.

6.6.2 Tar chemicals

Tar chemicals are mixture of condensable organic compounds and can vary from oxygenated products to heavy deoxygenated hydrocarbons and polycyclic aromatic compounds. Production of different species of tar depends on the operating parameters, feedstock composition, and process conditions such as temperature, pressure, type of oxidant, pressure, amount of oxidant, and feedstock residence time. One such example is the gasification of wood with higher amounts of stable aromatics results in higher tar concentration compared to coal [44].

Biomass tars undergo reactions when held at elevated temperatures at increased residence times. Reactions can even take place at room temperature if the tar is exposed to air.

Depending on the conditions, different reactions can occur:

1. Reactions between liquid/gaseous tar compounds and carbonaceous ash/char
2. Reactions between liquid tar compounds
3. Decomposition reactions in the gas phase in an inert atmosphere

A general reaction model for hydropyrolysis of aromatic compounds is developed by Nelson and Hüttinger and is shown in Fig. 25 [46].

Destabilization of the tar compound and radical information are two important steps that determine the reaction rate. Tar decomposition rate is not affected by hydrogen. When coal tars decompose, they generate reactive volatiles and radicals. Hydrogen can react with these volatiles and radicals to form stable tar compounds that do not decompose any further. This reaction with hydrogen prevents the reactive volatiles or radicals to react and to form carbon or high-molecular tar. If the pressure is increased, such subsequent reactions of radicals to form tar increase, taking place in the carbon particles, with the result that the volatile compounds cannot escape. Thus, hydrogasification reactions result in a reduced char yield and an increased yield of methane and other gases, while the tar concentration remains about the same or rises, compared with pyrolysis reactions under inert conditions [46].

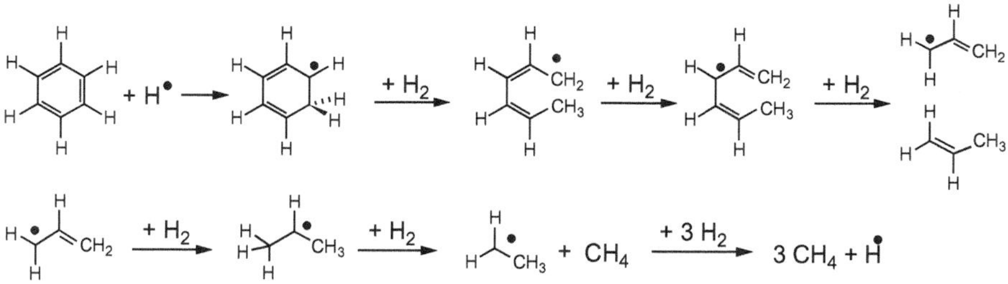

Fig. 25 General reaction model for hydropyrolysis of aromatic compounds is developed by Nelson and Hüttinger.
From B.J. Vreugdenhil, R.W.R. Zwart, Tar Formation in Pyrolysis and Gasification, Energy Research Center of the Netherlands, 2009.

References

[1] S.W. Snyder, Commercializing biobased products: opportunities, challenges, benefits, and, risks, Catalysis's Role in Bioproducts Update, RSC Publishing, Cambridge, UK, 2015, pp. 81 (Chapter 5).

[2] A. Gandini, Furans as offspring of sugars and polysaccharides and progenitors of a family of remarkable polymers: a review of recent progress, Polym. Chem. 1 (2010) 245–251.

[3] A. Gandini, M.N. Belgacem, Furans in polymer chemistry, Prog. Polym. Sci. 22 (1997) 1203–1379.

[4] A. Gandini, D. Coelho, M. Gomes, B. Reis, A. Silvestra, Materials from renewable resources based on furan monomers and furan chemistry: work in progress, J. Mater. Chem. 19 (2009) 8656–8664.

[5] M. Bicker, J. Hirth, H. Vogel, Dehydration of fructose to 5-hydroxymethylfurfural in sub- and supercritical acetone, Green Chem. 5 (2003) 280–284.

[6] Y. R-Leshkov, J.N. Chheda, J.A. Dumesic, Phase modifiers promote efficient production of hydroxymethylfurfural from fructose, Science 312 (2006) 1933–1937.

[7] C. Moreau, M.N. Belgacem, A. Gandini, Recent catalytic advances in the chemistry of substituted furans from carbohydrates and in the ensuing polymers, Top. Catal. 385 (2004) 1–13.

[8] Y.Y. Gorbanev, S.K. Klitgaard, J.M. Woodley, C.H. Christensen, A. Riisager, Gold-catalyzed aerobic oxidation of 5-hydroxymethylfurfural in water at ambient temperature, ChemSusChem 2 (7) (2009) 672–675.

[9] N.K. Gupta, S. Nishimura, A. Takagaki, K. Ebitani, Hydrotalcite-supported gold-nanoparticle-catalyzed highly efficient base-free aqueous oxidation of 5-hydroxymethylfurfural into 2,5-furandicarboxylic acid under atmospheric oxygen pressure, Green Chem. 12 (2011) 824–827.

[10] E. Taarning, I.S. Nielsen, K. Egeblad, R. Madsen, C.H. Christensen, Chemicals from renewables: aerobic oxidation of furfural and hydroxymethylfurfural over gold catalysts, ChemSusChem 1 (2008) 75–78.

[11] O. Casanova, S. Iborra, A. Corma, Biomass into chemicals: one pot-base free oxidative esterification of 5-hydroxymethyl-2-furfural into 2,5-dimethylfuroate with gold on nano-particulated ceria, J. Catal. 265 (2009) 109–116.

[12] H. Zhou, J.E. Holladay, H. Brown, C. Zhang, Metal chlorides in ionic liquid solvents convert sugars to 5-hydroxymethylfurfural, Science 316 (2007) 1597–1600.

[13] M.L. Ribeiro, U. Schuchardt, Cooperative effect of cobalt acetylacetonate and silica in the catalytic cyclization and oxidation of fructose to 2,5-furandicarboxylic acid, Catal. Commun. 4 (2003) 83–86.

[14] M. Kröger, U. Prüße, K.-D. Vorlop, A new approach for the production of 2,5-furandicarboxylic acid by in-situ oxidation of 5-hydroxymethylfurfural starting from fructose, Top. Catal. 13 (2000) 237–242.

[15] G.T. Tsao, N.J. Cao, J. Du, C.S. Gong, Production of multifunctional organic acids from renewable resources, Advances in Biochemical Engineering Biotechnology: Recent Progress in Bioconversion of Lignocellulosics, Springer, New York, NY, 1999, pp. 272.

[16] R. Goswami, V.K. Mishra, Sugar-derived industrially important C6 platform chemicals, Platform Chemical Biorefinery: Future Green Chemistry, Elsevier, Cambridge, MA, 2016, pp. 231–232.

[17] T.S. Moon, S.-H. Yoon, A.M. Lanza, J.D. R-Mayhew, K.L.J. Prather, Production of glucaric acid from a synthetic pathway in recombinant *Escherichia coli*, Appl. Environ. Microbiol. 75 (3) (2009) 589–595.

[18] Glutamic Acid: Structure, Formula, and Function. http://study.com/academy/lesson/glutamic-acid-structure-formula-function.html.
[19] Glutamic Acid, Tutorvista. http://chemistry.tutorvista.com/biochemistry/glutamic-acid.html.
[20] J.-J. Zhong, Developing a sustainable bioprocessing strategy based on a generic feedstock, Biomanufacturing, Springer, New York, NY, 2004, pp. 224.
[21] G. Najafpour, Industrial microbiology, Biochemical Engineering and Biotechnology, Elsevier, Waltham, MA, 2007, pp. 9.
[22] K. Karimi, Introduction to lignocellulose-based bioproducts, Lignocellulose-Based Bioproducts, Springer, Switzerland, 2015, pp. 24–25.
[23] R.A. Sheldon, I. Arends, U. Hanefeld, Chemicals from renewable raw materials, Green Chemistry and Catalysis, Wiley, Weinheim, Germany, 2007, pp. 368–370.
[24] Z.-X. Wang, J. Zhuge, H. Fang, B.A. Prior, Glycerol production by microbial fermentation: a review, Biotechnol. Adv. 19 (2001) 201–223.
[25] D-Sorbitol/Xylitol, Assay prcedure, Megazyme. https://secure.megazyme.com/files/Booklet/K-SORB_DATA.pdf.
[26] I.S.M. Rafiqul, A.M. Mimi Sakinah, Processes for the production of Xylitol-A—a review, Food Rev. Int. 29 (2013) 127–156. Taylor & Francis Group.
[27] F.D. Gunstone, R.J. Hamilton, Basic oleochemicals, Oleochemical Manufacture and Applications, CRC Press, Boca Raton, FL, 2001, pp. 4–8.
[28] What are PHAs, Renewable and Biodegradable Polymer for Engineered Solution, PolyFerm Canada. http://www.polyfermcanada.com/pha.html.
[29] J.M. Pachence, M.P. Bohrer, J. Kohn, Biodegradable polymers, in: B. Lanza, C. Langer, P. Vacanti (Eds.), Principles of Tissue Engineering, third ed., Elsevier, 2007, p. 323.
[30] A. Girdhar, M. Bhatia, S. Nagpal, A. Kanampalliwar, A. Tiwari, Process parameters for influencing polyhydroxyalkanoate producing bacterial factories: an overview, J. Pet. Environ. Biotechnol. 4 (5) (2013) 155–162.
[31] H.J. Endres, A. Siebert-Raths, Basics of PHA, Bioplast. Mag. 6 (2011) 42–45. http://f2.hs-hannover.de/fileadmin/media/doc/ifbb/Bioplastics_Magazine__03_11__Vol._6_S._43-45.pdf.
[32] M.V. Guettler, M.K. Jain, US Patent US5521075, 1996.
[33] E. Delwiche, A biotin function in succinic acid decarboxylation by Propionibacterium pentosaceum, J. Bacteriol. 59 (1950) 439–442.
[34] Succinic acid applications, Thirumalai Chemicals. https://www.thirumalaichemicals.com/images/sacapplications.pdf.
[35] L. Stols, M.I. Donnelly, Production of succinic acid through overexpression of NAD+-dependent malic enzyme in an Escherichia coli mutant, Appl. Environ. Microbiol. 63 (1997) 2695–2701.
[36] S. Vaswani, Bio-Based Succinic Acid, PEP Review 2010–2014, U. S. Department of Energy. https://www.ihs.com/pdf/RW2010-14_220240110917062932.pdf.
[37] J.W. Lee, H.U. Kim, S. Choi, J. Yi, S.Y. Lee, Microbial production of building block chemicals and polymers, Curr. Opin. Biotechnol. 22 (2011) 758–767.
[38] J.A. Maga, A.T. Tu, Food acidulants, Food Additive Toxicology, Marcel Dekker, New York, NY, 1994, pp. 39.
[39] M. Okabe, D. Lies, S. Kanamasa, E.Y. Park, Biotechnological production of itaconic acid and its biosynthesis in *Aspergillus terreus*, Appl. Microbiol. Biotechnol. 84 (2009) 597–606.
[40] T. Klement, J. Buchs, Itaconic acid—a biotechnological process in change, Bioresour. Technol. 135 (2013) 422–431.

[41] X. Zhang, W. Yang, C. Dong, Levoglucosan formation mechanisms during cellulose pyrolysis, J. Anal. Appl. Pyrolysis 104 (2013) 19–27. Elsevier.
[42] F. Shafizadeh, R.H. Furneaux, T.T. Stevenson, Some reactions of levoglucosenone, Carbohydr. Res. 71 (1) (1979) 169–191 Elsevier.
[43] G. Maschio, A. Lucchesi, G. Stoppato, Production of syngas from biomass, Bioresour. Technol. 48 (1994) 119–126.
[44] M.A. Yaqoob, Production of Syngas From Biomass and Its Purification, University of Engineering and Technology, Lahore, 2011.
[45] Q. Xie, S. Kong, Y. Liu, H. Zeng, Syngas production by two-stage method of biomass catalytic pyrolysis and gasification, Bioresour. Technol. 110 (2012) 603–609.
[46] B.J. Vreugdenhil, R.W.R. Zwart, Tar Formation in Pyrolysis and Gasification, Energy Research Center of the Netherlands, Petten, Netherlands, 2009.

Further reading

[1] C.H.R.M. Wilsens, Exploring the Application of 2,5-Furandicarboxylic Acid as a Monomer in High Performance Polymers: Synthesis, Characterization, and Properties, Technische Universiteit, Eindhoven University of Technology, Eindhoven, 2015.
[2] Sorbitol, Huakang Pharma. http://www.huakangpharma.com/en/index.php/product/1.
[3] S.A. Ashter, Types of biodegradable polymers, Introduction to Bioplastics Engineering, Elsevier, Waltham, MA, 2016, pp. 104–112 (Chapter 5).
[4] Y.H. Hui, G.G. Khachatourians, Microorganisms for organic acid production, Food Biotechnology: Microorganisms, Wiley, Hoboken, NJ, 2005, pp. 263.

Derivation of monomers from biomass

7

7.1 Introduction

There are different classes of biopolymers available for the production of chemicals or materials. Some of the examples include starch, lignin, cellulose, hemicellulose, and proteins. These polymers are converted into different chemicals and further chemically modified to yield other chemicals or materials [1–5]. The majority of biomass-derived polymers when broken down into their constituents consist of (a) cellulose-derived sugars of five or six carbon atoms, and (b) aromatic building blocks derived from lignin [6].

Gasification is a thermochemical process that transforms carbon material such as biomass into usable fuels or chemicals. It utilizes carbon dioxide during photosynthesis and does not add carbon to the atmosphere. This makes it a carbon neutral. During biomass gasification, a mixture of many gases, predominantly carbon monoxide and hydrogen, is released as the output product of the gasification process. Heating value of the product gas is dependent on:

- Raw material
- Gasifying agents
- Operating variables
- Design of gasifier
- Catalysts

Generated syngas can either be used for heating, power generation, or as a raw material for chemical feedstock production [7].

Fig. 1 shows overview of a typical gasification conversion route. The conversion process consists of two main stages after feed processing: (a) deconstruction and (b) transformation. Following are the steps for gasification conversion to biofuels and bioproducts:

- Feed processing
- Gasification
- Syngas cleanup and conditioning
- Product synthesis

This process, however, excludes steps such as separation and purification that may be needed to produce certain chemicals depending on the end finished product [8].

During gasification process, some or all of the following chemical reactions are observed [9]:

$$C + CO_2 \leftrightarrow 2CO$$

$$C + H_2O \leftrightarrow CH_4$$

Technology and Applications of Polymers Derived from Biomass. https://doi.org/10.1016/B978-0-323-51115-5.00007-4

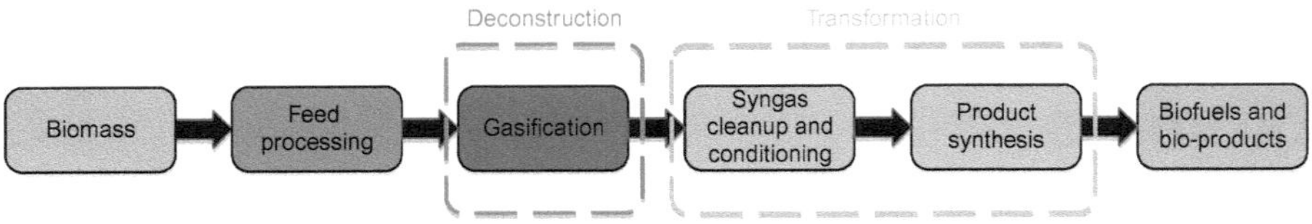

Fig. 1 Overview of gasification conversion route.
From J.A. Schaidle, M. Talmadge, M. Biddy, M. Nimlos, A. Bratis, Chemical Derived from Biomass Thermolysis and Gasification, Elsevier, 2017, p. 4.

$$C + 2H_2 \leftrightarrow CH_4$$

$$C + 0.5H_2 \leftrightarrow 2CO$$

$$C + O_2 \leftrightarrow CO_2$$

$$C + 0.5O_2 \leftrightarrow CO_2$$

$$CH_4 + 2O_2 \leftrightarrow CO_2 + 2H_2O$$

$$CO + H_2O \leftrightarrow CO_2 + H_2$$

$$2CO + 2H_2 \leftrightarrow CH_4 + CO_2$$

$$CO + 3H_2 \leftrightarrow CH_4 + H_2O$$

$$CO + 4H_2 \leftrightarrow CH_4 + 2H_2O$$

$$CH_4 + H_2O \leftrightarrow CO_2 + 3H_2$$

$$CH_4 + 0.5O_2 \leftrightarrow CO + 2H_2$$

The fermentation unit operation of biochemical conversion of biomass is the last process to turn biomass into final products through biochemical conversion. It consists of following four steps that are divided into solid-state fermentation and submerged fermentation [10]. In all these four steps, it is important to control conditions during fermentation to avoid decrease in the microbial growth.

1. Fractional hydrolysis fermentation process
2. Simultaneous saccharification and fermentation process
3. Simultaneous saccharification and cofermentation process
4. Consolidated bioprocessing process

7.2 Derivation of monomers by gasification

7.2.1 Polycyclic aromatic hydrocarbons

Polycyclic aromatic hydrocarbons (PAHs) consist of two or more single or fused aromatic rings with a pair of carbon atoms shared between rings in their molecules. They are classified as chemicals that occur naturally in coal, crude oil, and gasoline and can also be produced when coal, oil, gas, wood, garbage, and tobacco are burned. PAHs generated from these sources tend to bind and form small particles in the air [11].

Some of the physical properties of PAHs are as follows:

- High melting and boiling points
- Low vapor pressure
- Very low aqueous solubility
- Highly lipophilic

PAHs also manifest various functions such as light sensitivity, heat resistance, conductivity, emittability, corrosion resistance, and physiological action.

The lower molecular weight PAHs, such as naphthalene, acenaphthalene, and phenanthrene, are found almost in gas phase, while the higher molecular weight PAH such as benzo[a]pyrene is almost exclusively in particle phase [11]. The molecular structures of sixteen parent PAH compounds are listed in Fig. 2 [11]

The process of converting lignin into aromatic hydrocarbons is shown in Fig. 3. Catalytic hydrodeoxygenation of uncondensed pyrolysis vapors coupled with lignin pyrolysis results in the production of PHAs. Use of the vapor-phase hydrotreatment

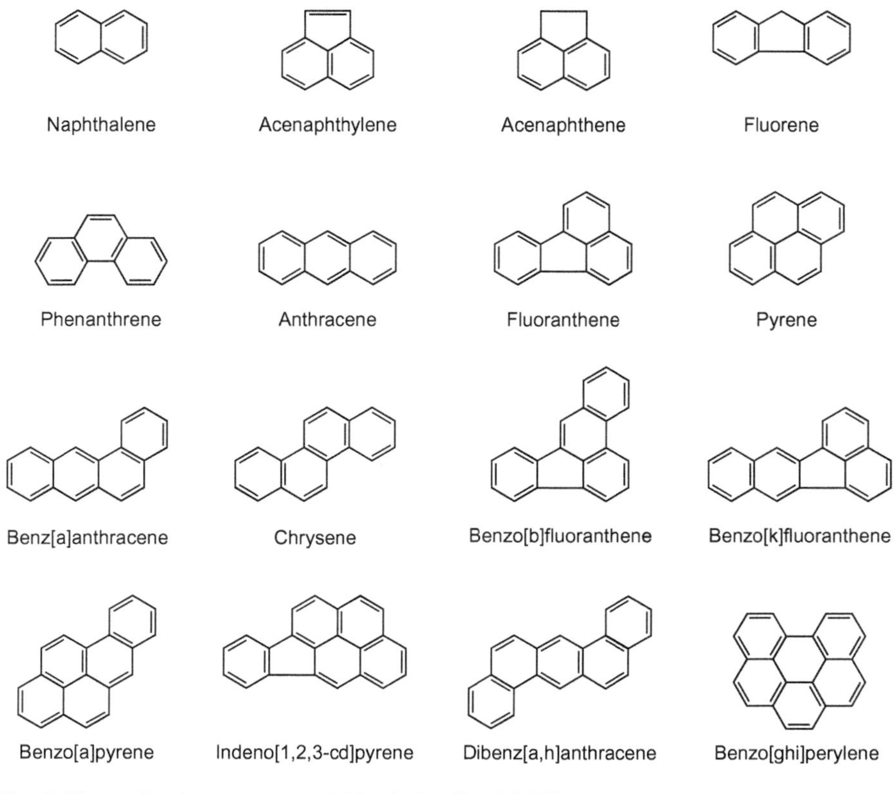

Fig. 2 The molecular structures of 16 priority-listed PAHs.
From H. Shen, Research Background, Polycyclic Aromatic Hydrocarbons: Their Global Athmospheric Emissions, Transport, and Lung Cancer Risk (Springer Thesis), Springer-Verlag, 2016, p. 8 (Chapter 2).

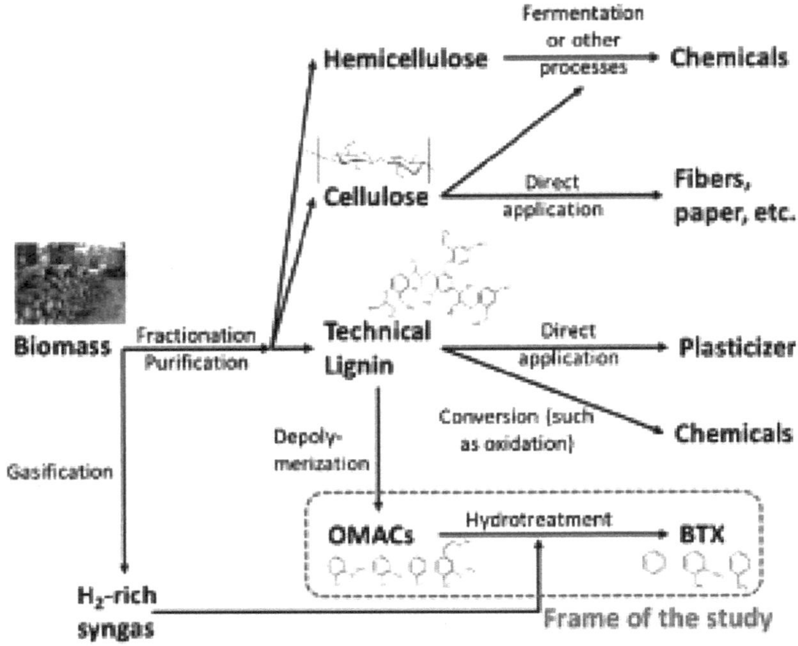

Fig. 3 Lignin to aromatic hydrocarbon conversion process.
From A. Dufour, Bioenergy Chain and Process Scales, Thermochemical Conversion of Biomass for the Production of Energy and Chemicals, Wiley, 2016, pp. 11–12.

of pyrolysis vapors over iron-based catalysts is preferred as it reduces the problem of handling the reactive lignin oils and promotes heat recovery. This process is more versatile in comparison to direct catalytic pyrolysis because of the low partial pressure and the separating pyrolysis and catalytic reactors. A second hydrotreatment refining step can be used on condensed bio-oils after the first catalytic hydrodeoxygenation of vapors [12].

7.2.2 Dioxins and furans

Polychlorinated dibenzo-p-dioxins (PCDD) and polychlorinated dibenzofuran (PCDF) collectively are termed as dioxins. They are chlorinated aromatic hydrocarbons similar in structure and chemical properties. They consist of two benzene rings that are interconnected by either two oxygen bridges, in case of PCDD or one oxygen bridge, and a single C—C bond in case of PCDF as shown in Fig. 4 [13]. The PCDDs and the PCDFs are pollutants that are produced from thermal and combustion operations.

The number and position of chlorine atoms bound to the two aromatic rings dictate the chemical and toxicological properties of these compounds [14,15]. Certain dioxin isomers are well known for their toxicological properties, e.g., exert carcinogenic and mutagenic effects. For example, the dioxin isomer with chlorine substituents as shown

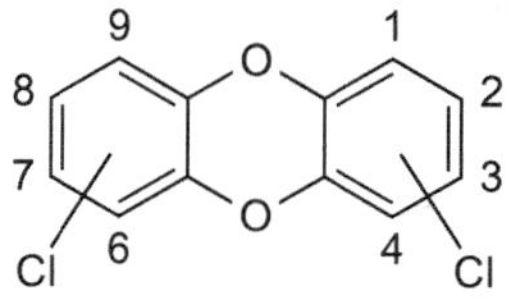

Polychlorinated dibenzo-p-dioxins (PCDD)

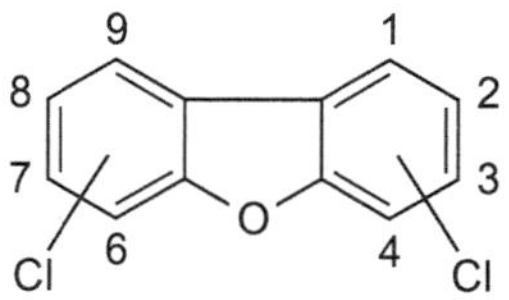

Polychlorinated dibenzofurans (PCDF)

Fig. 4 PCDD and PCDF structures.
From Q. Gao, Dioxins and Dioxin-like Compounds in Thermochemical Conversion of Biomass (Doctoral Thesis), Department of Chemistry, Umea University, Sweden, 2016.

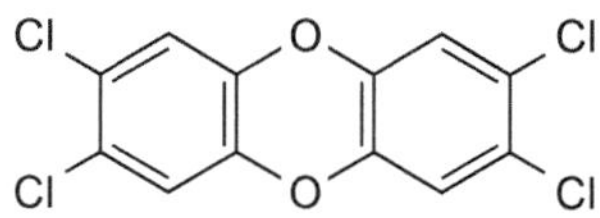

2, 3, 7, 8-tetrachlorodibenzo-p-dioxin

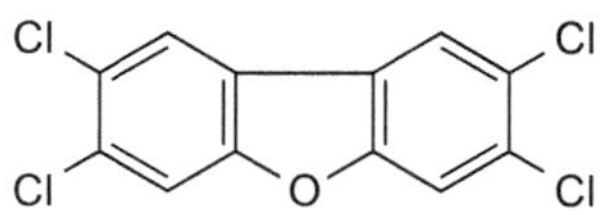

2, 3, 7, 8-tetrachlorodibenzofuran

Fig. 5 Molecular structure of 2, 3, 7, 8-tetra isomers.
From H.K. Chagger, A. Kendall, A. McDonald, M. Pourkashanian, A. Williams, Formation of dioxins and other semi-volatile organic compounds in biomass combustion, Appl. Energy 60 (1998) 101–114; E.R. Altwicker, Sci. Total Environ. 104 (1991) 47; H. Vogg, L. Stieglitz, Chemosphere, 15 (1986) 1373; L. Stieglitz, G. Zwick, J. Beck, W. Roth, H. Vogg, Chemosphere 18 (1989) 1216.

in Fig. 5 in positions 2, 3, 7, and 8, or 2, 3, 7, 8-tetrachlorodibenzo-p-dioxin, is one of the toxic compounds [16–20].

Although, incineration of municipal, chemical, and hospital wastes is the major source of dioxins, combustion of oils, wood, gasoline, smelting of copper and scrap metals, recovery of plastic-coated (PVC) wire, and natural combustion such as forest fires results in the formation of dioxins [21,22].

There are two ways a dioxin can be formed by:

1. De Novo Synthesis
2. Undergoing multistep reaction

De novo synthesis which postulates heterogeneous catalytic assembly of chlorinated dioxin structures from a carbon, oxygen, and chlorine source at a temperature window of 300–325°C in the postcombustion zone [18–20,23]. The second mechanism for the formation of PCDD/PCDF is a multistep reaction in the postcombustion zones [24–26]:

- Aromatization of aliphatic compounds
- Chlorination
- Catalytic reaction of chlorinated aromatic precursors [27,28]
- Gas-phase coupling reactions of chlorinated precursors
- Adsorption on the organic phase [23,29–31].

Fig. 6 shows a moving grit gasifier to evaluate dioxins and furan emission during gasification [18]. The equipment was fed with municipal solid waste (MSW), to generate and subsequently burn the synthesis gas. The device was fed manually; the combustion system was controlled manually, as well. The gasified MSW was provided by a landfill that receives waste collected from 14 small municipalities of the Brazilian states of Santa Catarina and Paraná, and air was used as gasifying agent [18].

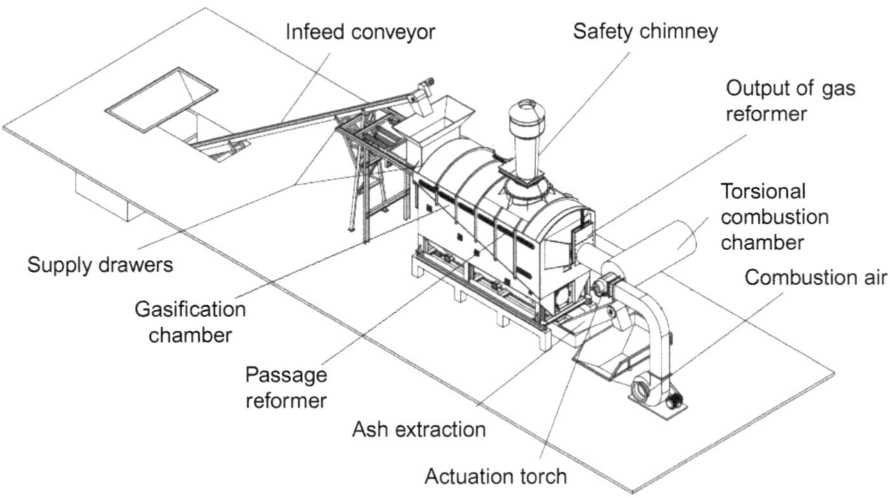

Fig. 6 Gasifier used in the process of MSW gasification.
From H.K. Chagger, A. Kendall, A. McDonald, M. Pourkashanian, A. Williams, Formation of dioxins and other semi-volatile organic compounds in biomass combustion, Appl. Energy 60 (1998) 101–114.

7.3 Derivation of monomers by fermentation

7.3.1 4-Hydroxyphenyllactic acid

4-Hydroxyphenyllactic acid (HPLA) is a tyrosine metabolite as shown in the structure in Fig. 7. The level of L-hydroxyphenyllactic acid is elevated in patients with a deficiency of the enzyme p-hydroxyphenylpyruvate oxidase. The D-form of hydroxyphenyllactate is of bacterial origin and is also found in individuals with bacterial overgrowth or unusual gut microflora [32].

Depending on the origin of the bacteria, hydroxyphenyllactic acid can have two different forms:

- *Bacillus proteus* will lead to production of D-*p*-hydroxyphenyllactic acid
- *Bacillus subtilis* produces L-*p*-hydroxyphenyllactic acid

B. proteus transforms the DL-tyrosine into the D-hydroxyacid, whereas *B. subtilis* transforms DL-tyrosine into L-hydroxyacid. Fig. 8 shows how different stereochemistries of tyrosine can lead to different forms of hydroxyphenyllactic acid [33].

Production of HPLA is strain dependent and can be produced by a wide range of lactic acid bacteria species. It also exists as metabolites in lactic acid bacteria strains through phenylalanine and tyrosine degradation, respectively [34–36].

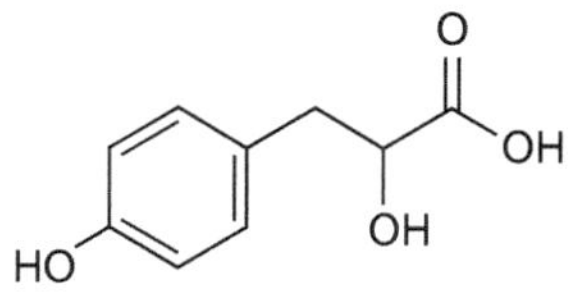

Fig. 7 Structure of 4-hydroxyphenyllactic acid (HPLA).
From http://www.sigmaaldrich.com/catalog/product/aldrich/h3253?lang=en&region=US.

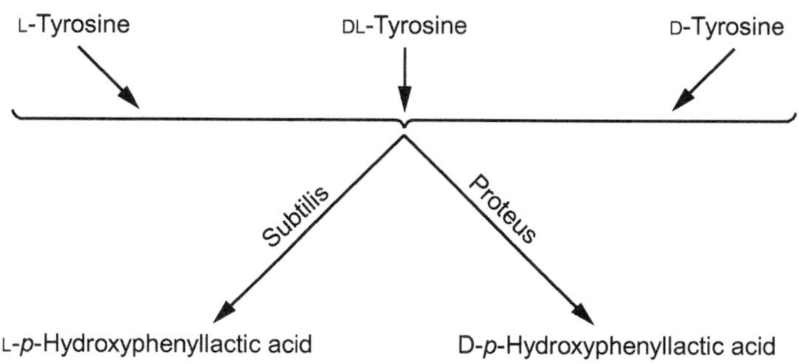

Fig. 8 Different stereochemistries of tyrosine leading to different forms of hydroxyphenyllactic acid.
From T. Sasaki, I. Otsuka, The stereochemistry of the bacterial decomposition of albumin, J. Biol. Chem. 32–33 (1917) 533.

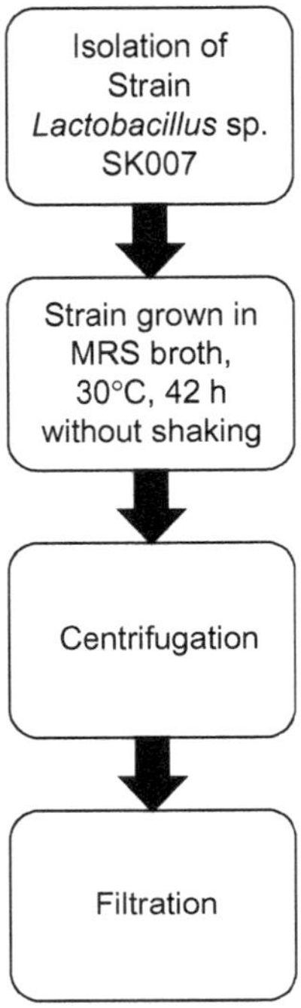

Fig. 9 Schematics for producing 4-hydroxyphenyllactic acid by *Lactobacillus* sp. SK007 fermentation [37].
From Y.T. Shah, Cellulosic ethanol, Water for Energy and Fuel Production, CRC Press, 2014, p. 255.

Mu et al. successfully produced 4-hydroxyphenyllactic acid by *Lactobacillus* sp. SK007 fermentation as indicated in Fig. 9 [37]:

7.3.2 Microbial monomers

7.3.2.1 Lignin monomers

Lignin monomers are complex amorphous phenolic polymers that are not sugar based and fermentable. The phenol in lignins may be either a guaiacyl or a syringyl unit. These units are bonded by alpha- or beta-ether linkages. A variety of C—C linkages may also be present and these are less common. The distribution of linkages in lignin is random and highly resistant to chemical, enzymatic, and microbial hydrolysis due to extensive cross-linking [38]. Structures of three lignin precursors are shown in Fig. 10 [39].

A typical lignin monomer can be isolated from extractives-free wood as an insoluble after hydrolytic removal of polysaccharides. It can be hydrolyzed and extracted from wood or converted to a soluble derivative. There are different methods to remove lignin from the polysaccharides [39].

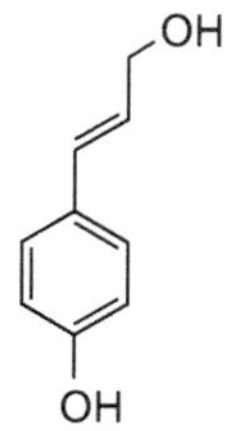

p-Coumaryl alcohol

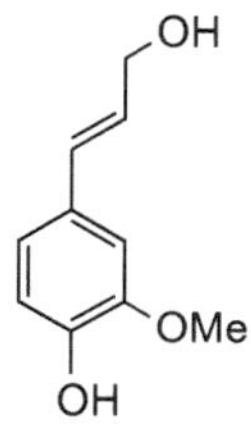

Coniferyl alcohol

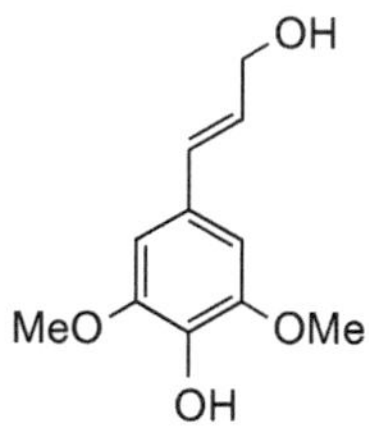

Sinapyl alcohol

Fig. 10 Lignin precursors.
From http://web.nchu.edu.tw/pweb/users/taiwanfir/lesson/10476.pdf.

1. Klason lignin: Obtained after removing the polysaccharides from extractives-free wood by hydrolysis with 72% sulfuric acid.
2. Cellulolytic enzyme lignin: Polysaccharide is removed by enzymes from finely divided wood meal. Although, this method is cumbersome, the original structure of lignin remains unchanged.
3. Milled wood lignin also known as Bjorkman lignin: Best known preparation approach. The wood meal is ground in a ball mill either dry or in the presence of nonswelling solvents such as toluene. In this process, the cell structure of the wood is destroyed and a small portion of lignin is obtained from the suspension by extraction with dioxane–water mixture.
4. Artificially prepared lignin: Lignin precursors are treated with peroxide in the presence of peroxidase enzyme to produce artificial lignin, also known as dehydrogenation polymer or DHP. In a different approach, suspension cultures of spruce wood cells are isolated as secretion products. These types of lignin are called released suspension culture lignin and represents carbohydrate-free coniferous lignin.

References

[1] C. Brunschwig, W. Moussavou, J. Blin, Use of bioethanol for biodiesel production, Prog. Energy Combust. Sci. 38 (2012) 283–301.

[2] J.T. Dam, U. Hanefeld, Renewable chemicals: dehydroxylation of glycerol and polyols, ChemSusChem 4 (2011) 1017–1034.

[3] I. Bechthold, K. Bretz, S. Kabasci, R. Kopitzky, A. Springer, Succinic acid: a new platform chemical for biobased polymers from renewable resources, Chem. Eng. Technol. 37 (5) (2008) 647–654.

[4] K.-K. Cheng, X.-B. Zhao, J. Zeng, J.-A. Zhang, Biotechnological production of succinic acid: current state and perspectives, Biofuels Bioprod. Biorefin. 6 (2012) 302–318.

[5] R.T. Mathers, How well can renewable resources mimic commodity monomers and polymers? Polym. Chem. 50 (2011) 1–15.

[6] http://www.ndsuresearchfoundation.org/rft478.

[7] V.S. Sikarwar, M. Zhao, Biomass Gasification, Elsevier, Cambridge, UK, 2017, p. 2.

[8] J.A. Schaidle, M. Talmadge, M. Biddy, M. Nimlos, A. Bratis, Chemical Derived from Biomass Thermolysis and Gasification, Elsevier, Cambridge, UK, 2017, p. 4.

[9] S. Kumar, S.K. Shukla, A review on recent gasification methods for biomethane gas production, Int. J. Energy Eng. 6 (1A) (2016) 32–43.

[10] H. Chen, L. Wang, Pandect of practice unit and process engineering of biomass biochemical conversion, Technologies for Biochemical Conversion of Biomass, Academic Press, London, UK, 2017, p. 16.

[11] H. Shen, Research Background, Polycyclic Aromatic Hydrocarbons: Their Global Athmospheric Emissions, Transport, and Lung Cancer Risk, (Springer Thesis)Springer-Verlag, Berlin, 2016, p. 8 (Chapter 2).

[12] A. Dufour, Bioenergy chain and process scales, Thermochemical Conversion of Biomass for the Production of Energy and Chemicals, Wiley, Weinheim, Germany, 2016, pp. 11–12.

[13] Q. Gao, Dioxins and Dioxin-like Compounds in Thermochemical Conversion of Biomass, (Doctoral Thesis), Umea University, Umea, Sweden, 2016.

[14] M. Altarawneh, B.Z. Dlugogorski, E.M. Kennedy, J.C. Mackie, Mechanisms for formation, chlorination, dechlorination and destruction of polychlorinated dibenzo-p-dioxins and dibenzofurans (PCDD/Fs), Prog. Energy Combust. Sci. 35 (3) (2009) 245–274.

[15] E.J. Lopes, L.A. Okamura, C.I. Yamamoto, Formation of dioxins and furans during municipal solid waste gasification, Braz. J. Chem. Eng. 32 (1) (2015) 87–92.

[16] H. Huang, A. Buekens, De novo synthesis of polychlorinated dibenzo-p-dioxins and dibenzofurans proposal of a mechanistic scheme, Sci. Total Environ. 193 (2) (1996) 121–141.

[17] J.V. Assunção, C.R. Pesquero, Dioxins and furans: origins and risks, J. Public Health 33 (5) (1999) 7.

[18] H.K. Chagger, A. Kendall, A. McDonald, M. Pourkashanian, A. Williams, Formation of dioxins and other semi-volatile organic compounds in biomass combustion, Appl. Energy 60 (1998) 101–114.

[19] E.R. Altwicker, Some laboratory experimental designs for obtaining dynamic property data on dioxins, Sci. Total Environ. 104 (1991) 47.

[20] H. Vogg, L. Stieglitz, Thermal behavior of PCDD/PCDF in fly ash from municipal incinerators, Chemosphere 15 (1986) 1373.

[21] N. Steisel, R. Morris, M.J. Clarke, The impact of the dioxin issue on resource recovery in the United States, J. Waste Manage Res. 5 (1987) 381.
[22] H. Tosine, K. Choudhary, Rappe (Eds.), Chlorinated Dioxins and Dibenzofurans in the Total Environment, Butterworth, London, 1983.
[23] L. Stieglitz, G. Zwick, J. Beck, W. Roth, H. Vogg, On the de-novo synthesis of PCDD/PCDF on fly ash of municipal waste incinerators, Chemosphere 18 (1989) 1216.
[24] L. Stieglitz, H. Vogg, On formation conditions of PCDD/PCDF in fly ash from municipal waste incinerators, Chemosphere 16 (1987) 1917.
[25] H.M. Hagenmaier, H. Kraft, R. Brunner, R. Haag, Catalytic effects of fly ash from waste incineration facilities on the formation and decomposition of polychlorinated dibenzo-p-dioxins and polychlorinated dibenzofurans, Environ. Sci. Technol. 21 (1987) 1080.
[26] E. Benefanti, G. Mariani, R. Fancelli, S. Zucotti, "De novo" synthesis of PCDD, PCDF, PCB, PCN and PAH in a pilot incinerator, Chemosphere 22 (1991) 1045.
[27] O. Hutzinger, M.J. Blumlich, M.V.D. Berg, K. Olie, Sources and fate of PCDDs and PCDFs: an overview, Chemosphere 14 (1985) 581.
[28] F.W. Karasek, L.C. Dickson, Model studies of polychlorinated dibenzo-p-dioxin formation during municipal refuse incineration, Science 237 (1987) 754.
[29] W.M. Shaub, W. Tsang, Dioxin formation in incinerators, Environ. Sci. Technol. 17 (1983) 721.
[30] L. Stieglitz, H. Vogg, G. Zwick, J. Beck, H. Bautz, On formation conditions of organohalogen compounds from particulate carbon of fly ash, Chemosphere 23 (1991) 255.
[31] K. Jay, L. Stieglitz, On the mechanism of formation of polychlorinated aromatic compounds with copper (II) chloride, Chemosphere 22 (1991) 987.
[32] http://www.hmdb.ca/metabolites/HMDB00755.
[33] T. Sasaki, I. Otsuka, The stereochemistry of the bacterial decomposition of albumin, J. Biol. Chem. 32–33 (1917) 533.
[34] M. Yvon, S. Thirouin, L. Rijnen, D. Fromentier, J.C. Gripon, An aminotransferase from Lactococcus lactis initiates conversion of amino acids to cheese flavor compounds, Appl. Environ. Microbiol. 63 (1997) 414–419.
[35] M. Yvon, S. Berthelot, J.C. Gripon, Adding α-ketoglutarate to semi-hard cheese curd highly enhances the conversion of amino acids to aroma compounds, Int. Diary J. 8 (1998) 889–898.
[36] A. Kieronczyk, S. Skeie, T. Langsrud, M. Yvon, Cooperation between Lactococcus lactis and nonstarter lactobacilli in the formation of cheese aroma from amino acids, Appl. Environ. Microbiol. 69 (2003) 734–739.
[37] W. Mu, Y. Yang, J. Jia, T. Zhang, B. Jiang, Production of 4-hydroxyphenyllactic acid by Lactobacillus sp. SK007 fermentation, J. Biosci. Bioeng. 109 (4) (2010) 369–371.
[38] Y.T. Shah, Cellulosic ethanol, Water for Energy and Fuel Production, CRC Press, Boca Raton, FL, 2014, p. 255.
[39] http://web.nchu.edu.tw/pweb/users/taiwanfir/lesson/10476.pdf.

Further reading

[1] http://www.sigmaaldrich.com/catalog/product/aldrich/h3253?lang=en&region=US.
[2] P. Laverymicocca, F. Valerio, A. Evidente, S. Lazzaroni, A. Corsett, M. Gobbetti, Purification and characterization of novel antifungal compounds from the sourdough lactobacillus plantarum strain 21B, Appl. Environ. Microbiol. 66 (2000) 4084–4090.

Polymerization of biomass-based-monomers

8

8.1 Introduction

With the increase in cost of fossil fuels, higher cost, and concern over the environmental impact, there has been a push to develop polymers and chemicals from renewable and sustainable source. In addition, rising cost of fossil fuels and continuous decrease in the reserves has pushed the chemical industry to look for alternative feedstock. Although, they are not considered cheap, biomass is abundant in nature and offers sustainable solution.

Biomass require breakdown of organic compounds that constitute biomass in order to convert to monomer that can be tailored to any number of material types and applications. Monomers then undergo polymerization to produce biofuels and useful chemicals.

Some of the polymerization techniques include (a) free radical polymerization, (b) anionic polymerization, and (c) cationic polymerization. Within each polymerization approach, monomer can undergo a set of independent reactions. For example, furan monomers can undergo the following substitution reactions:

- Alkylation
- Halogenation
- Sulfonation
- Nitration

This chapter will be focused on providing in-depth understanding on polymerization of different biomass-derived monomers

8.2 Furans

Furan is a volatile colorless liquid with an aromatic scent and is one of the major representatives of the five-member unsaturated heterocycle family, which also includes pyrrole and thiophene as well as several other less widespread homologs. The general structures of furan, pyrrole, and thiophene are shown in Fig. 1 [1,2].

Resonance-contributing structures [A–D] for the furan heterocycle as determined by molecular orbital calculations are shown in Fig. 2 [1]. There are four substitution reactions that occur in furan:

1. Alkylation
2. Halogenation
3. Sulfonation
4. Nitration

Technology and Applications of Polymers Derived from Biomass. https://doi.org/10.1016/B978-0-323-51115-5.00008-6

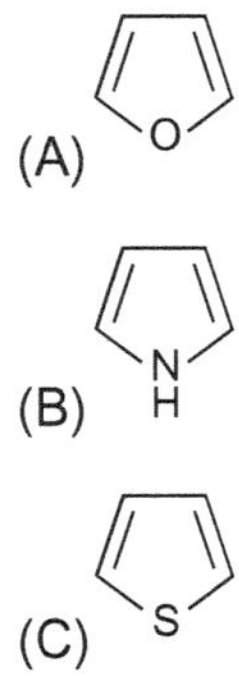

Fig. 1 General structure of (A) furan, (B) pyrrole, and (C) thiophene.
From http://www.sigmaaldrich.com/catalog/product/aldrich/f19558?lang=en&region=US; http://www.sigmaaldrich.com/catalog/product/mm/807492?lang=en&region=US; http://www.sigmaaldrich.com/catalog/product/aldrich/t31801?lang=en&region=US.

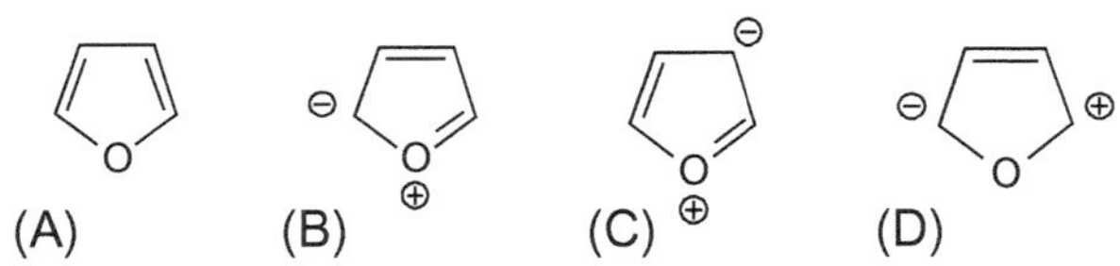

Fig. 2 Resonance-contributing structures for furan heterocycle determined by molecular orbital calculations.
From A. Gandini, M.N. Belgacem, Furan Derivatives and Furan Chemistry Service of Macromolecular Materials, Monomers, Polymers, and Composites from Renewable Resources, Elsevier Publishers, p. 116 (Chapter 6).

Substitution reaction in furan (A) leads to structure B, C, and D. Substitution reactions occur at C2 and/or C5 when the carbon positions are not substituted. This suggests that structure B is dominant.

8.2.1 Furfural

Agricultural to forestry residues resources are the main source of furfural. For example, the pentoses found in by-products of the food industry as corn cobs, oat and rice hulls, sugarcane bagasse, cotton seeds, olive husks and stones, as well as wood chips can be used to produce furfural.

Furfural is produced by acid-catalyzed dehydrolytic depolymerization of the hemicellulose followed by dehydration of the ensuing aldopentoses. Most of the furfural is converted into furfural alcohol as a precursor to a variety of resins. The generalized structure of furfural alcohol is shown in Fig. 3 [3].

The second derivative of furan is hydroxymethylfurfural (HMF) which is prepared from C6 polysaccharides or sugars. In situ conversion of HMF produces very stable aldehyde derivatives—2,5-furandicarboxydialdehyde (FCDA) and

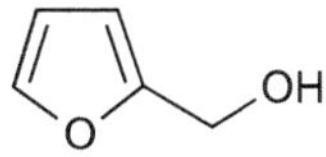

Fig. 3 Structure of furfural alcohol.
From http://www.sigmaaldrich.com/catalog/product/aldrich/185930?lang=en&region=US.

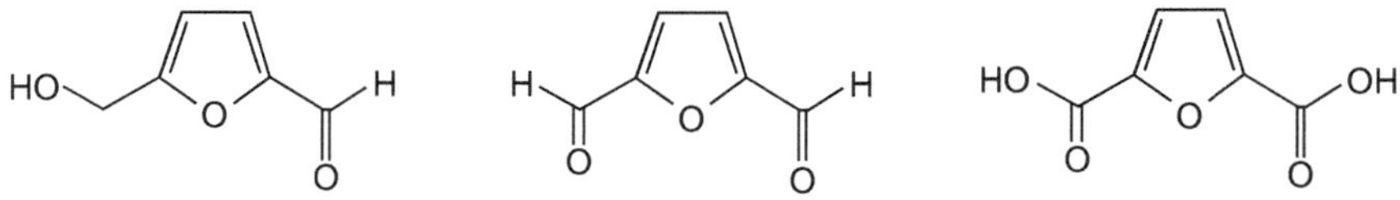

Fig. 4 Derivatives of furan.
From A. Gandini, M.N. Belgacem, Furan Derivatives and Furan Chemistry Service of Macromolecular Materials, Monomers, Polymers, and Composites from Renewable Resources, Elsevier Publishers, p. 116 (Chapter 6).

2,5-furandicarboxylic acid (FDCA). The generalized structures of these derivatives are shown in Fig. 4 [1].

Furfural (F) and 5-methylfurfural (MF) are precursors and can be polymerized using chain reaction mechanisms. Structures of products derived from furfural and 5-methylfurfural are shown in Fig. 5. The exception to this approach is 2-hydroxymethylfuran (FA) which instead is produced by catalytic reduction of furfural [4–8].

These products are prone to chain reaction polymerization. Furan monomers undergo three different types of polymerization reactions:

1. Free radical polymerization
2. Cationic polymerization
3. Anionic polymerization

8.2.2 Free radical polymerization

Furan monomers bearing an alkenyl moiety deprived of any stabilizing environment after its transformation into a free radical are not expected to polymerize to any sensible extent because the preferred free radical structure would be that involving the much more stable furyl moiety [7,8].

Fig. 6 shows the generic formula for free radical initiation that is utilized to synthesize novel inhibitors [7,8].

Fig. 7 shows a highly stabilized structure generated by trapping of a free radical. This makes it a highly stabilized structure in which unpaired electron is delocalized over the entire conjugated moiety. These structures have a very high efficiency in stopping free radical-based reactions such as photochemical and oxidation processes [7,9].

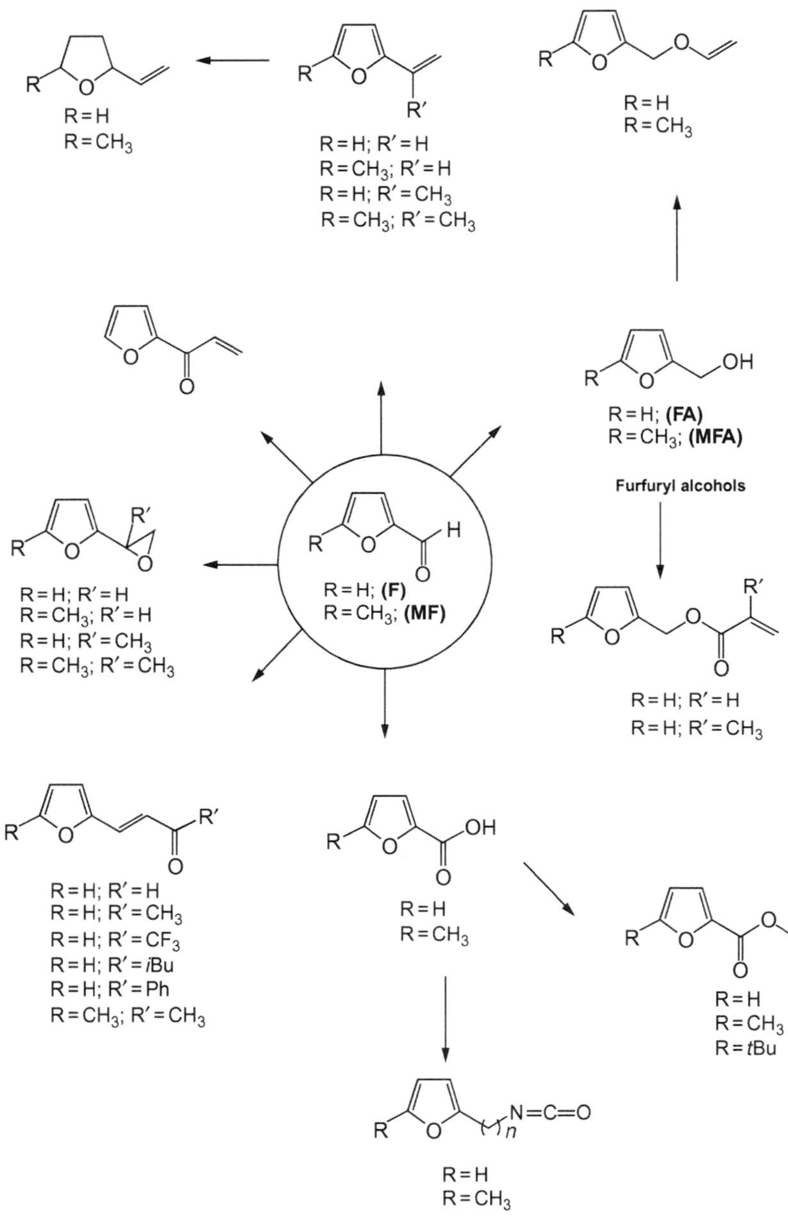

Fig. 5 Structures of products derived from furfural and 5-methylfurfural.
From A. Gandini, Adv. Polym. Sci. 25 (1977) 47; A. Gandini, Encyclopedia of Polymer Science and Engineering, in: H.F. Mark, N.M. Bikales, C.G. Overberger, G. Menges (Eds.), second ed., vol. 7, John Wiley, New York, 1988, p. 454; W.J. McKillip, ACS Symp. Ser. 385 (1989) 408; A. Gandini, M.N. Belgacem, Prog. Polym. Sci. 22 (1997) 1203; C. Moreau, A. Gandini, M.N. Belgacem, Topics Catal. 27 (2004) 11.

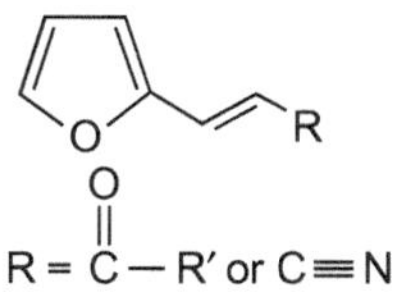

Fig. 6 Generic formula for free radical initiation.
From A. Gandini, M.N. Belgacem, Prog. Polym. Sci. 22 (1997) 1203; C. Moreau, A. Gandini, M.N. Belgacem, Topics Catal. 27 (2004) 11.

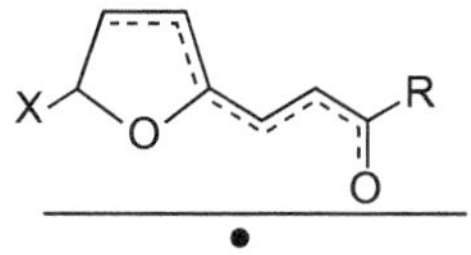

Fig. 7 Highly stabilized structure generated by trapping of free radical.
From C. Moreau, A. Gandini, M.N. Belgacem, Topics Catal. 27 (2004) 11; A. Gandini, J. Rieumont, Tetrahedron Lett. 25 (1976) 2101.

8.2.3 Cationic polymerization

Furan and its methylated group of monomers readily undergo cationic polymerization initiated by both Lewis and Bronsted acids. These branched products with a degree of polymerization ranging between 50 and 300 have complex structures that are made up of both dihydro- and tetrahydrofuran rings [4,7].

Fig. 8 shows structures of some 2-alkenylfurans that are readily activated by cationic initiators [7].

Some monomers produce polymers bearing different units. An example to this category is 2-vinylfuran. These types of monomers undergo two types of side reactions accompanied by propagation through the vinyl unsaturation. Fig. 9 shows a structure generated as a result of alkylation reaction of monomer molecules at C5.

Other furan monomers which undergo cationic polymerization include 2-furfuryl vinyl ether, 2-vinyl furoate, furfural, and 5-methylfurfural as comonomers in conjunction with substituted styrenes and vinyl ethers, as well as 2-furfurylidene methyl ketone [7].

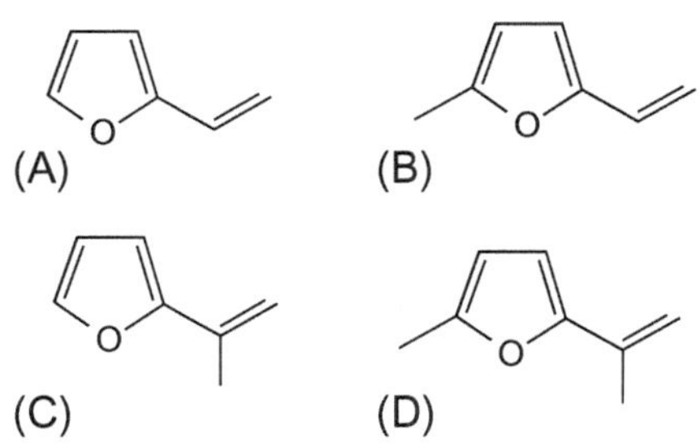

Fig. 8 General structures of (A) 2-vinylfuran, (B) 2-vinyl-5-methylfuran, (C) 2-isopropenylfuran, and (D) 2-isopropenyl-5-methylfuran.
From A. Gandini, M.N. Belgacem, Prog. Polym. Sci. 22 (1997) 1203.

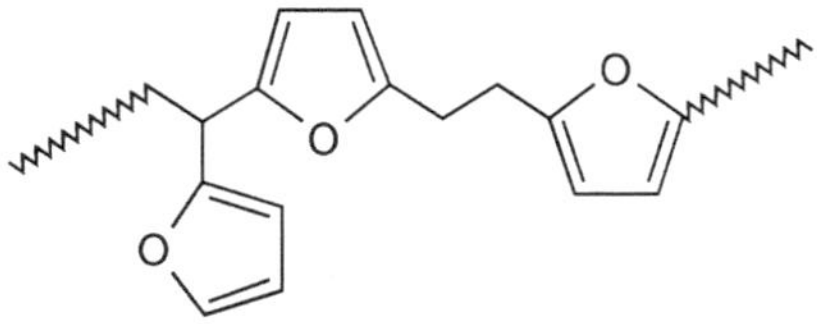

Fig. 9 Structure generated as a result of alkylation reaction.
From A. Gandini, M.N. Belgacem, Prog. Polym. Sci. 22 (1997) 1203.

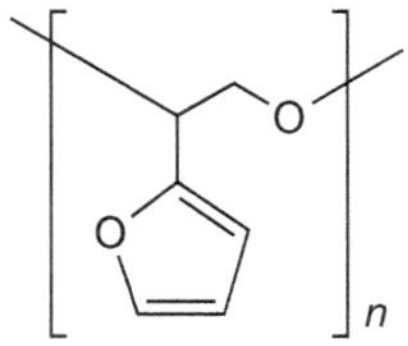

Fig. 10 Irregular polymer structures formed by conventional initiators.
From A. Gandini, M.N. Belgacem, Prog. Polym. Sci. 22 (1997) 1203.

8.2.4 Anionic polymerization

2-Alkenylfurans do not polymerize under the effect of conventional anionic initiators. 2-vinylfuran is the only monomer which can produce a modest yield of polymer when introduced with sodium biphenyl at room temperature. However, conversion higher than 10% cannot be achieved.

2-Furyloxirane is very prone to anionic initiation through the opening of the epoxide ring, even with such mild nucleophiles as alcohols, and gives the corresponding polyether [7,8]. Irregular polymer structures as shown in Fig. 10 were formed when conventional initiators such as tBuOK induced the opening of the ring. This produced regular macromolecules with appreciable tacticity [7].

8.3 Rosin

Rosin is a nonvolatile residue obtained after the distillation of volatiles from the resin. They are made up of 90%–95% of diterpenic monocarboxylic acids, or also referred to as resin acids. The general formula of resin acids is $C_{19}H_{29}COOH$. The most common resin acids found in pine rosin are derived from the three basic tricyclic carbon skeletons as shown in Fig. 11 [10,11]:

- Abietane
- Pimarane
- Isopimarane
- Labdane

The abietane skeleton is shared by four resin acids, namely, abietic, neoabietic, levopimaric, palustric, and dehydroabietic acids as shown in Fig. 12 [11]. The structures of these compounds differ only in the position of the conjugated double

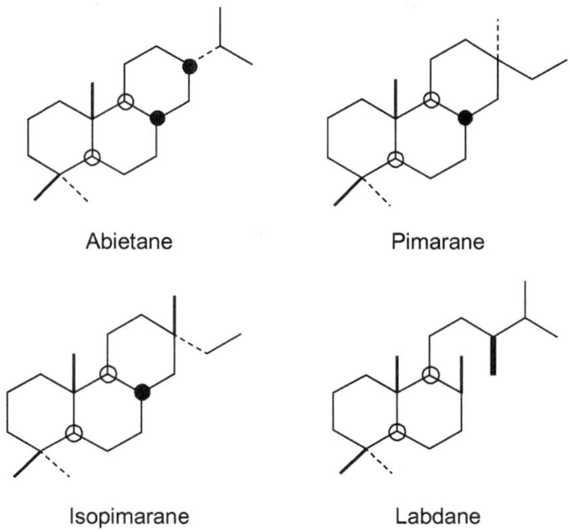

Fig. 11 Diterpene carbon skeletons found in the most common resin acids.
From A.J.D. Silvestre, A. Gandini, Rosin: Major Sources, Properties and Applications, Monomers, Polymers and Composites From Renewable Resources, Elsevier, 2008, pp. 67–88 (Chapter 4).

bond. They are together known as abietadienoic acids. The double bond influences the chemical reactivity of each compound as well as the applications of the products [11].

The most common pimarane-type acids as shown in Fig. 13 are [11] as follows:

- Pimaric
- Isopimaric
- Sandaracopimaric

The differences between pimarane and the abietic acids are as follows:

- Basic skeleton
- The double bond system is not conjugated in pimarane and thereby reduces significantly the possible chemical exploitation of these compounds

Rosin-derived monomers are prepared by vinyl interchange of resin acids with vinyl acetate. The monomer is synthesized by using hydrogenated resin acid such as dehydroabietic acid (DA). The chemical structure of rosin-derived monomers with vinyl ester group is shown in Fig. 14 [12].

Tang and group developed a series of rosin-derived monomers with an acrylate group by reaction of DA and various acrylate with a hydroxyl group. The synthetic strategy is shown in Fig. 15 [12].

Tang and group also used controlled or living radical polymerization (CRP) approach to fabricate the rosin-based homopolymers. This approach synthesizes well-defined polymers with controlled molar mass and narrow molecular weight distribution. The three most widely used CRP methods are [12] as follows:

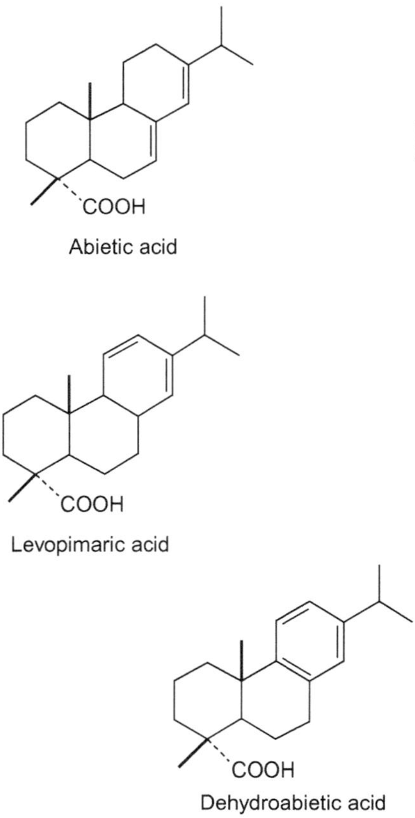

Fig. 12 Structures of the most common conjugated resin acids.
From A.J.D. Silvestre, A. Gandini, Rosin: Major Sources, Properties and Applications, Monomers, Polymers and Composites From Renewable Resources, Elsevier, 2008, pp. 67–88 (Chapter 4).

- Atom transfer radical polymerization (ATRP)
- Reversible addition fragmentation chain-transfer (RAFT)
- Nitroxide-mediated polymerization (NMP)

ATRP is a robust technique to precisely control the chemical composition and architecture of polymers as well as the uniform growth of polymer chains, while tolerating a wide range of monomers.

8.4 Terpenes

Terpenes are secondary metabolites synthesized primarily by plants, but also by a limited number of insects, marine micro-organisms, and fungi [13]. Most terpenes share isoprene (2-methyl-1,4-butadiene) as a common carbon skeleton building block. Terpenes can be classified according to the number of isoprene units (Table 1) [14].

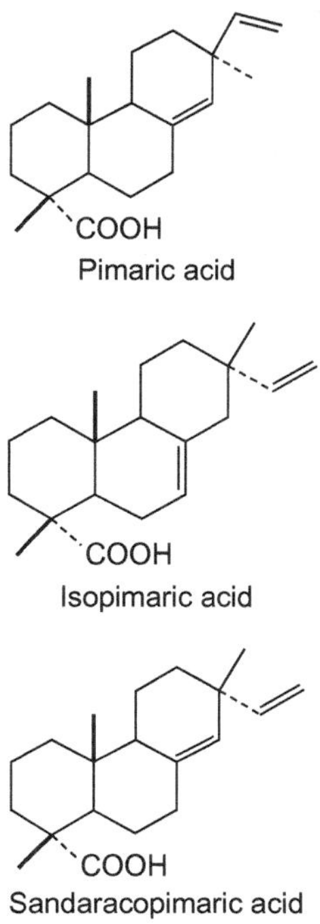

Fig. 13 Structures of the most common nonconjugated resin acids.
From A.J.D. Silvestre, A. Gandini, Rosin: Major Sources, Properties and Applications, Monomers, Polymers and Composites From Renewable Resources, Elsevier, 2008, pp. 67–88 (Chapter 4).

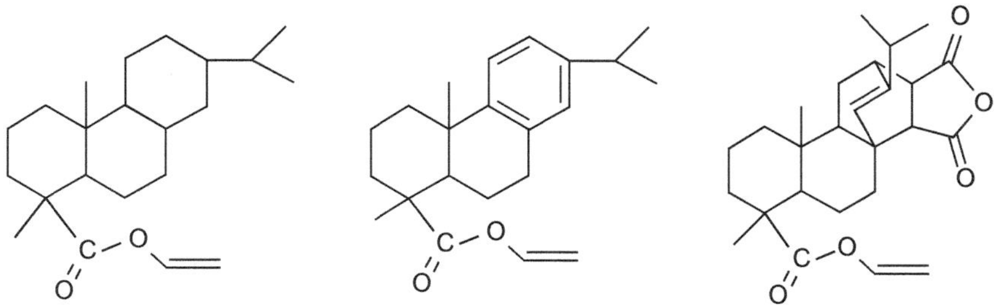

Fig. 14 Chemical structure of rosin-derived monomers with vinyl ester group.
From C. Tang, C.Y. Ryu, Rosin-derived monomers and their progress in polymer applications, Sustainable Polymers From Biomass, Wiley-VCH, p. 112.

There are two ways terpenes can be obtained:

1. Condensation
2. Cyclization and/or rearrangement of a defined number of precursors sharing a common biosynthetic pathway

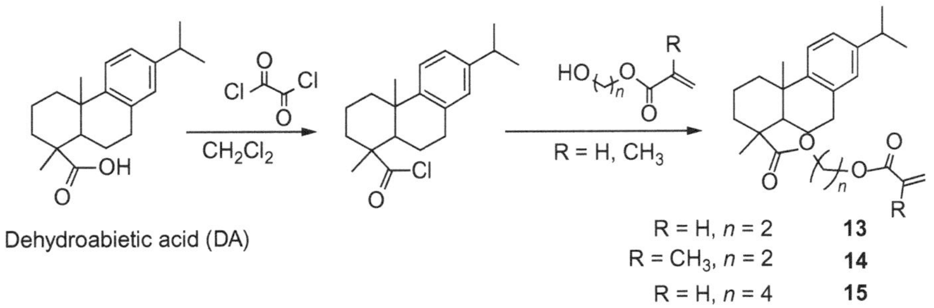

Fig. 15 Synthesis of rosin-derived monomers with acrylate group.
From C. Tang, C.Y. Ryu, Rosin-derived monomers and their progress in polymer applications, Sustainable Polymers From Biomass, Wiley-VCH, p. 112.

Table 1 Major group of terpene compounds according to the number of isoprenic units

Classification	Isoprene units
Hemiterpenes	1
Monoterpenes	2
Sesquiterpenes	3
Diterpenes	4
Sesterpenes	5
Triterpenes	6
Tetraterpenes	8
Polyterpenes	n

From A.J.D. Silvestre, A. Gandini, Terpenes: Major Sources, Properties and Applications, Monomers, Polymers and Composites from Renewable Resources, Elsevier, 2008, pp. 17–38.

The key precursor for monoterpenes is geranyl pyrophosphate, which, upon elimination of the pyrophosphate moiety, forms a *p*-menthane-type carbenium ion. Most common monoterpenes are shown in Fig. 16 [14].

Terpenes undergo four different types of polymerization:

1. Conventional cationic polymerization
2. Living cationic polymerization
3. Living cationic-radical copolymerization with synthetic monomers
4. Radical polymerization of monoterpenes

8.4.1 Conventional cationic polymerization

8.4.1.1 β-Pinene

The cationic polymerization of β-pinene in toluene has been evaluated with different Lewis acid metal halides as initiators [15]. These metal halides include $AlCl_3$,

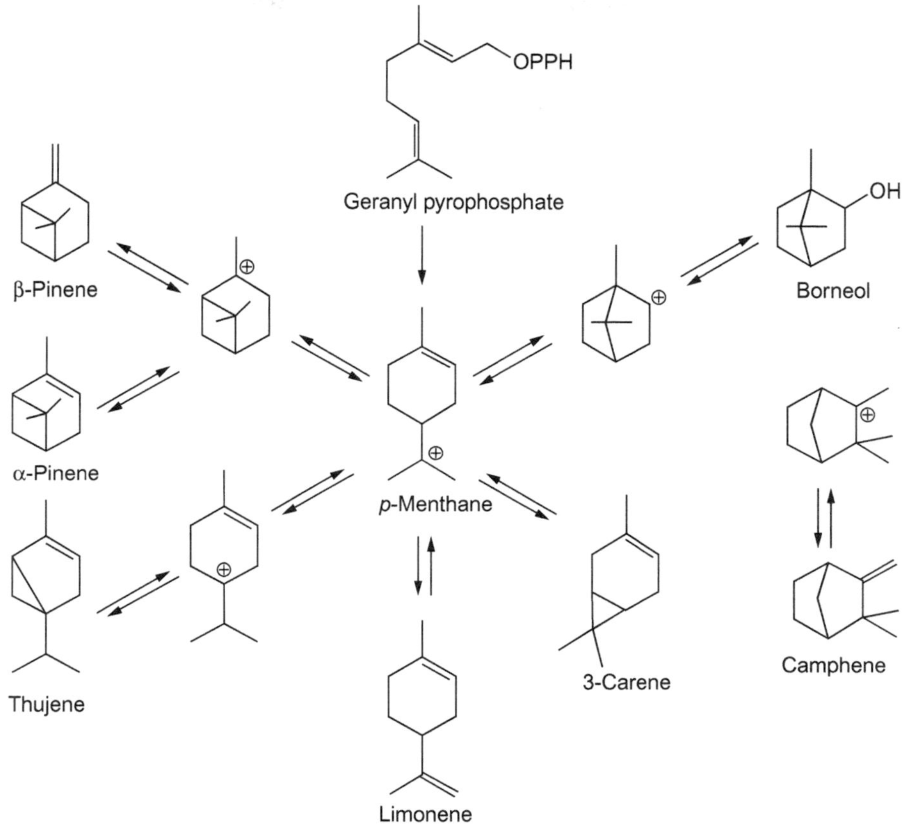

Fig. 16 General outline of the biosynthetic pathway to the formation of main monoterpene skeletons.
From A.J.D. Silvestre, A. Gandini, Terpenes: Major Sources, Properties and Applications, Monomers, Polymers and Composites From Renewable Resources, Elsevier, 2008, pp. 17–38.

$AlBr_3$, and $ZrCl_4$. The driving force of this reaction is the strain release caused by the opening of the fused cyclobutane ring and the formation of a stable tertiary carbonium ion.

Fig. 17 shows proposed polymerization mechanism involving a reaction between Lewis acid and water, followed by the protonation of the pinene molecule. The carbenium ion rearranges to form a *p*-menthane-type carbocation responsible for the propagation reaction [15,16]. This process is extremely rapid and highly exothermic.

The temperature during polymerization is kept low to reduce the kinetic energy of the contribution of the transfer reaction with respect to chain propagation. This produces polymers with molecular weights of few thousand [15,16].

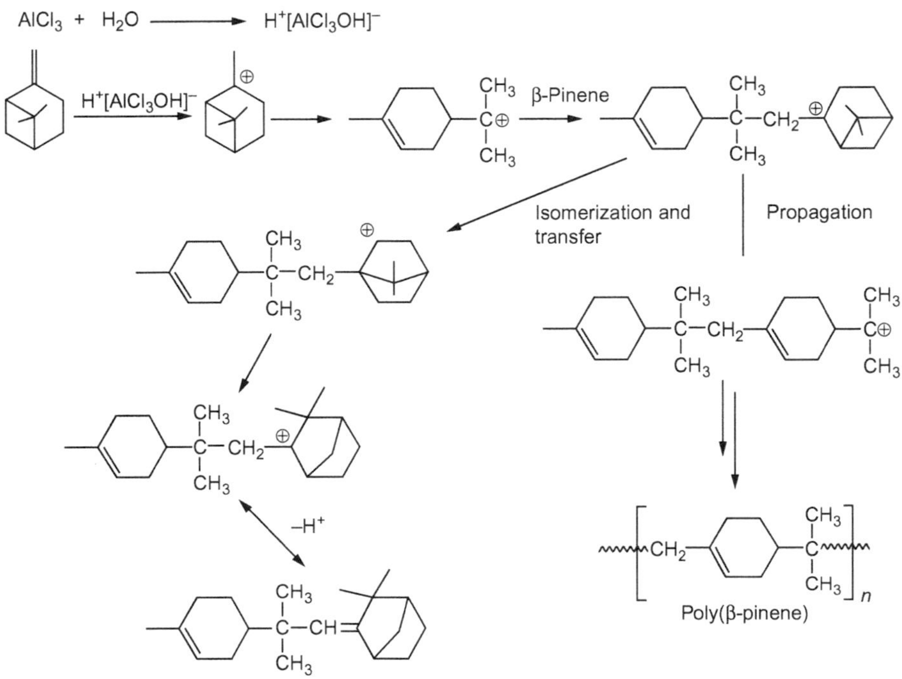

Fig. 17 Proposed mechanism of the cationic polymerization of β-pinene catalyzed by Lewis acid.
From A.J.D. Silvestre, A. Gandini, Terpenes: Major Sources, Properties and Applications, Monomers, Polymers and Composites From Renewable Resources, Elsevier, 2008, pp. 17–38; W.J. Roberts, A.R. Day, A study of the polymerization of alpha-pinene and beta-pinene with Friedel Crafts type catalysts, J. Am. Chem. Soc. 72(3) (1950) 1226–1230.

8.4.1.2 α-Pinene

There are two mechanisms where α-pinene undergoes cationic polymerization as shown in Fig. 18 [14]. Here, α-pinene is readily protonated to form a tertiary carbocation, which can then rearrange to an unsaturated *p*-menthane isomer. The steric hindrance limits the attack of the endocyclic double bond to the isopropenyl cationic site. There is an alternate α-pinene polymerization pathway that involves the rearrangement of the tertiary pinane carbocation to a saturated secondary bornane isomer.

8.4.2 Living cationic polymerization

8.4.2.1 Homopolymerization of β-pinene

Fig. 19 shows the proposed polymerization mechanism involving the rearrangement of the pinane skeleton with the formation of a *p*-menthane-type repeating unit [17,18]. It follows that the strain release resulting from the cyclobutane ring opening is still a key driving force in this propagation mechanism.

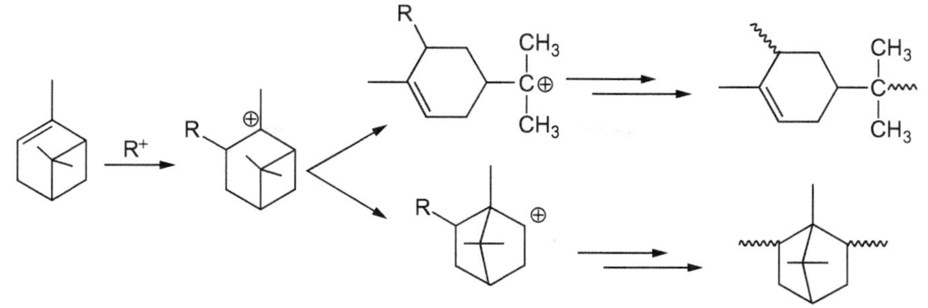

Fig. 18 Alternative mechanism of α-pinene cationic polymerization.
From A.J.D. Silvestre, A. Gandini, Terpenes: Major Sources, Properties and Applications, Monomers, Polymers and Composites From Renewable Resources, Elsevier, 2008, pp. 17–38.

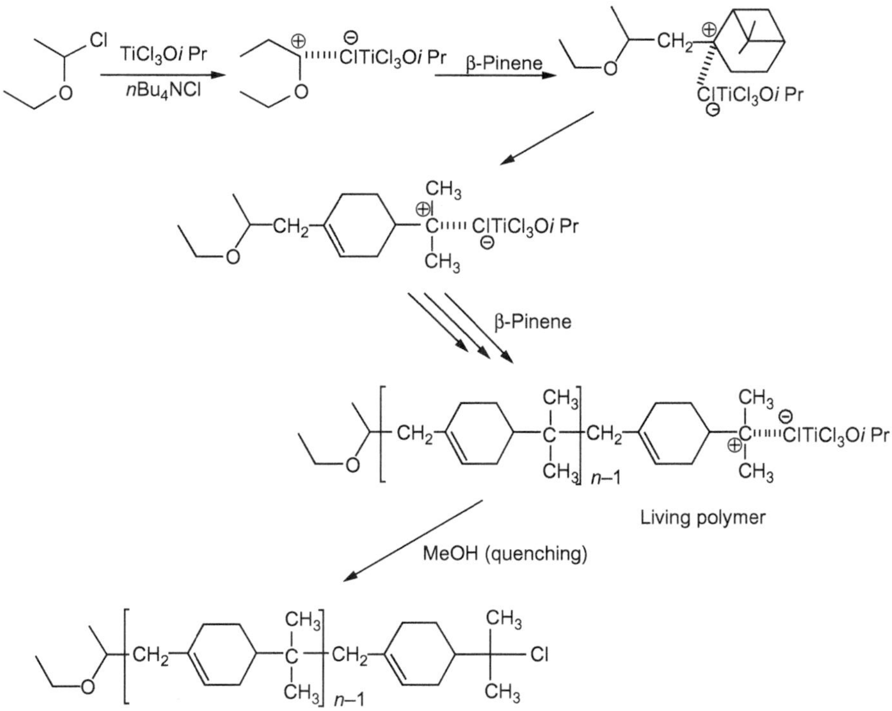

Fig. 19 Synthesis of star-shaped poly(βPIN) with three chlorine end groups.
From J. Lu, M. Kamigaito, M. Sawamoto, T. Higashimura, Y.X. Deng, Living cationic isomerization polymerization of betapinene, Initiation with HCl-2-chloroethyl vinyl ether adduct TiC13(OiPr) in conjunction with nBu(4)NCl, Macromolecules, 30(1) (1997) 22–26; J. Lu, M. Kamigaito, M. Sawamoto, T. Higashimura, Y.X. Deng, Living cationic isomerization polymerization of betapinene, Synthesis of block and random copolymers with styrene or p-methylstyrene, Macromolecules 30(1) (1997) 27–31.

This novel initiating system is capable of providing controlled cationic polymerizations with respect to quasi-living conditions (LCP). Best conditions involved the use of the HCl adduct of 2-chloroethylvinyl ether, isopropoxytitanium chloride in the presence of tetra-*n*-butylammonium chloride in dichloromethane at −40°C to −78°C through a system previously optimized for the LCP of styrene [19].

8.4.3 Living cationic radical copolymerization with synthetic monomers

Fig. 20 shows the mechanism in which the terminal reactive moiety for both homo- and copolymers of β-pinene prepared by LCP is a methacrylic group, from which the free radical polymerization of methyl methacrylate (MMA) is carried out to give block copolymers with interesting properties [20].

Fig. 21 shows an approach that combines cationic and radical mechanisms. Allylic brominated poly(βPIN) with Br/β-pinene unit ratios of 1.0 and 0.5 were obtained by treatment with *N*-bromosuccinimide/AIBN and then used as macroinitiators in conjunction with CuBr and 2,2-bipyridine for the atom transfer radical polymerization of acrylic monomers [21,22].

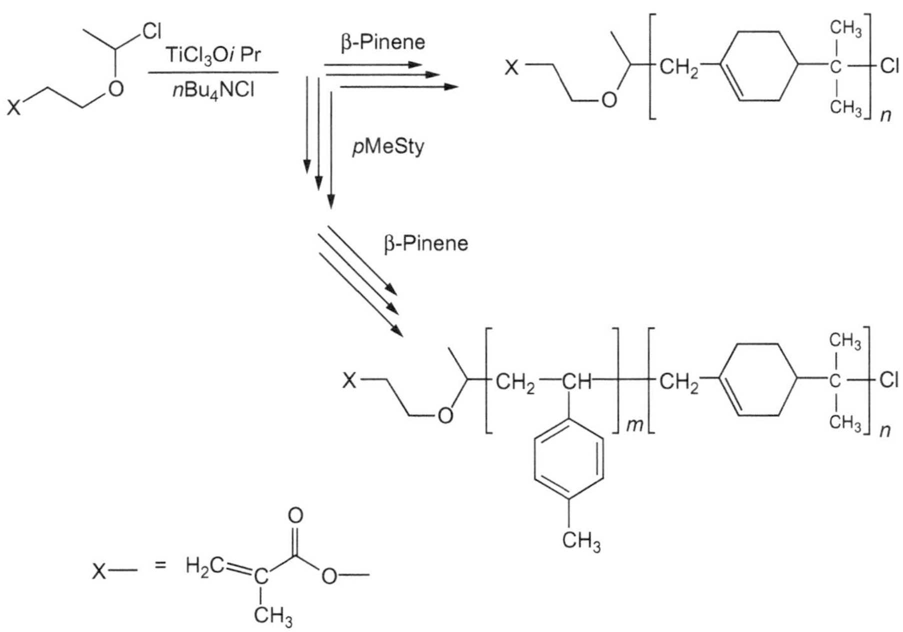

Fig. 20 Synthesis of methacrylic end-functionalized poly(βPIN) and poly(pMeSt)-block-poly (βPIN) by LCP.
From J. Lu, M. Kamigaito, M. Sawamoto, T. Higashimura, Y.X. Deng, Living cationic isomerization polymerization of beta-pinene, synthesis of end-functionalized polymers and graft copolymers, J. Polym. Sci. A Polym. Chem. 35(8) (1997) 1423–1430.

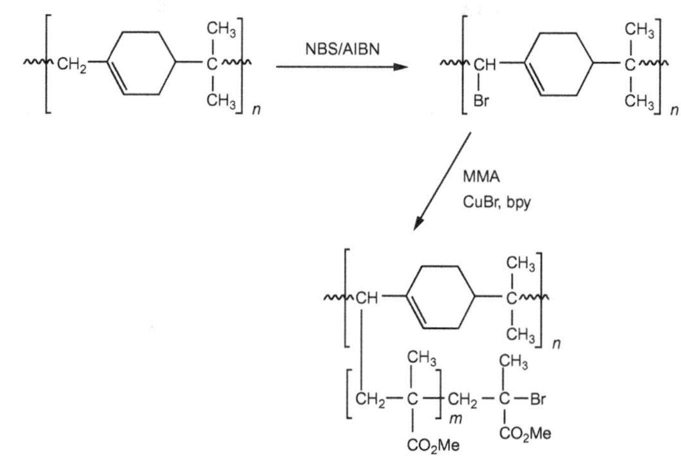

Fig. 21 Synthesis of poly(βPIN)-g-poly(MMA).
From J. Lu, H. Liang, W. Zhang, Q. Cheng, Synthesis of poly(beta-pinene)-g poly(meth) acrylate by the combination of living cationic polymerization and atom transfer radical polymerization, J. Polym. Sci. A Polym. Chem. 41(9) (2003) 1237–1242; A.L. Li, H. Liang, J. Lu, Synthesis of copolymers of beta-pinene by the combination of living cationic polymerization and ATRP, Acta Polym. Sin. (1) (2006) 151–155.

8.4.4 Radical polymerization of monoterpenes

8.4.4.1 Copolymerization using pinenes

Fig. 22 shows a proposed mechanism for the radical copolymerization of β-pinene and acrylonitrile using AIBN as the initiator [23]. The addition of Lewis acids resulted in an increased tendency to alternate. Reduction in the reactivity ratio of acrylonitrile to approximately 0.11 results in an increase in the incorporation of β-pinene levels. This leads to the formation of nearly alternated copolymer structures [23].

8.4.4.2 Copolymerization of limonene

The radical copolymerization of limonene with MMA, AN, St, and *N*-vinylpyrrolidone gave alternating copolymers because both reactivity ratios were close to zero. Fig. 23 shows the mechanism proposed for this process in the case of MMA [24–28].

8.4.4.3 Copolymerization of monoterpene alcohols

Most monoterpene alcohols do not undergo radical homopolymerization, but intervene in copolymerization with conventional monomers. Fig. 24 depicts the mechanism proposed for the system involving MMA as comonomer [29]. α-Terpineol readily copolymerizes with MMA and BMA. In all these systems, the reactivity ratios were found to be very low, which favored a high degree of alternation in the ensuing materials [29–31].

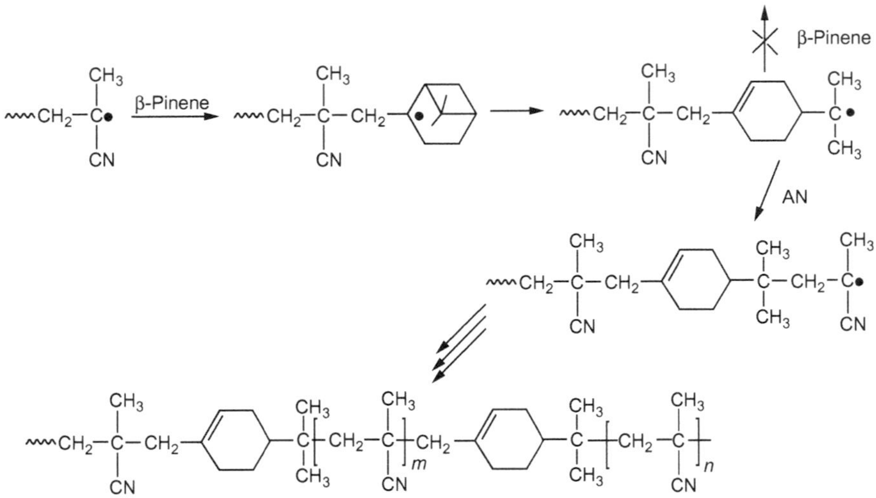

Fig. 22 Proposed mechanism for the radical copolymerization of β-pinene and acrylonitrile. From A.L. Li, Y. Wang, H. Liang, J. Lu, Controlled radical copolymerization of beta-pinene and acrylonitrile, J. Polym. Sci. A Polym. Chem. 44(8) (2006) 2376–2387.

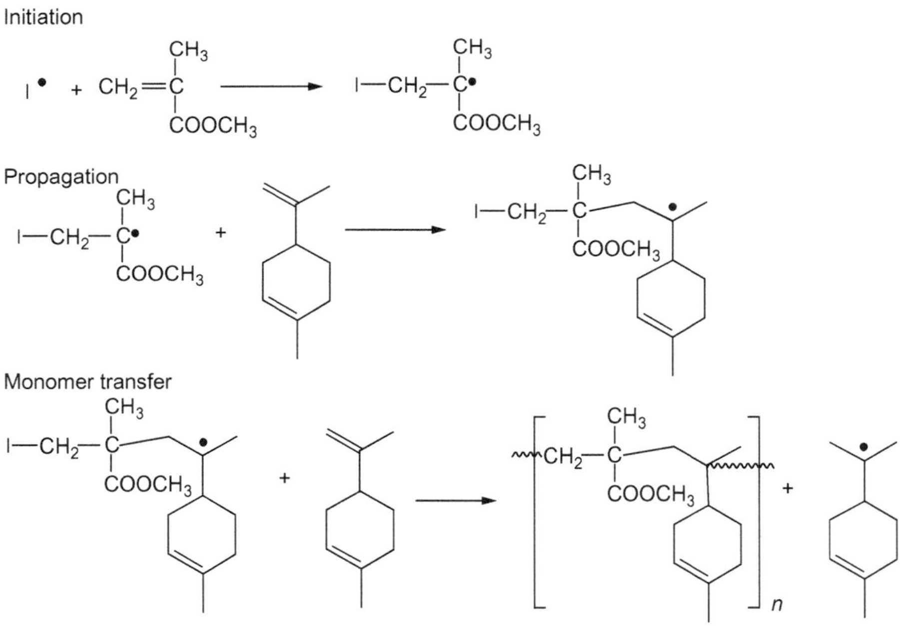

Fig. 23 Proposed mechanism for the radical copolymerization of limonene and MMA. From S. Sharma, A.K. Srivastava, Alternating copolymers of limonene with methyl meth acrylate: kinetics and mechanism, J. Macromol. Sci. A Pure Appl. Chem. 40(6) (2003) 593–603; S. Sharma, A.K. Srivastava, Free radical copolymerization of limonene with butyl methacrylate: synthesis and characterization, Indian J. Chem. Technol. 12(1) (2005) 62–67; S. Sharma, A.K. Srivastava, Radical copolymerization of limonene with acrylonitrile: kinetics and mechanism, Polym. Plast. Technol. Eng. 42(3) (2003) 485–502; S. Sharma, A.K. Srivastava, Synthesis and characterization of copolymers of limonene with styrene initiated by azobi-sisobutyronitrile, Eur. Polym. J. 40(9) (2004) 2235–2240; S. Sharma, A.K. Srivastava, Radical co-polymerization of limonene with N-vinyl pyrrolidone: synthesis and characterization, Des. Monomers Polym. 9(5) (2006) 503–516.

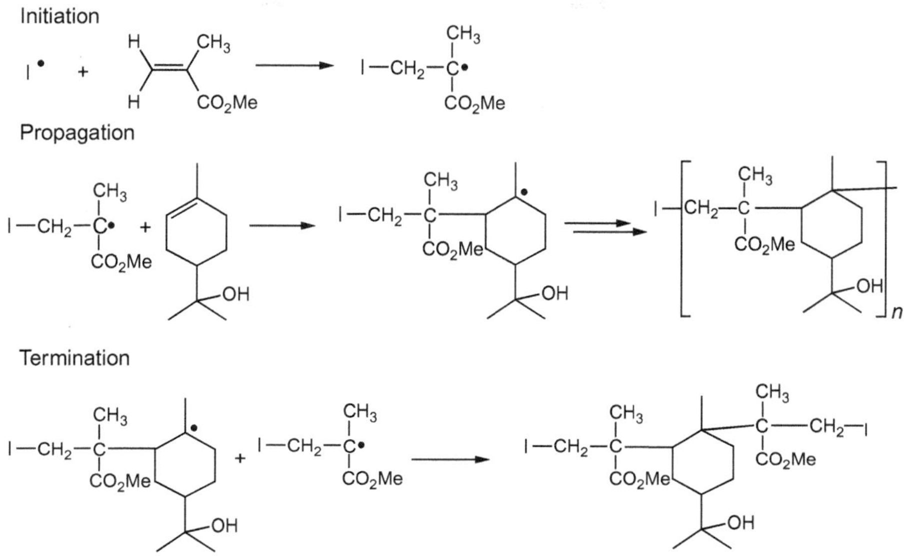

Fig. 24 Proposed mechanism for the radical copolymerization of α-terpineol with MMA. From S. Yadav, A.K. Srivastava, Copolymerization of alpha-terpineol with styrene: synthesis and characterization, J. Polym. Sci. A Polym. Chem. 41(11) (2003) 1700–1707; S. Yadav, A.K. Srivastava, Kinetics and mechanism of copolymerization of alpha-terpineol with methylmethacrylate in presence of azobisisobutyronitrile as an initiator, J. Polym. Res. Taiwan 9(4) (2002) 265–270; S. Yadav, A.K. Srivastava, Synthesis of functional and alternating copolymer of alpha-terpineol with butylmethacrylate, Polym. Plast. Technol. Eng. 43(4) (2004) 1229–1243.

8.5 Glycerol

Fig. 25 shows structure of glycerol [32]. Glycerol contains three hydrophilic alcoholic hydroxyl groups that are responsible for its solubility in water and its hygroscopic nature. It is a highly flexible molecule forming both intra- and intermolecular hydrogen bonds. There are approximately 126 conformations of glycerol. In condensed phases, glycerol is characterized by a high degree of association due to hydrogen bonding [33].

Mixture of linear and branched oligomers is produced by the self-condensation of glycerol catalyzed by both homogeneous and heterogeneous basic catalysis. The linear oligomers are produced with homogeneous catalysis while heterogeneous catalysis results in branched oligomers [10].

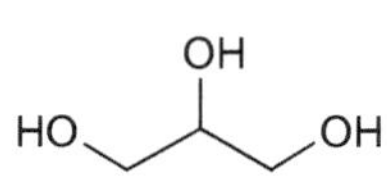

Fig. 25 Structure of glycerol.
From http://www.sigmaaldrich.com/catalog/product/sigma/g5516?lang=en&region=US.

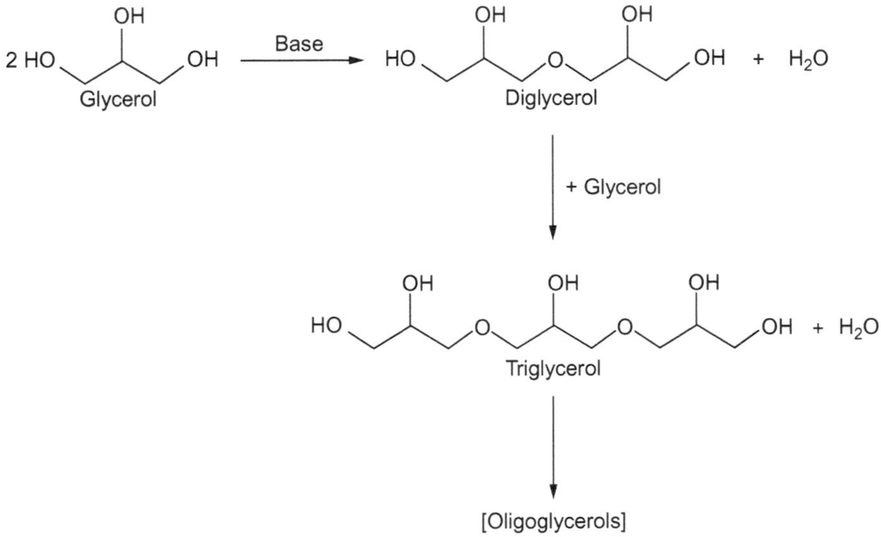

Fig. 26 Schematic illustration of the linear growth.
From K. Loos, Monomers and macromonomers from renewable resources, Biocatalysis in Polymer Chemistry, Wiley-VCH, 2011.

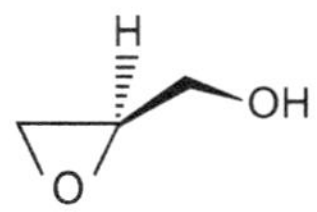

Fig. 27 Glycidol, a glycerol derivative.
From M. Pagliaro, M. Rossi, Glycerol: properties and production, The Future of Glycerol: New Uses of a Versatile Raw Material, RSC Publishing, 2008, p. 2.

Fig. 26 shows a schematic illustration of the linear growth, involving only the primary hydroxyl group [10].

Controlled etherification of glycerol to form polyglycerols occurring via anionic polymerization of glycidol is shown in Fig. 27. This anionic polymerization happens through a rapid cation exchange equilibrium, resulting in a narrow molecular weight distribution. Partially deprotonated 1,1,1-tris(hydroxymethyl)propane is used as an initiator for the anionic polymerization carried out under slow addition conditions, when polymerization or cyclization is minimized in the absence of the initiator [33].

8.6 Vegetable Oils

Vegetable oil is a triglyceride formed by polycondensation of one molecule of glycerol with three molecules of fatty acids or their derivatives. Fig. 28 shows the structure of a triglyceride, where R represents the hydrocarbon parts of different fatty

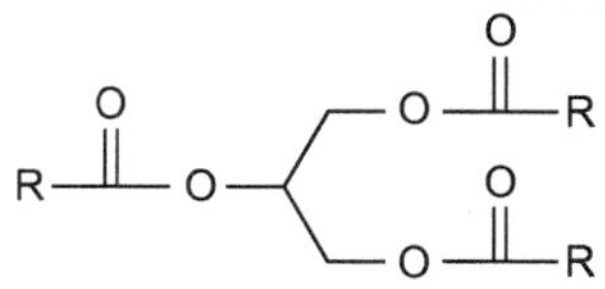

Fig. 28 Structure of a triglyceride, where R represents the hydrocarbon parts of different fatty acids of the oil.
From N. Karak, Vegetables oils and their derivatives, Vegetable Oil-Based Polymers: Properties, Processing, and Applications, Woodhead Publishing, 2012, p. 55 (Chapter 3).

acids of the oil. The saturated fatty acids have no double bond but unsaturated acids have one or more double bonds. These double bonds may contain an alternative single and double bond structures [34].

Some of other facts about vegetable oils are as follows:

- Nonvolatile but are unstable at high temperatures
- May contain phospholipids, free sterols, tocols, triterpene alcohols, hydrocarbons, and fat soluble vitamins

Vegetable oil-based polyesters also known as alkyds are prepared by the polycondensation of polyhydric alcohol and polybasic acid or ester, at least one reactant of which must be obtained from vegetable oil [35].

There are two techniques to produce oil-modified polyesters:

1. Fusion process
2. Solvent process

8.6.1 Polymerization approaches

8.6.1.1 Fusion process

This process involves all reactants to be processed together at the reaction temperature in a stream of nitrogen or argon gas. This initiates an esterification reaction which removes the water and prevents oxidation [35,36]. Volatile reactants are released at high reaction temperatures producing polyester. This process can also be used to produce polyester with high percentage of oil content. This process is also highly desirable because of its relatively low cost and requires no water condenser [35].

8.6.1.2 Solvent process

Solvent process involves polycondensation of one molecule of oil and two molecules of glycerol in the presence of a solvent. It is important that this solvent be immiscible in water and is capable of forming an azeotropic mixture. Xylene is the commonly used solvent and the amount can range from 3% to 10% of the batch. The method is typically for producing short oil polyester resin [35].

Fig. 29 shows synthesis of polyester resin by alcoholysis process. In this process, polyester resin was produced based on *Mesua ferrea* L. seed oil with different

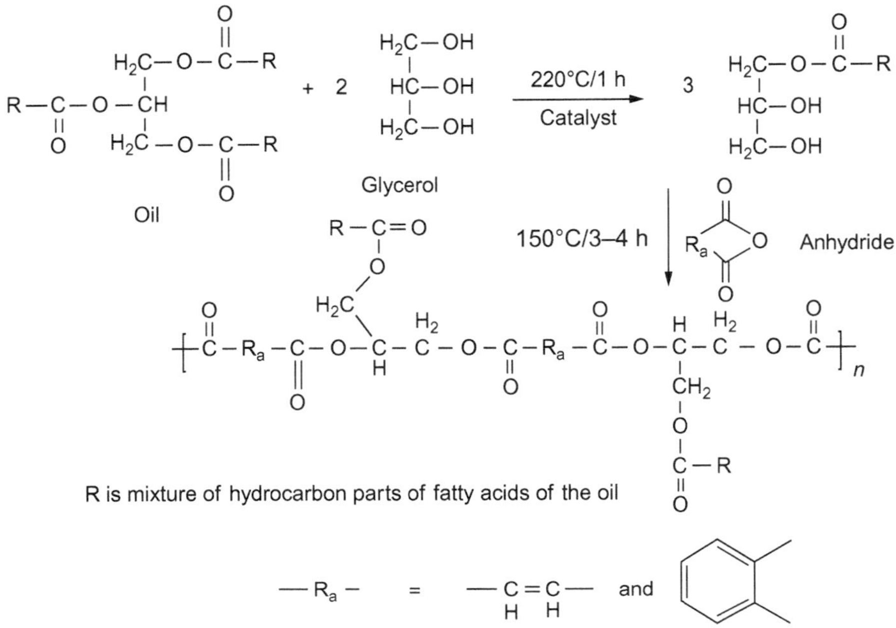

Fig. 29 Reaction scheme for synthesis of polyester resin by an alcoholysis process.
From N. Karak, Vegetables oil-based polyesters, Vegetable Oil-Based Polymers: Properties, Processing, and Applications, Woodhead Publishing, 2012, p. 96 (Chapter 4); N. Dutta, N. Karak, S.K. Dolui, Synthesis and characterization of polyester resins based on Nahar seed oil, Prog. Org. Coat. 49 (2004) 146–152.

compositions of phthalic and maleic anhydrides using a minimum amount of solvent [35,37].

This process has advantages over other process:

- Short reaction time
- Extent and rate of reaction can be determined easily
- Batch-to-batch reproducibility
- Uniform molecular distribution
- Resist attack by atmospheric oxygen

Linear and cross-linked structures of oleic acid-based polyesters prepared by the chemo-enzymatic route, using lipase immobilized on an acrylic resin are shown in Fig. 30 [35,38].

8.7 Tannins

Tannins can be derived from different sources including trees and shrubs, black wattle or black mimosa bark, quebracho wood, oak bark, chestnut wood, mangrove wood, *Terminalia* and *Phyllantus* tree species, divi-divi, algarobilla chilena, tara, and the bark of several species of pines and firs.

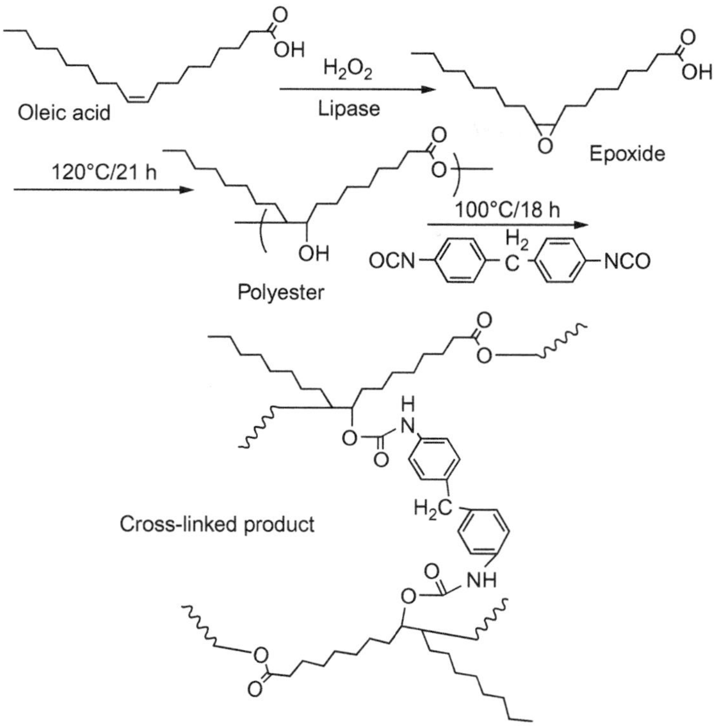

Fig. 30 Linear and cross-linked polyesters prepared by the chemo-enzymatic route [38]. From N. Karak, Vegetables oil-based polyesters, Vegetable Oil-Based Polymers: Properties, Processing, and Applications, Woodhead Publishing, 2012, p. 96 (Chapter 4); N. Dutta, N. Karak, S.K. Dolui, Synthesis and characterization of polyester resins based on Nahar seed oil, Prog. Org. Coat. 49 (2004) 146–152.

The term natural vegetable tannins is used loosely to define two broad classes of chemical compounds of mainly phenolic nature, namely, condensed or polyflavonoid tannins and hydrolysable tannins. The recognized oligomeric nature of condensed tannins [39–42] contrasts with the allegedly nonpolymeric nature of hydrolysable tannins [40–42].

8.7.1 Polymerization

Tannins can be produced from two types of polymerization approaches [43]:

1. Hydrolysable tannins
2. Condensed tannins

8.7.1.1 Hydrolyzable tannins

Fig. 31 shows different low molecular weight fraction of hydrolyzable tannins [43]. These are mixture of phenols and esters of sugar. The carbohydrates have a covalent

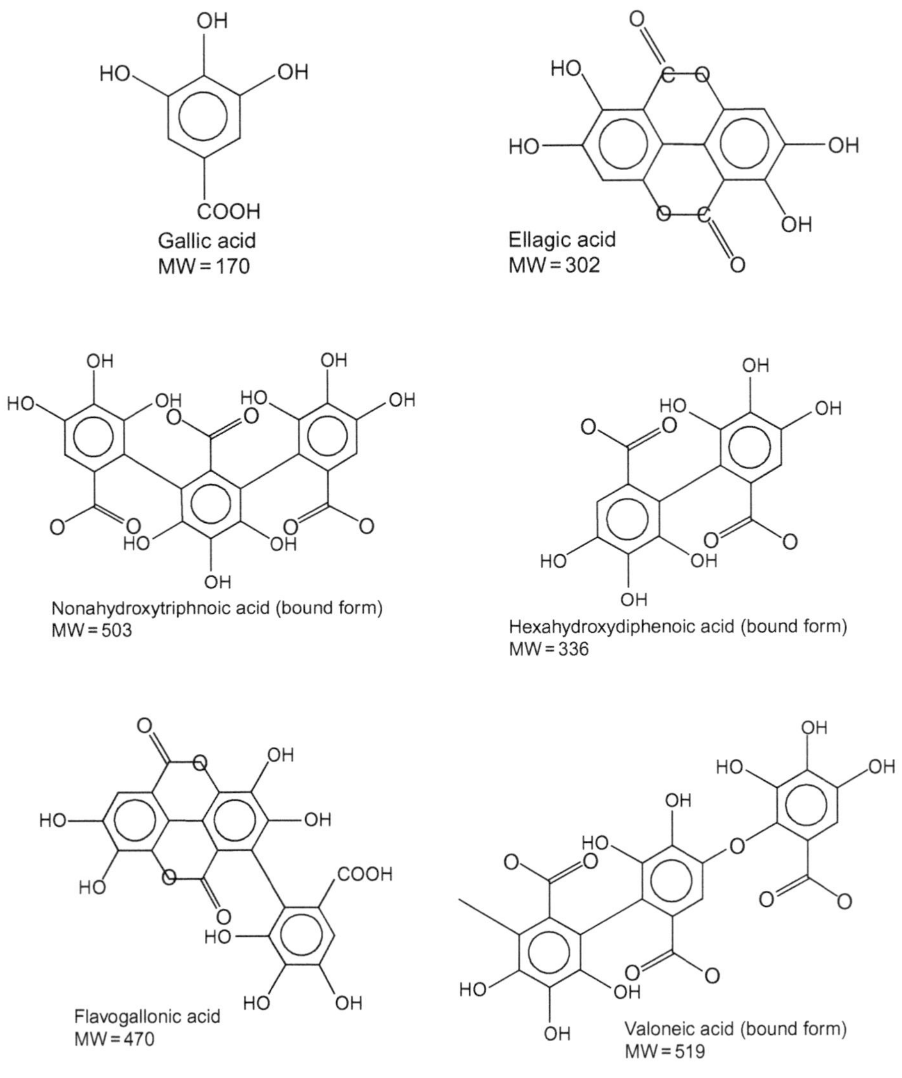

Fig. 31 Different low molecular weight fraction of hydrolysable tannins.
From A. Pizzi, Tannins: major sources, properties and applications, in: M.N. Belgacem, A. Gandini (Eds.), Monomers, Polymers and Composites From Renewable Resources, Elsevier Publishers, pp. 179–199 (Chapter 8).

bond linked to the phenolic moieties in the structure forming complex structures. Several studies have shown that chestnut tannin extract has positional isomers. A total of 30.4% castalagin and vescalagin, 6.6% of castalin and vescalin, 6% of gallic acid, and 3% of pentagalloyl monomer exist as positional isomers [39,40]. Fig. 32 shows positional isomers (a) castalagin and vescalagin, and (b) castalin and vescalin.

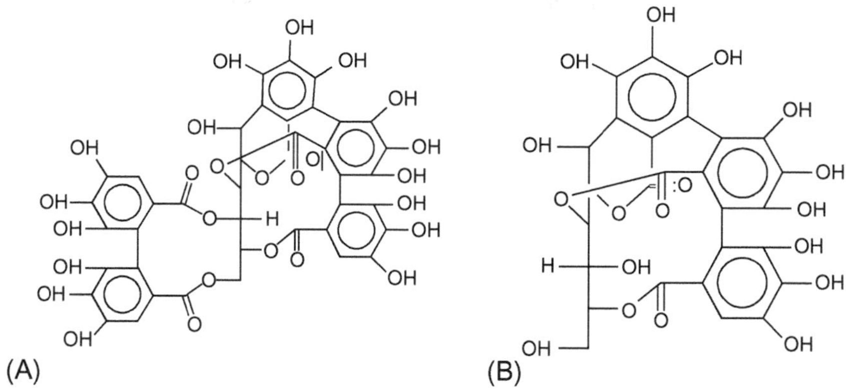

Fig. 32 Positional isomers (A) castalagin and vescalagin, and (B) castalin and vescalin. From D. Fengel, G. Wegener, Wood, Chemistry, Ultrastructure, Reactions, Walther DeGruyter, Berlin, 1984; H.R. Tang, R.A. Hancock, A.D. Covington, studies on commercial tannin extracts, in: XXI IULTCS (International Union of Leather Trades Chemists), Proceedings, Barcelona, Spain, September 1991, 1991, pp. 1503–1527.

Some of the hydrolyzable tannins are as follows:

- Chestnut (*Castanea sativa*)
- Myrabolans (*Terminalia* and *Phyllantus* tree species)
- Divi-divi (*Caesalpinia coriaria*)
- Tara
- Algarrobilla
- Valonea
- Oak

There are several commercial tannin extracts available including [43]:

- Gallic acid
- Ellagic acids
- Nonahydroxytriphnoic acid
- Hexahydroxydiphenoic acid
- Flavogallonic acid
- Valoneic acid

8.7.1.2 Condensed or polyflavonoid tannins

Source of condensed tannins are typically found in the wood and bark of various trees. These include:

- *Acacia* (wattle or mimosa bark extract)
- *Schinopsis* (quebracho wood extract)
- *Tsuga* (hemlock bark extract)
- *Rhus* (sumach extract)
- *Pinus* (bark extract)

These tannins are highly reactive toward aldehydes and other reagents. This makes them potential candidate in the production of adhesives, resins, and other

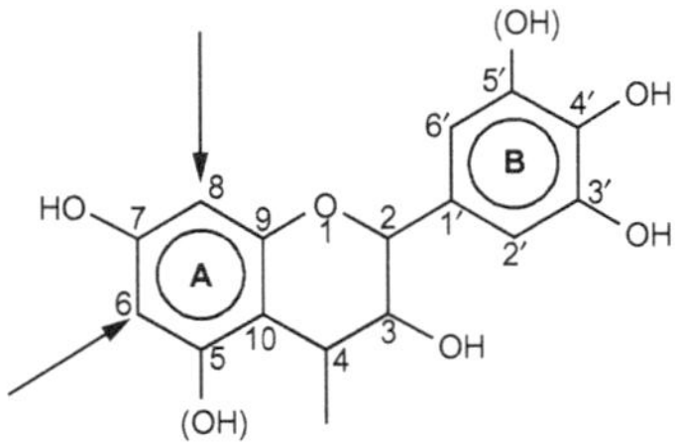

Fig. 33 Structure of flavonoid constituting condensed tannins.
From A. Pizzi, Tannins: major sources, properties and applications, in: M.N. Belgacem, A. Gandini (Eds.), Monomers, Polymers and Composites From Renewable Resources, Elsevier Publishers, pp. 179–199 (Chapter 8).

applications. Fig. 33 represents structure of the flavonoid constituting condensed tannins [43].

This flavonoid unit is repeated 2–11 times in mimosa and quebracho tannins [44,45], with an average degree of polymerization of 4–5, and up to 30 times for pine tannins, with an average degree of polymerization of 6–7 for their soluble extract fraction [46]. The nucleophilic centers on the A-ring of a flavonoid unit tend to be more reactive in aromatic substitution than those found on the B-ring. This is due to the vicinal hydroxyl substituents, which cause general activation in the B-ring, without any localized effects such as those found in the A-ring [44].

Fig. 34 shows an illustration of two tannin structures with three-unit fragments [44–46]:

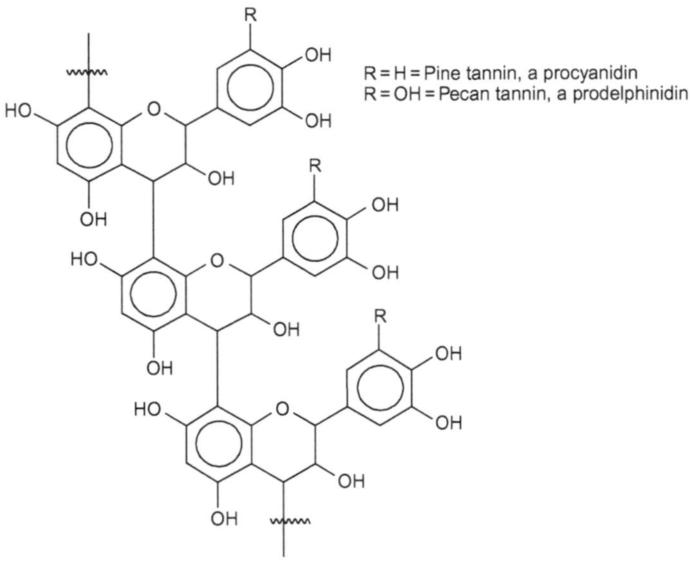

Fig. 34 Illustration of three-unit fragment of two tannin structures.
From A. Pizzi, Wood adhesives—chemistry and technology, Volume I, in: A. Pizzi (Ed.), J. Macromol. Sci. Chem., Ed., vol. C18(2), 1980, 247 (Chapter 4); H. Pasch, A. Pizzi, K. Rode, Polymer 42(18) (2001) 7531–7539; A. Pizzi, Advanced Wood Adhesives Technology, Dekker, New York, NY, 1994.

8.8 Lignin

Lignin is an organic substance binding the cells, fibers, and vessels which constitute wood and the lignified elements of plants, as in straw. Fig. 35 shows the biosynthetic route toward production of monolignols *p*-coumaryl, coniferyl, and sinapyl alcohol [47,48].

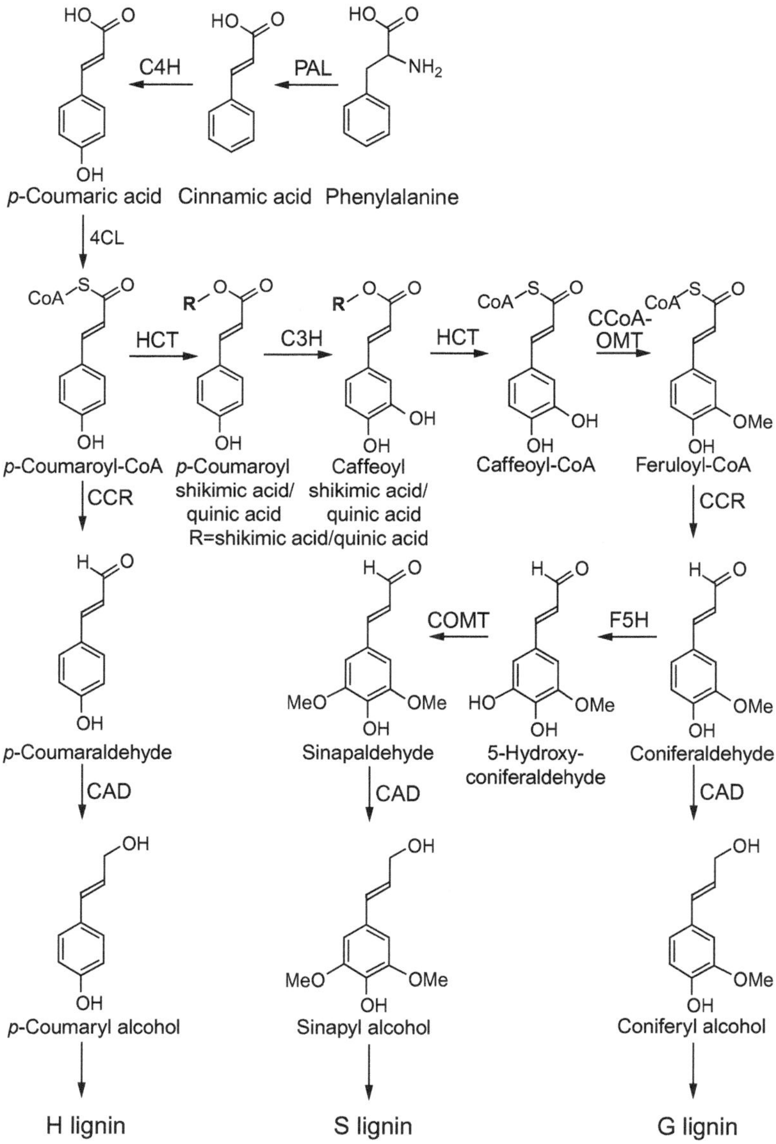

Fig. 35 The main biosynthetic route toward production of monolignols *p*-coumaryl, coniferyl, and sinapyl alcohol.
From R. Vanholme, B. Demedts, K. Morreel, J. Ralph, W. Boerjan, Lignin biosynthesis and structure, Plant Physiol. 153(9) (2010) 895–905; W. Boerjan, J. Ralph, M. Baucher, Lignin biosynthesis, Annu. Rev. Plant Biol. 54 (2003) 519–546.

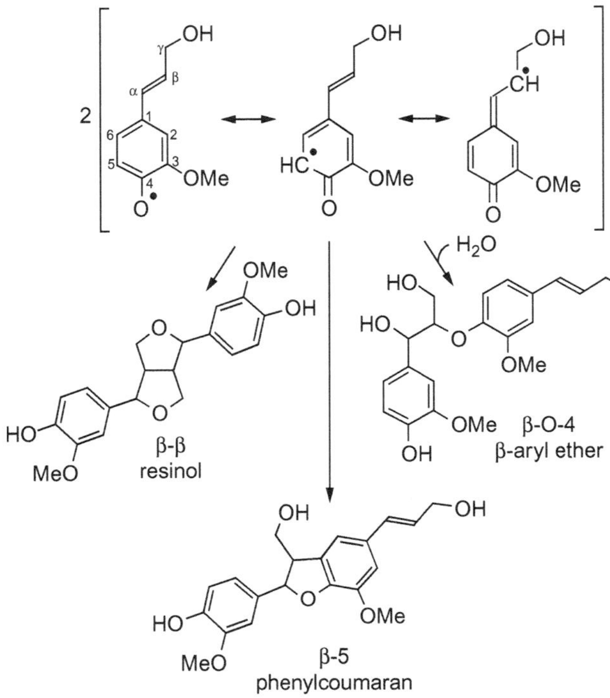

Fig. 36 Dimerization of two dehydrogenated coniferyl alcohol monomers.
From R. Vanholme, B. Demedts, K. Morreel, J. Ralph, W. Boerjan, Lignin biosynthesis and structure, Plant Physiol. 153(9) (2010) 895–905; W. Boerjan, J. Ralph, M. Baucher, Lignin biosynthesis, Annu. Rev. Plant Biol. 54 (2003) 519–546.

Typical lignin polymerization happens in two stages:

1. Oxidative radicalization of phenols
2. Combinatorial radical coupling

During the oxidative radicalization of phenols, the monolignol phenol undergoes dehydrogenation resulting in phenolic radical. This radical is relatively stable due to delocalization of the unpaired electron in the conjugated system. Two monomer radicals may couple to form a dimer, thereby establishing a covalent bond between both subunits. Monolignol radicals favor coupling at their β positions, resulting essentially in only the β-β, β-O-4, and β-5 dimers. Fig. 36 shows dimerization of two dehydrogenated coniferyl alcohol monomers [47,48].

8.9 Suberin

Suberin is complex polyester built from polyfunctional long-chain fatty acids such as suberin acids and glycerol. The structural formula of the main suberin monomers are shown in Fig. 37 [49].

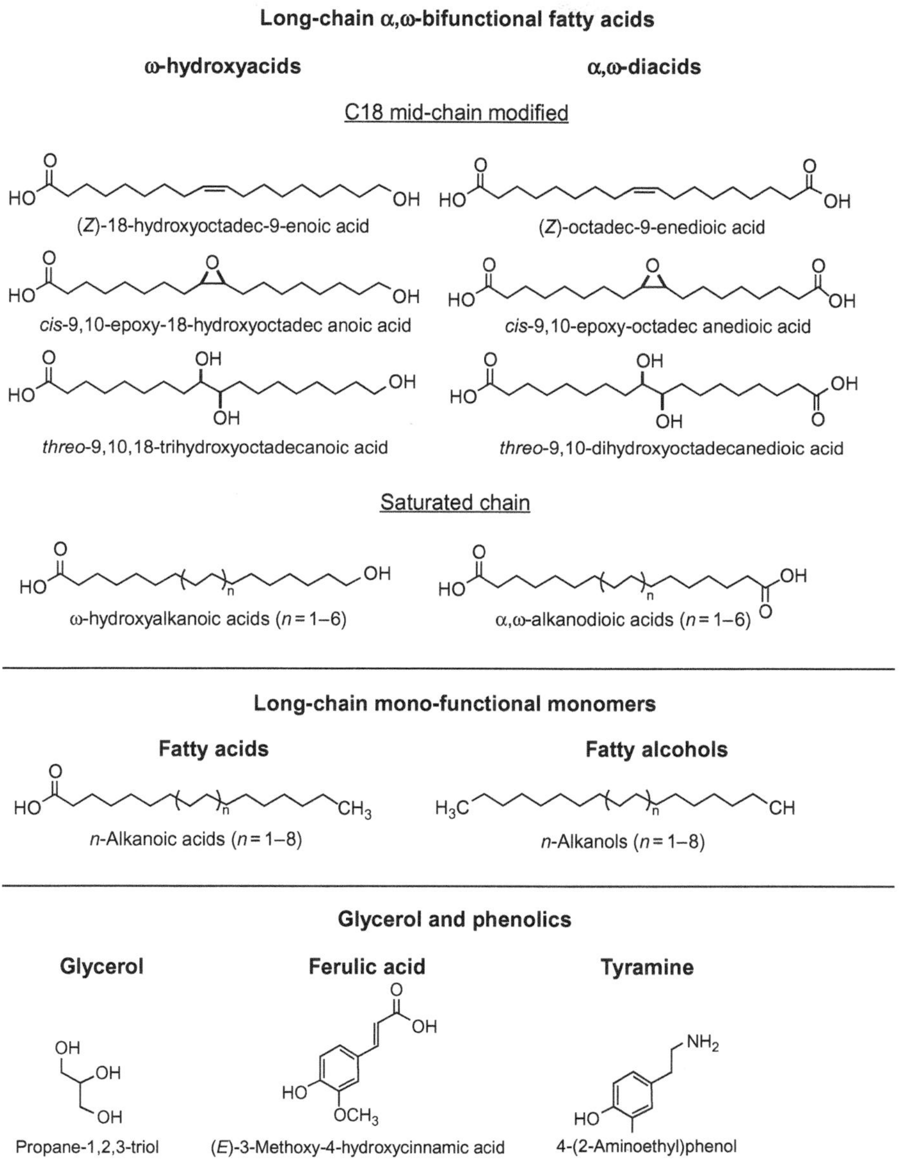

Fig. 37 Structural formula of the main suberin monomers.
From J. Graca, Suberin: the biopolyester at the frontier of plants, Front. Chem. 3(62) (2015).

The oligomers and monomers of suberin macromolecule are obtained by the partial depolymerization of the suberin polyester, using mild or partial hydrolysis conditions. The dominant oligomer species from the partial depolymerization of suberin are glycerides having as acyl moieties all types of suberin acids, namely, α-ω-diacids, ω-hydroxyacids, and alkanoic acids [49].

There are three species of acylglycerol at all the glycerol positions in suberin oligomers [50]:

- 1(3)- and 2-monoacylglycerols
- 1,2- and 2,3-diacylglycerols
- Triacylglycerols

Partial depolymerization of suberins produces suberin acid as oligomer fragments which are linked to glycerol. These linear aliphatic esters are included mainly as monomer residues ω-hydroxyacids, since the ω-hydroxyl group is necessary to make the ester bond [51].

The other significant ester structure found in the oligomers released from the partial depolymerization of suberins is ferulates of ω-hydroxyacids. All ω-hydroxyacids are esterified though their ω-hydroxyl group to ferulic acid as oligomer species [52–54]. There is a high proportion of ω-hydroxyacids found as suberin monomers and the relative abundance of their ferulates found as suberin oligomers.

It is important to understand that when the suberin polyester is depolymerized by any ester breaking technique, small amounts of phenolics are cosolubilized together with the main aliphatic suberin monomers. Soluble phenolics such as the hydroxycinnamic ferulic acid, and aminated phenolics such as tyramine, both in native form induce potato periderm and is an important element in the overall structure of suberized cell walls [55,56].

References

[1] A. Gandini, M.N. Belgacem, Furan Derivatives and Furan Chemistry Service of Macromolecular Materials, Monomers, Polymers, and Composites from Renewable Resources, Elsevier Publishers, 2008, p. 116 (Chapter 6).

[2] I. Delidovich, P.J.C. Hausoul, L. Deng, R. Pfützenreuter, M. Rose, R. Palkovits, Alternative monomers based on lignocellulose and their use for polymer production, Chem. Rev. 116 (2016) 1540–1599. ACS Publications.

[3] http://www.sigmaaldrich.com/catalog/product/aldrich/185930?lang=en&region=US.

[4] A. Gandini, The behavior of furan derivatives in polymerization reactions, Adv. Polym. Sci. 25 (1977) 47.

[5] A. Gandini, Encyclopedia of Polymer Science and Engineering, in: H.F. Mark, N.M. Bikales, C.G. Overberger, G. Menges (Eds.), second ed., vol. 7, John Wiley, New York, NY, 1988, p. 454.

[6] W.J. McKillip, Chemistry of furan polymers, ACS Symp. Ser. 385 (1989) 408.

[7] A. Gandini, M.N. Belgacem, Furans in polymer chemistry, Prog. Polym. Sci. 22 (1997) 1203.

[8] C. Moreau, A. Gandini, M.N. Belgacem, Recent catalytic advances in the chemistry of substituted furans from carbohydrates and in the ensuing polymers, Top. Catal. 27 (2004) 11.

[9] A. Gandini, J. Rieumont, The reaction of furan derivatives with azo-bis-isobutyronitrile, Tetrahedron Lett. 25 (1976) 2101.

[10] K. Loos, Monomers and macromonomers from renewable resource, Biocatalysis in Polymer Chemistry, Wiley-VCH, Weinheim, Germany, 2011.

[11] A.J.D. Silvestre, A. Gandini, Rosin: major sources, properties and applications, Monomers, Polymers and Composites From Renewable Resources, Elsevier, Amsterdam, 2008, pp. 67–88 (Chapter 4).

[12] C. Tang, C.Y. Ryu, Rosin-derived monomers and their progress in polymer applications, Sustainable Polymers from Biomass, Wiley-VCH, Weinheim, Germany, 2017, p. 112.

[13] W.F. Erman, Chemistry of Monoterpenes, Marcel Dekker, New York, NY, 1985.
[14] A.J.D. Silvestre, A. Gandini, Terpenes: major sources, properties and applications, Monomers, Polymers and Composites from Renewable Resources, Elsevier, Amsterdam, The Netherlands, 2008, pp. 17–38.
[15] W.J. Roberts, A.R. Day, A study of the polymerization of alpha-pinene and beta-pinene with Friedel Crafts type catalysts, J. Am. Chem. Soc. 72 (3) (1950) 1226–1230.
[16] E.R. Ruckel, H.J. Arlt, D.F. Zinkel, J. Russel (Eds.), Polyterpene Resins, Pulp Chemical Association, New York, NY, 1989, pp. 510–530.
[17] J. Lu, M. Kamigaito, M. Sawamoto, T. Higashimura, Y.X. Deng, Living cationic isomerization polymerization of betapinene, Initiation with HCl-2-chloroethyl vinyl ether adduct $TiCl_3(OiPr)$ in conjunction with nBu(4)NCl, Macromolecules 30 (1) (1997) 22–26.
[18] J. Lu, M. Kamigaito, M. Sawamoto, T. Higashimura, Y.X. Deng, Living cationic isomerization polymerization of betapinene, synthesis of block and random copolymers with styrene or p -methylstyrene, Macromolecules 30 (1) (1997) 27–31.
[19] T. Hasebe, M. Kamigaito, M. Sawamoto, Living cationic polymerization of styrene with $TiCl_3(OiPr)$ as a Lewis acid activator, Macromolecules 29 (19) (1996) 6100–6103.
[20] J. Lu, M. Kamigaito, M. Sawamoto, T. Higashimura, Y.X. Deng, Living cationic isomerization polymerization of beta-pinene, synthesis of end-functionalized polymers and graft copolymers, J. Polym. Sci. A Polym. Chem. 35 (8) (1997) 1423–1430.
[21] J. Lu, H. Liang, W. Zhang, Q. Cheng, Synthesis of poly(beta-pinene)- *g* poly(meth)acrylate by the combination of living cationic polymerization and atom transfer radical polymerization, J. Polym. Sci. A Polym. Chem. 41 (9) (2003) 1237–1242.
[22] A.L. Li, H. Liang, J. Lu, Synthesis of copolymers of beta-pinene by the combination of living cationic polymerization and ATRP, Acta Polym. Sin. 1 (2006) 151–155.
[23] A.L. Li, Y. Wang, H. Liang, J. Lu, Controlled radical copolymerization of beta-pinene and acrylonitrile, J. Polym. Sci. A Polym. Chem. 44 (8) (2006) 2376–2387.
[24] S. Sharma, A.K. Srivastava, Alternating copolymers of limonene with methyl meth acrylate: Kinetics and mechanism, J. Macromol. Sci. Pure Appl. Chem. A40 (6) (2003) 593–603.
[25] S. Sharma, A.K. Srivastava, Free radical copolymerization of limonene with butyl methacrylate: synthesis and characterization, Indian J. Chem. Technol. 12 (1) (2005) 62–67.
[26] S. Sharma, A.K. Srivastava, Radical copolymerization of limonene with acrylonitrile: kinetics and mechanism, Polym. Plast. Technol. Eng. 42 (3) (2003) 485–502.
[27] S. Sharma, A.K. Srivastava, Synthesis and characterization of copolymers of limonene with styrene initiated by azobi- sisobutyronitrile, Eur. Polym. J. 40 (9) (2004) 2235–2240.
[28] S. Sharma, A.K. Srivastava, Radical co-polymerization of limonene with *N*-vinyl pyrrolidone: synthesis and characterization, Des. Monomers Polym. 9 (5) (2006) 503–516.
[29] S. Yadav, A.K. Srivastava, Copolymerization of alpha-terpineol with styrene: synthesis and characterization, J. Polym. Sci. A Polym. Chem. 41 (11) (2003) 1700–1707.
[30] S. Yadav, A.K. Srivastava, Kinetics and mechanism of copolymerization of alpha-terpineol with methylmethacrylate in presence of azobisisobutyronitrile as an initiator, J. Polym. Res. Taiwan 9 (4) (2002) 265–270.
[31] S. Yadav, A.K. Srivastava, Synthesis of functional and alternating copolymer of alpha-terpineol with butylmethacrylate, Polym. Plast. Technol. Eng. 43 (4) (2004) 1229–1243.
[32] http://www.sigmaaldrich.com/catalog/product/sigma/g5516?lang=en&region=US.
[33] M. Pagliaro, M. Rossi, Glycerol: properties and production, The Future of Glycerol: New Uses of a Versatile Raw Material, RSC Publishing, Cambridge, UK, 2008, 2.
[34] N. Karak, Vegetables oil-based polyesters, Vegetable Oil-Based Polymers: Properties, Processing, and Applications, Woodhead Publishing, Cambridge, UK, 2012, 55, (Chapter 3).
[35] N. Karak, Vegetables oil-based polyesters, Vegetable Oil-Based Polymers: Properties, Processing, and Applications, Woodhead Publishing, Cambridge, UK, 2012, 96, (Chapter 4).

[36] V.C. Malshe, M. Sikchi, Basics of Paint Technology, Part I, UICT, Mumbai, 2004.
[37] N. Dutta, N. Karak, S.K. Dolui, Synthesis and characterization of polyester resins based on Nahar seed oil, Prog. Org. Coat. 49 (2004) 146–152.
[38] S. Miao, S. Zhang, Z. Su, P. Wang, Chemo-enzymatic synthesis of oleic acid-based polyesters for use as highly stable biomaterials, J. Polym. Sci. A Polym. Chem. 46 (2008) 4243–4248.
[39] D. Fengel, G. Wegener, Wood, Chemistry, Ultrastructure, Reactions, Walther DeGruyter, Berlin, 1984.
[40] H.R. Tang, R.A. Hancock, A.D. Covington, Studies on commercial tannin extracts, in: XXI IULTCS (International Union of Leather Trades Chemists), Proceedings, Barcelona, Spain, September 1991, 1991, pp. 1503–1527.
[41] E. Haslam, The Chemistry of Vegetable Tannins, Academic Press, London, 1966.
[42] D.G. Roux, Reflections on the chemistry and affinities of the major commercial condensed tannins, in: R.W. Hemingway, P.E. Laks (Eds.), Plant Polyphenols, vol. 1, Plenum, New York, NY, 1992, pp. 27–39.
[43] A. Pizzi, Tannins: major sources, properties and applications, in: M.N. Belgacem, A. Gandini (Eds.), Monomers, Polymers and Composites From Renewable Resources, Elsevier Publishers, 2011, pp. 179–199 (Chapter 8).
[44] A. Pizzi, Wood adhesives–chemistry and technology, volume I, in: A. Pizzi (Ed.), J. Macromol. Sci. Chem., Ed., C18 (2) Marcel Dekker, New York, NY, 1980, p. 247, (Chapter 4).
[45] H. Pasch, A. Pizzi, K. Rode, MALDI-TOF mass spectrometry of polyflavonoid tannins, Polymer 42 (18) (2001) 7531–7539.
[46] A. Pizzi, Advanced Wood Adhesives Technology, Dekker, New York, NY, 1994.
[47] R. Vanholme, B. Demedts, K. Morreel, J. Ralph, W. Boerjan, Lignin biosynthesis and structure, Plant Physiol. 153 (9) (2010) 895–905.
[48] W. Boerjan, J. Ralph, M. Baucher, Lignin biosynthesis, Annu. Rev. Plant Biol. 54 (2003) 519–546.
[49] J. Graca, Suberin: the biopolyester at the frontier of plants, Front. Chem. 3 (62) (2015) 1–12.
[50] J. Graça, V. Cabral, S. Santos, P. Lamosa, O. Serra, M. Molinas, Partial depolymerization of genetically modified potato tuber periderm reveals intermolecular linkages in suberin polyester, Phytochemistry 117 (2015) 209–219.
[51] J. Graça, S. Santos, Linear aliphatic dimericesters from cork suberin, Biomacromolecules 7 (2006) 2003–2010.
[52] J. Graça, H. Pereira, Feruloylestersofomega-hydroxyacidsincork suberin, J. Wood Chem. Technol. 18 (1998) 207–217.
[53] J. Graça, H. Pereira, Glyceryl-acyl and aryl-acyldimers in Pseudotsuga menziesii bark suberin, Holzforschung 53 (1999) 397–402.
[54] J. Graça, H. Pereira, Methanolysisofbarksuberins:analysis of glycerol and acid monomers, Phytochem. Anal. 11 (2000) 45–51.
[55] J. Graça, Hydroxycinnamates in suberin formation, Phytochem. Rev. 9 (2010) 85–91.
[56] O. Borgolivier, B. Monties, Lignin, suberin, phenolic-acids and tyramine in the suberized, wound-induced potato periderm, Phytochemistry 32 (1993) 601–606.

Further reading

[1] http://www.sigmaaldrich.com/catalog/product/aldrich/f19558?lang=en&region=US.
[2] http://www.sigmaaldrich.com/catalog/product/mm/807492?lang=en&region=US.
[3] http://www.sigmaaldrich.com/catalog/product/aldrich/t31801?lang=en&region=US.

Direct derivatives of polymers from biomass

9

9.1 Introduction

Sugars are made from a wide variety of woody material, wastes, and other residues. Many of the routes onward from sugar are then expected to use microorganisms to generate finished products or useful intermediates. There are many option and combinations of feedstock, pretreatment options, sugars, and conversion technologies that can be followed as viable pathways to produce biofuels. [1].

Large number of polymers produced from sugar is further converted into biofuels for commercial use. Type of conversion route dictates the type of conversion. For example, selective oxidation of levulinic acid (LA) produces succinic and acrylic acid while 3-Hydroxypropionic acid (3-HPA) is converted to 1,3-propanediol (1,3-PDO) by undergoing hydrogenation/reduction reactions.

In this chapter, direct derivatives of polymers produced from biomass will be discussed. It will also include derivatives obtained from microbial polymers.

9.2 Epoxy resin from biomass

Epoxy resins are one of the most important members of the thermoset family because of their use in adhesives, composites, and elastomers [2]. There has been a lot of research studying different type of lignin-based epoxy resins [3,4]. One of such research was reported by Malutan et al. They reported epoxy resins derived from hydroxymethylated lignin [5]. Belgacem et al. [6] reported saccharide-based epoxy resins and furan derivatives with epoxy groups from pentoses and hexoses. There are several other studies that reported new types of epoxy resins with polyester chains derived from saccharides, lignin, and glycerol [7–13].

As shown in Fig. 1, epoxy resins are typically prepared by dissolving saccharides and lignin in polyols such as ethylene glycol and glycerol [14]. These polyols acts as solvent in the system. Equal amounts of biomass components and polyols by weight are added. Amount of polyol can be reduced however; this leads to high viscous polyol-biomass solution.

Derivatives of polycarboxylic acid from biomass components can be prepared by allowing primary alcoholic groups to react with the phenolic hydroxyl group on saccharides, lignin, and glycerol under optimum conditions. All the hydroxyl groups are converted to half esters of dicarboxylic acid anhydrides in the presence of amine catalyst such as dimethylbenzylamine.

Fig. 2 shows the synthetic scheme of epoxy resins containing lignin as a biomass component. Other polymer derivative obtained from biomass include ethylene glycol poly acid derivative (EGPA) that is prepared from ethylene glycol [14].

Technology and Applications of Polymers Derived from Biomass. https://doi.org/10.1016/B978-0-323-51115-5.00009-8

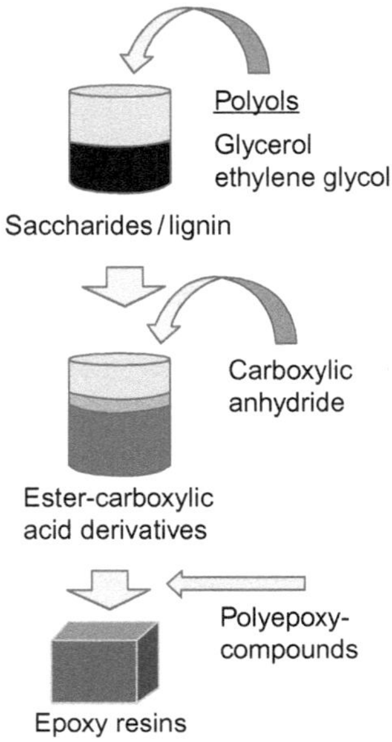

Fig. 1 Preparation procedure of epoxy resins.
From S. Hirose, T. Hatakeyama, H. Hatakeyama, Novel epoxy resins derived from biomass components, Procedia Chem. 4 (2012) 26–33.

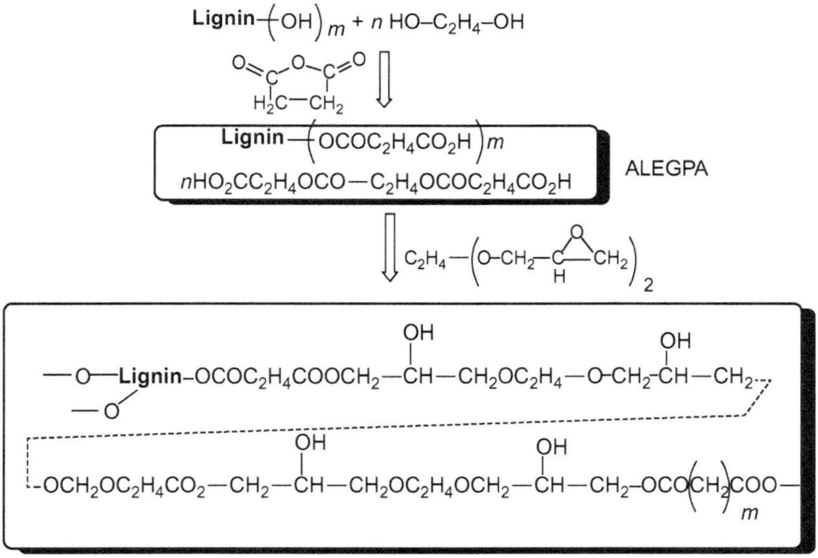

Fig. 2 Reaction scheme for epoxy resins derived from lignin.
From S. Hirose, T. Hatakeyama, H. Hatakeyama, Novel epoxy resins derived from biomass components, Procedia Chem. 4 (2012) 26–33.

9.3 Six-carbon compounds

Hydrolysis of cellulose fraction to lignocellulosic biomass to produce glucose allows access to six carbon sugar (C6) platform chemicals. Chemical processing is another way to convert glucose into useful chemical building blocks. Hydrolysis of hemicellulose results in mixed six- and five-carbon platforms. These carbon platforms (C5 and C6) can undergo selective dehydration, hydrogenation, and oxidation reactions to give useful products such as sorbitol, furfural, LA, and 2,5-furan dicarboxylic acid (FDCA). Fermentation and chemical transformation of these streams can provide a range of useful molecules as well [15].

9.3.1 *Sorbitol*

Sorbitol is produced by the hydrogenation of glucose. Derivatives of sorbitol include isosorbide, propylene glycol, ethylene glycol, glycerol, lactic acid, anhydrosugars, and branched polysaccharides. The derivatives and their uses are described in Fig. 3 [15,16].

Due to lack in the development of process conditions and dehydration of catalysts that can provide high yield pose significant challenge in producing isosorbide from sorbitol. Although, yields have been reported around 76%, it would be desirable to

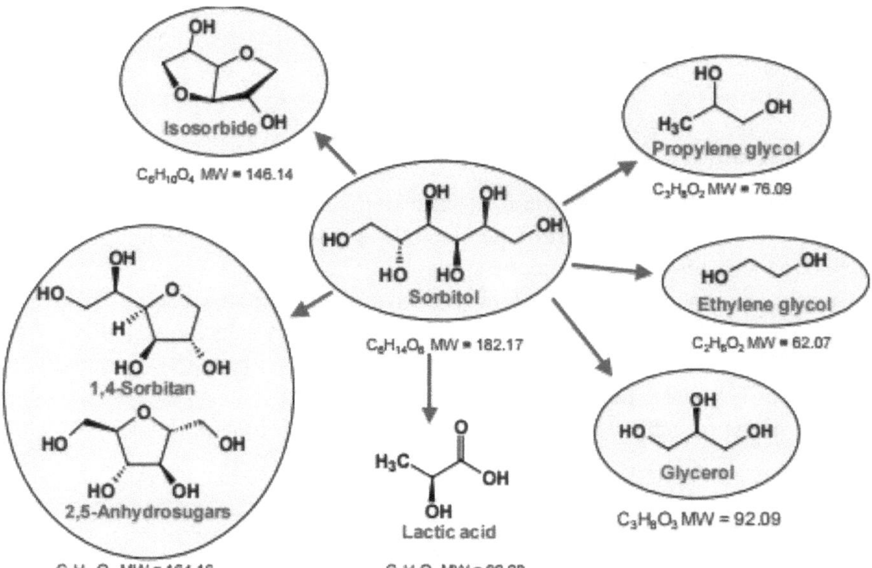

Fig. 3 Derivatives of sorbitol and their uses.
From A. Pandey, T. Bhaskar, M. Stocker, R. Sukumaran, Advances in thermochemical conversion of biomass—introduction, Recent Advances in Thermochemical Conversion of Biomass, Elsevier, 2015, p. 22; http://www.nrel.gov/docs/fy04osti/35523.pdf.

have increased yields to 90%. Higher yield not only help in reducing the production cost but also reduces the recovery and purification costs. Copolymerization with other glycols in the unsaturated polyester resin market would be a major opportunity [16].

9.3.2 2,5-Furan dicarboxylic acid

FDCA is a member of the furan family and is formed by an oxidative dehydration of glucose. The production process uses either oxygen or electrochemistry. The conversion can also be carried out by oxidation of 5-hydroxymethylfurfural, which is an intermediate in the conversion of six-carbon sugars into LA.

There are numerous routes to its production, including dehydration of hexose derivatives, oxidation of 2,5-disubstituted furans, and catalytic conversion of furan derivatives as shown in Fig. 4 [1].

New kind of polyesters can be produced from FDCA derivatives. When these polyesters are combined with FDCA, it leads to completely new biomass-derived products. For example, new nylons are produced when FDCA reacts with diamines, or when FDCA is converted to 2,5-bis(aminomethyl)-tetrahydrofuran [17]. Some of the possible derivatives of FDCA are (Fig. 5) [16]:

1. Succinic acid
2. 2,5-furan dicarbaldehyde
3. 2,5-dihydroxymethylfuran
4. 2,5-dihydroxymethyltetrahydrofuran
5. 2,5-bis(aminomethyl)tetrahydrofuran

9.3.3 LA derivatives

LA is formed when C6 sugar obtained from starch or lignocellulosic is treated with acid. Similarly, sugar derived from hemicelluloses can be converted to LA by adding a reduction step after acid treatment.

Current conversion yields of LA to its derivatives are approximately 70%. Yields can be increased by developing new catalysts as well as by selective dehydration process. As an example, it is important to analyze use of oxidants during the selective oxidation of LA that produces succinic and acrylic acid.

Some of the possible derivatives of LA are (Fig. 6) [16] as follows:

1. Levulinate esters
2. 1,4-Pentanediol
3. Angelilactones
4. G-Valerolactone
5. 2-Methyl-THF
6. Acrylic acid
7. d-Aminolevulinate
8. Diphenolic acid
9. b-Acetylacrylic acid

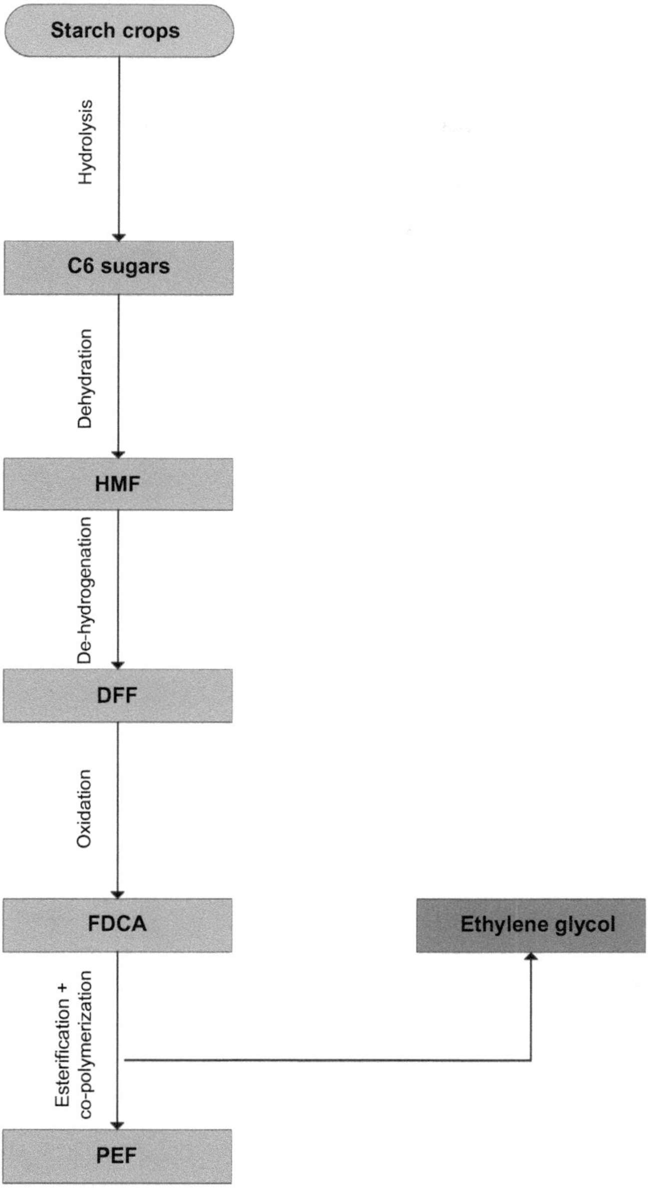

Fig. 4 Derivatives of FDCA produced by catalytic conversion of furan derivatives. From From the Sugar Platform to biofuels and biochemicals, European Commission Directorate-General Energy, No. ENER/C2/423-2012/S12.673791, April 2015.

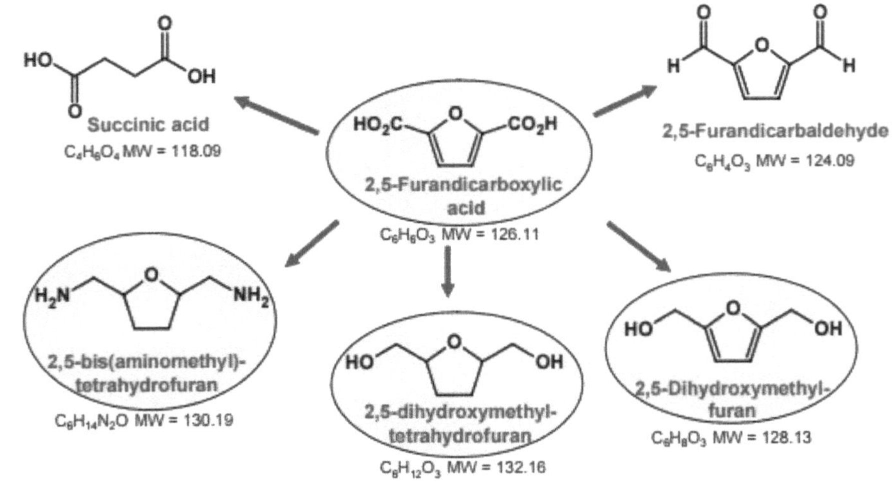

Fig. 5 Derivatives of 2,5-furan dicarboxylic acid (FDCA).
From http://www.nrel.gov/docs/fy04osti/35523.pdf.

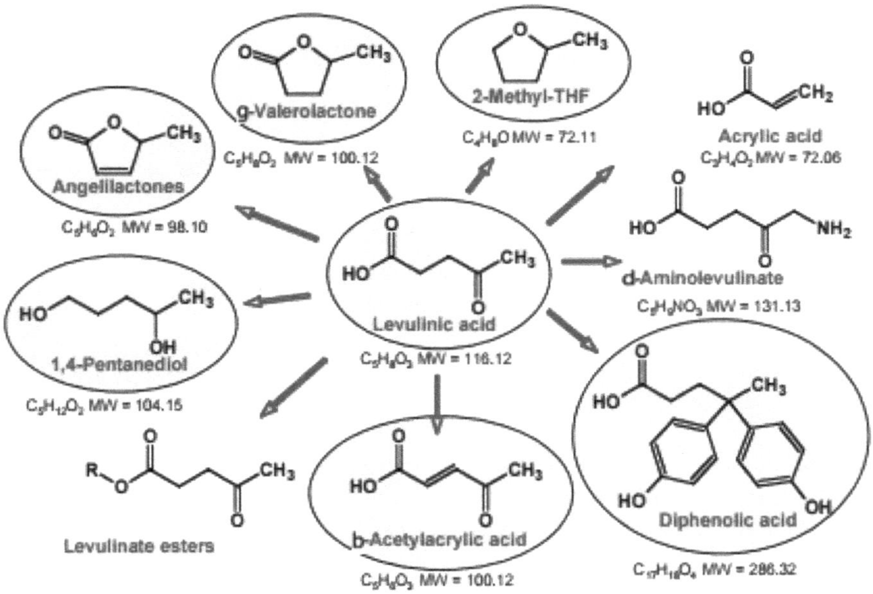

Fig. 6 Derivatives of levulinic acid.
From http://www.nrel.gov/docs/fy04osti/35523.pdf.

9.4 3-HPA derivatives

3-HPA is a C3 building block producing specific derivatives when it undergoes hydrogenation, reduction, esterification, or dehydration reactions. 3-HPA is converted to 1,3-propanediol (1,3-PDO) by undergoing hydrogenation/reduction reactions. Development of new catalyst systems capable of performing direct reduction of carboxylic acid groups to alcohol is important when converting 3-HPA to 1,3-PDO.

Similarly, development of new acid catalyst system with high selectivity is required for the dehydration of 3-HPA to the family of acrylates including acrylic acid and acrylamide. During this dehydration reaction, there is a potential for polymerization and must be prevented [16].

Some of the derivatives of 3-HPA are shown in Fig. 7 [16].

9.5 Glucaric acid derivatives

Glucaric acid is a member of a much larger family of materials known as oxidized sugars and can be produced from glucose by oxidation with nitric acid. Converting cheap glucose into glucaric acid arises from two features:

- Glucaric acid can serve as a starting point in the production of wide range of products with applicability in high volume markets.

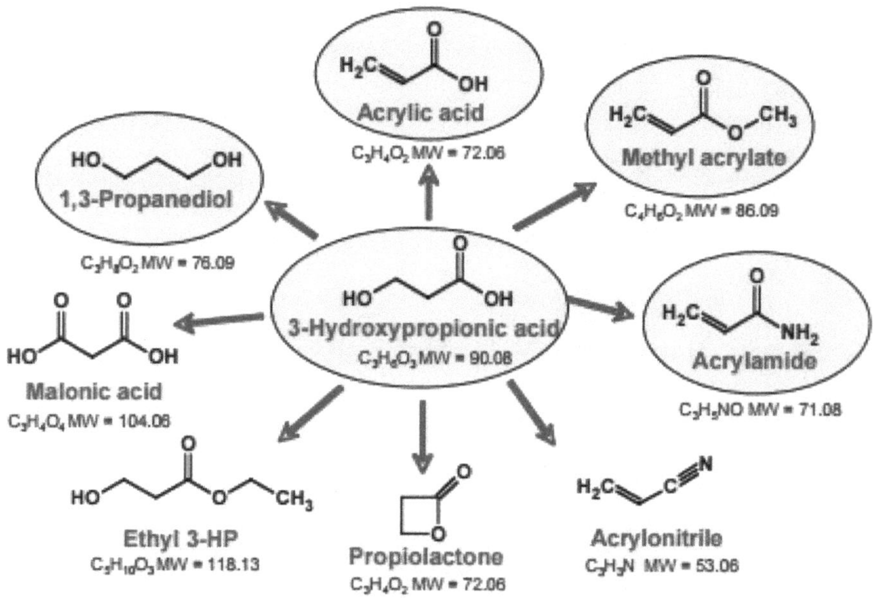

Fig. 7 Derivatives of 3-Hydroxypropionic acid (3-HPA). From http://www.nrel.gov/docs/fy04osti/35523.pdf.

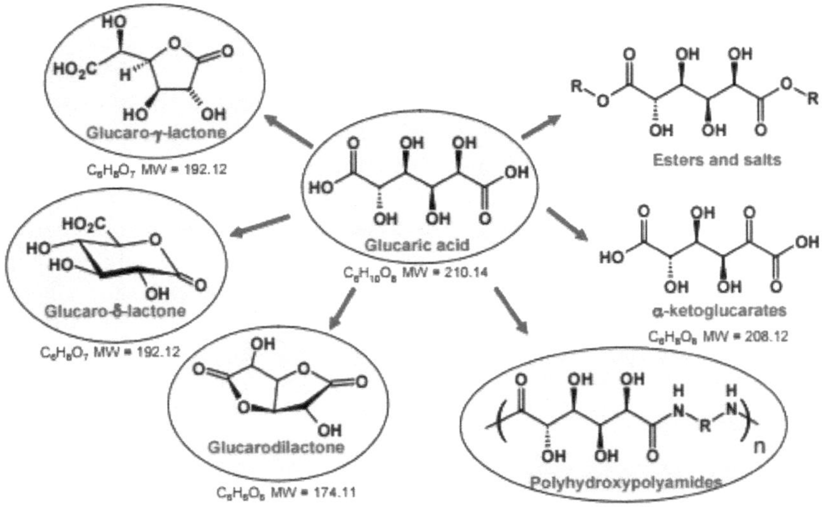

Fig. 8 Derivatives of 3-Hydroxypropionic acid (3-HPA).
From http://www.nrel.gov/docs/fy04osti/35523.pdf.

- Development of efficient processes for production of glucaric acid will also be applicable to efficient oxidation of other inexpensive sugars studied in this evaluation, such as xylose or arabinose.

Some of the derivatives of glucaric acid (Fig. 8) are [16] as follows:

1. Glucaro-ϒ-lactone
2. Esters and salts
3. α-ketoglucarates
4. Polyhydroxypolyamides
5. Glucarodilactone
6. Glucaro-δ-lactone

9.6 Derivatives from microbial polymers

There are two main transformations that are required to convert biomass into hydrocarbon fuels using microbes:

(1) Deoxygenation
(2) Carbon chain extension

Since most of the biomass is made from hexoses and pentoses available in cellulose and hemicellulose, they contain six or less carbon structures.

There are three different routes which can be used to decrease oxygen content of the biomass intermediates:

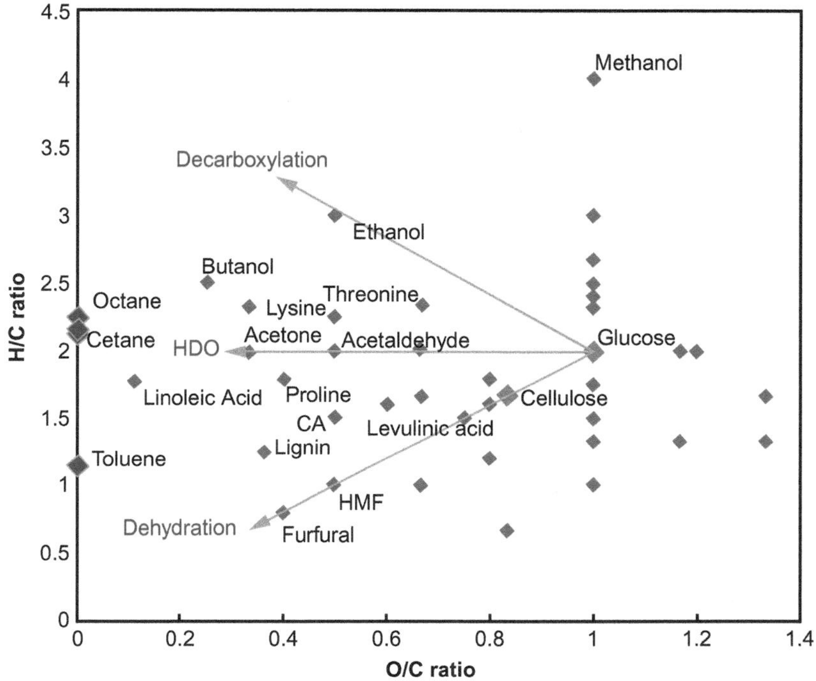

Fig. 9 van Krevelen diagram of glucose and cellulose, bioproducts, and hydrocarbon fuel components.
From H. Pilath, A. Mittal, L. Moens, T.B. Vinzant, W. Wang, D.K. Johnson, A route from biomass to hydrocarbons via depolymerization and decarboxylation of microbially produced polyhydroxybutyrate, Direct Microbial Conversion of Biomass to Advanced Fuels, Elsevier Publishers, 2015, pp. 383–394.

(1) Dehydration
(2) Hydrodeoxygenation
(3) Decarboxylation

Fig. 9 shows van Krevelen diagram of glucose and cellulose, bioproducts, and hydrocarbon fuel components. The arrows indicate the direction in which the ratios would change for the three deoxygenation reactions [18].

Decarboxylation route is typically pathway for converting glucose into polyhydroxybutyrate (PHB) into propene and then into hydrocarbon as shown in Fig. 10 [18].

The overall reaction pathway for converting glucose to hydrocarbons via PHB and CA is shown in Fig. 11 [18].

Pathway to convert polyhydroxyalkanoates (PHA) to alkenes involves several microorganisms. There are more than 140 different kinds of Eubacteria with alkyl chains ranging from 3 to 16 carbons long. These microorganisms synthesize PHAs and deposit these polyesters as cytoplasmic granules [18].

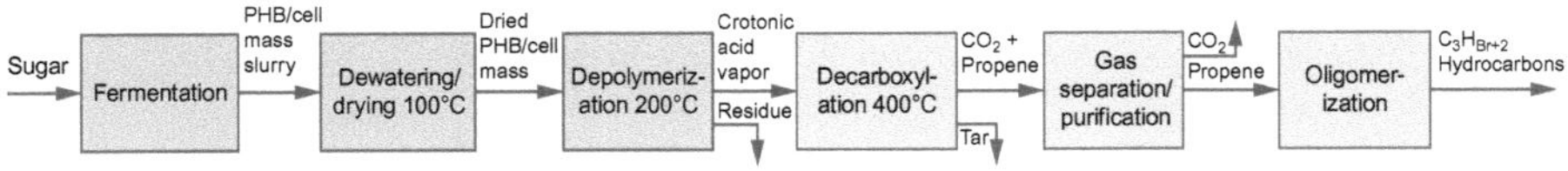

Fig. 10 Proposed process flow diagram for converting sugars into hydrocarbons via PHB. From H. Pilath, A. Mittal, L. Moens, T.B. Vinzant, W. Wang, D.K. Johnson, A route from biomass to hydrocarbons via depolymerization and decarboxylation of microbially produced polyhydroxybutyrate, Direct Microbial Conversion of Biomass to Advanced Fuels, Elsevier Publishers, 2015, pp. 383–394.

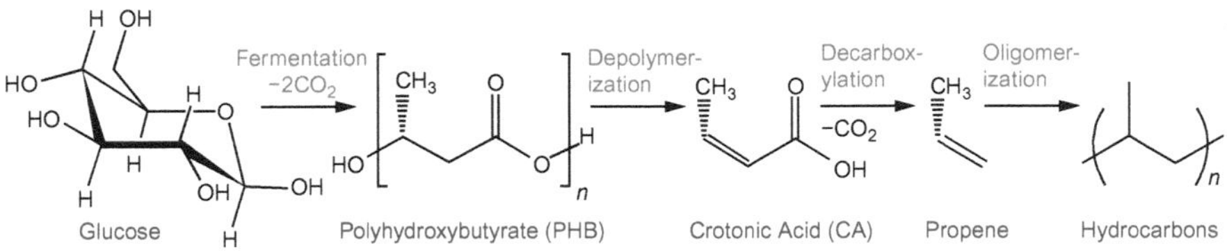

Fig. 11 Proposed route for transforming sugars into PHB and then into a hydrocarbon fuel component.
From H. Pilath, A. Mittal, L. Moens, T.B. Vinzant, W. Wang, D.K. Johnson, A route from biomass to hydrocarbons via depolymerization and decarboxylation of microbially produced polyhydroxybutyrate, Direct Microbial Conversion of Biomass to Advanced Fuels, Elsevier Publishers, 2015, pp. 383–394.

9.7 Xylan derivatives

Xylans are the most common hemicelluloses and are composed of a backbone of β-(1 → 4)-linked anhydroxylose units (AXU) [19–22]. Xylans are not only available in wood but also in plants such as grasses, cereals, and herbs [23]. Derivatives of xylan recently gained importance as new biopolymeric-based materials and functional polymers [24–27].

Fig. 12 shows reaction path for the preparation of xylan derivatives via etherification in aqueous system or esterification in DMF/LiCl DMSO [28]. By exposing to the respective moieties, anionic and cationic xylan derivatives can be produced. Anionic xylan derivatives based on GX obtained from birch can be produced by the introduction of carboxylmethyl (CM) moieties. Similarly, cationic xylan derivatives from birch wood GX are produced by the functionalization with hydroxypropyltrimethylammonium (HPMA) groups [28].

In order to access the functional groups for the reagents, it is important to activate xylan. This activation can be carried out heterogeneously or starting from the dissolved polymer. The heterogeneous activation starts with a stirred suspension of the xylan in the slurry medium, which is usually an alcohol or 1,2-dimethoxyethane (DME). Although, the addition of aqueous sodium hydroxide leads to the activation of the xylan, it creates an aggregated system. In order to prevent aggregation, xylan should be dissolved in aqueous sodium hydroxide and precipitated by addition of 2-propanol to yield a uniform suspension and the activated xylan [28].

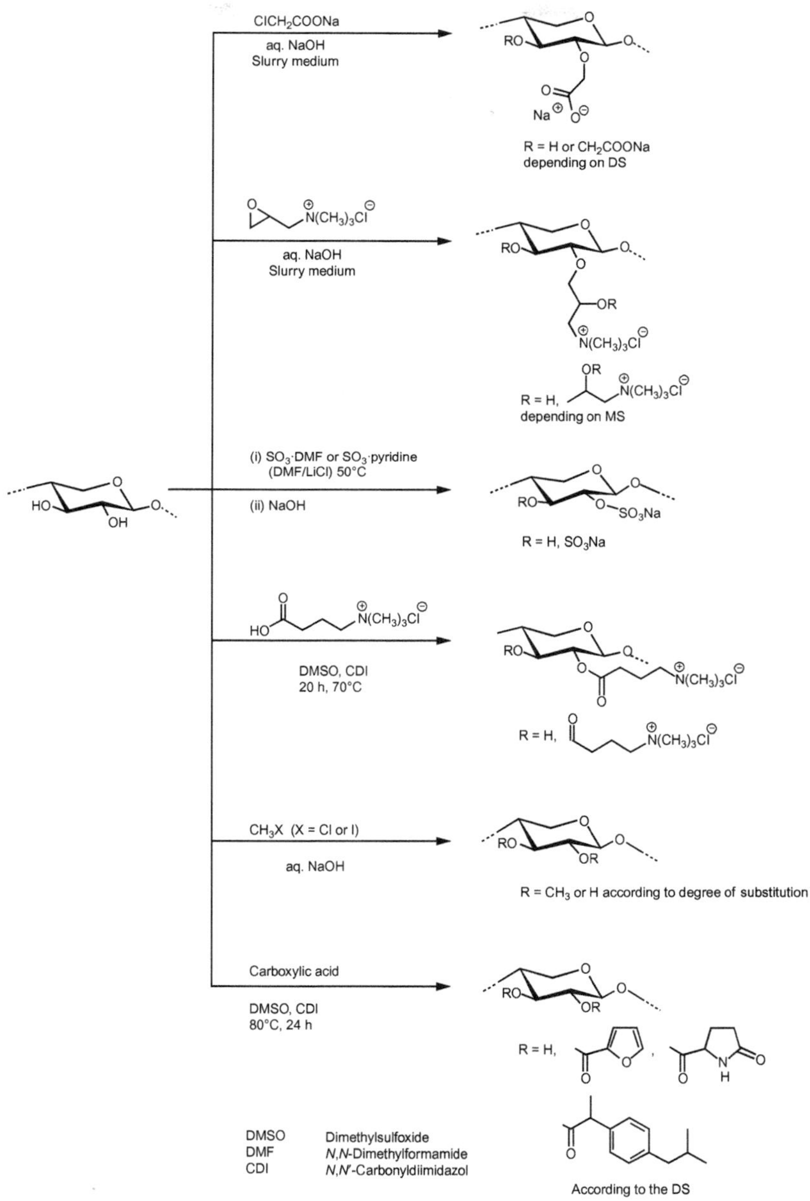

Fig. 12 Reaction path for the preparation of xylan derivatives via etherification in aqueous systems or esterification in DMF/LiCl DMSO.
From K. P-Welcke, K. Schwikal, S. Daus, T. Heinze, Xylan derivatives and their application potential—mini-review of own results, Carbohydr. Polym. 100 (2014) 80–88.

References

[1] From the Sugar Platform to biofuels and biochemicals, European Commission Directorate-General Energy, No. ENER/C2/423-2012/S12.673791, April 2015.

[2] J.-P. Pascault, R.J.J. Williams, General concepts about epoxy polymers, in: J.-P. Pascault, R.J.J. Williams (Eds.), Epoxy Polymers, Wiley-VCH, Weinheim, 2010, pp. 1–12.

[3] D. Feldman, D. Banu, Kinetic data on the curing of an epoxy polymer in the presence of lignin, J. Polym. Sci. A Polym. Chem. 26 (1988) 973–983.

[4] K. Hofmann, W. Glasser, Engineering plastics from lignin, 23. Network formation of lignin-based epoxy resins, Macromol. Chem. 195 (1994) 65–80.

[5] T. Malutan, R. Nicu, V.I. Popa, Lignin modification by epoxidation, Bioresources 3 (2008) 1371–1376.

[6] S. Boufi, M.N. Belgacem, A. Gandini, 2-Furyloxiranes III. Chain extension with different polyols, Polym. J. 29 (1997) 479–486.

[7] S. Hirose, M. Kobayashi, H. Kimura, H. Hatakeyama, Synthesis of lignin-based polyester-epoxy resins, in: J.F. Kennedy, G.O. Phillips, P.A. Williams, H. Hatakeyama (Eds.), Recent Advances in Environmentally Compatible Polymers, Woodhead, Cambridge, 2001, pp. 73–78.

[8] S. Hirose, T. Hatakeyama, H. Hatakeyama, Synthesis and thermal properties of epoxy resins from ester-carboxylic acid derivative of alcoholysis lignin, Macromol. Symp. 197 (2003) 157–169.

[9] S. Hirose, T. Hatakeyama, H. Hatakeyama, Thermal properties of epoxy resins from ester-carboxylic acid derivatives of mono-and disaccharides, Netsu Sokutei 30 (2003) 154–160.

[10] S. Hirose, T. Hatakeyama, H. Hatakeyama, Curing and glass transition of epoxy resins from ester-carboxylic acid derivatives of mono- and disaccharides, and alcoholysis Lignin, Macromol. Symp. 224 (2005) 343–353.

[11] S. Hirose, T. Hatakeyama, H. Hatakeyama, Glass transition and thermal decomposition of epoxy resins from the carboxylic acid system consisting of ester-carboxylic acid derivatives of alcoholysis lignin and ethylene glycol with various dicarboxylic acids, Thermochim. Acta 431 (2005) 76–80.

[12] T.N.M.T. Ismail, H.A. Hassan, S. Hirose, Y. Taguchi, T. Hatakeyama, H. Hatakeyama, Synthesis and thermal properties of ester-type crosslinked epoxy resins derived from lignosulfonate and glycerol, Polym. Int. 59 (2010) 181–186.

[13] S. Hirose, New epoxy resins from oil palm components, J. Oil Palm Res. 23 (2011) 1110–1114.

[14] S. Hirose, T. Hatakeyama, H. Hatakeyama, Novel epoxy resins derived from biomass components, Procedia Chem. 4 (2012) 26–33.

[15] A. Pandey, T. Bhaskar, M. Stocker, R. Sukumaran, Advances in thermochemical conversion of biomass—introduction, Recent Advances in Thermochemical Conversion of Biomass, Elsevier, Waltham, MA, 2015, 22.

[16] http://www.nrel.gov/docs/fy04osti/35523.pdf.

[17] D. Sengupta, R.W. Pike, Chemicals from biomass: integrating bioprocesses into chemical production complexes for sustainable development, in: Chemicals from Biomass, CRC Press, Boca Raton, FL, 2013, pp. 102–103 (Chapter 3).

[18] H. Pilath, A. Mittal, L. Moens, T.B. Vinzant, W. Wang, D.K. Johnson, A route from biomass to hydrocarbons via depolymerization and decarboxylation of microbially produced polyhydroxybutyrate, in: Direct Microbial Conversion of Biomass to Advanced Fuels, Elsevier Publishers, Amsterdam, Netherlands, 2015, pp. 383–394.

[19] A. Cunha, A. Gandini, Turning polysaccharides into hydrophobic materials: a critical review. Part 2. hemicelluloses, chitin/chitosan, starch, pectin and alginates, Cellulose 17 (2010) 1045–1065.
[20] A. Ebringerová, Structural diversity and application potential of hemicelluloses, Macromol. Symp. 232 (2005) 1–12.
[21] A. Ebringerová, T. Heinze, Xylan and xylan derivatives—biopolymers with valuable properties. 1. Naturally occurring xylans: structures, isolation, procedure and properties, Macromol. Rapid Commun. 21 (2000) 542–556.
[22] I. Spiridon, V.I. Popa, Hemicelluloses: major sources, properties and applications, in: M. N. Belgacem, A. Gandini (Eds.), Monomers, Polymers and Composites From Renewable Resources, Elsevier, Amsterdam, 2008, pp. 289–304.
[23] A.M. Stephen, G.O. Aspinall (Ed.), The Polysaccharides, Academic, San Diego, CA, 1983, pp. 98–193.
[24] T. Heinze, A. Koschella, A. Ebringerova, Chemical functionalization of xylan: a short review, in: P. Gatenholm, M. Tenkanen (Eds.), Hemicelluloses: Science and Technology, ACS Symposium Series, Vol. 864, 2004, pp. 312–325.
[25] M.S. Lindblad, A.-C. Albertsson, Chemical modification of hemi-celluloses and gums, in: S. Dumitriu (Ed.), Polysaccharides: Structural Diversity and Functional Versatility, second ed., Marcel Dekker, Boca Raton, FL, 2004, pp. 491–508.
[26] A.B. Samuelsen, I. Lund, J.M. Djahromi, B.S. Paulsen, J.K. Wold, S.H. Knutsen, Structural features and anti-complementary activity of some heteroxylan polysaccharide fractions from the seeds of Plantago major L, Carbohydr. Polym. 38 (1999) 133–143.
[27] M. Bukovsky, A. Kardosová, H. Koscova, D. Kostálová, Immunomodu- lating activity of 4-O-methyl-d-glucurono-d-xylan from *Rudbeckia fulgida* var. *sullivantii*, Biologia 53 (1998) 771–775.
[28] K. P-Welcke, K. Schwikal, S. Daus, T. Heinze, Xylan derivatives and their application potential—mini-review of own results, Carbohydr. Polym. 100 (2014) 80–88.

Processing of biomass-derived polymers

10

10.1 Introduction

The conversion of biomass to polymers has considerable economic and environmental value. Range of advanced methodologies to produce polymers has been investigated. This allowed redesigning of new value chains starting from the desired product properties and resulting in a great potential for sustainable production processes and products. Lignin, hemicellulose, and cellulose are the three most important plant-derived raw materials since their waste streams can be utilized. Depending on the type, range of processing technologies such as fractionation, gasification, and fermentation can be implemented as standalone or as a combination to convert monomers into polymers.

In recent times, the concept of biorefinery has simplified bioconversion of biomass. A concept of biorefinery came into existence in which a biomass is treated and converted into its end products. This approach supports the development of technology that can minimize or eliminate waste resulting from biomass processing, and thus converting each waste stream into a by-product stream. Fig. 1 shows schematics of a biorefinery [1].

A biorefinery approach involves multistep processes [1]:

1. Step 1: involves treating the precursor-containing biomass to make it more amenable for further processing. This step is conventionally referred as pretreatment.
2. Step 2: Following pretreatment, the biomass components are subject to a combination of biological and/or chemical treatments.
3. Step 3: The outputs from this pretreatment step (specialty chemicals or reducing sugars) could be further converted to chemical building blocks for further processing uses.

10.2 Production using heterogeneous and homogeneous catalysis

10.2.1 *Lignocellulosic biomass*

Lignocellulosic biomass is the most abundant and biorenewable biomass on earth and has a critical importance. It has been projected as an abundant carbon-neutral renewable source, which can decrease CO_2 emissions and atmospheric pollution. Thus it is a promising alternative to limit crude oil, which can be utilized to produce biofuels, biomolecules, and biomaterials. Furthermore, the major component of lignocellulosic biomass, cellulose, is considered as the strongest potential candidate for the substitution of petroleum-based polymers owing to its ecofriendly properties like renewability, biocompatibility, and biodegradability [2].

Technology and Applications of Polymers Derived from Biomass. https://doi.org/10.1016/B978-0-323-51115-5.00010-4

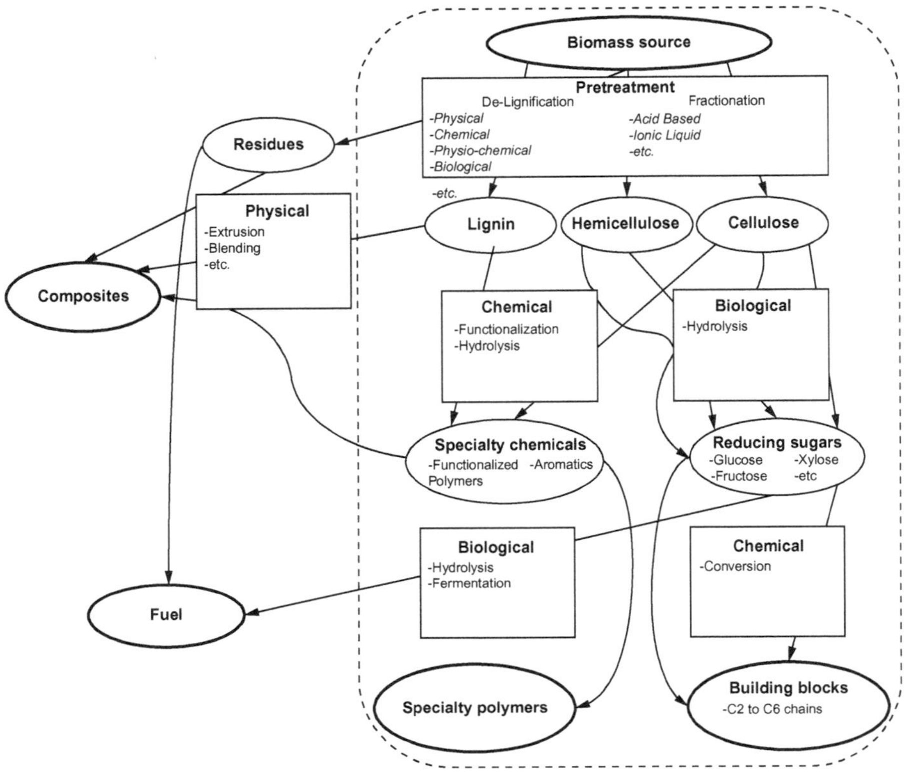

Fig. 1 Schematics of a biorefinery.
From M. Fitzpatrick, P. Champagne, M.F. Cunningham, R.A. Whitney, A biorefinery processing perspective: treatment of lignocellulosic materials for the production of value-added products, Bioresour. Technol. 101 (2010) 8915–8922.

Lignocellulosic biomass is mainly composed of three polymers as shown in Fig. 2:

1. Cellulose
2. Hemicellulose
3. Lignin together

Depending on the type of lignocellulosic biomass, these polymers are organized in complex nonuniform three-dimensional structures to different degrees and varying relative composition [2].

Cellulose: It is the major component of lignocellulosic biomass. Disaccharide cellobiose is the repeating unit of the cellulose chain and its structure consists of extensive intramolecular and intermolecular hydrogen bonding networks that binds the glucose units.

Hemicellulose: It is the second most abundant polymer. The heteropolymers of hemicellulose are composed of different C5 and C6 monosaccharide units such as pentoses, hexoses, and acetylated sugars. Hemicelluloses are imbedded in the plant cell

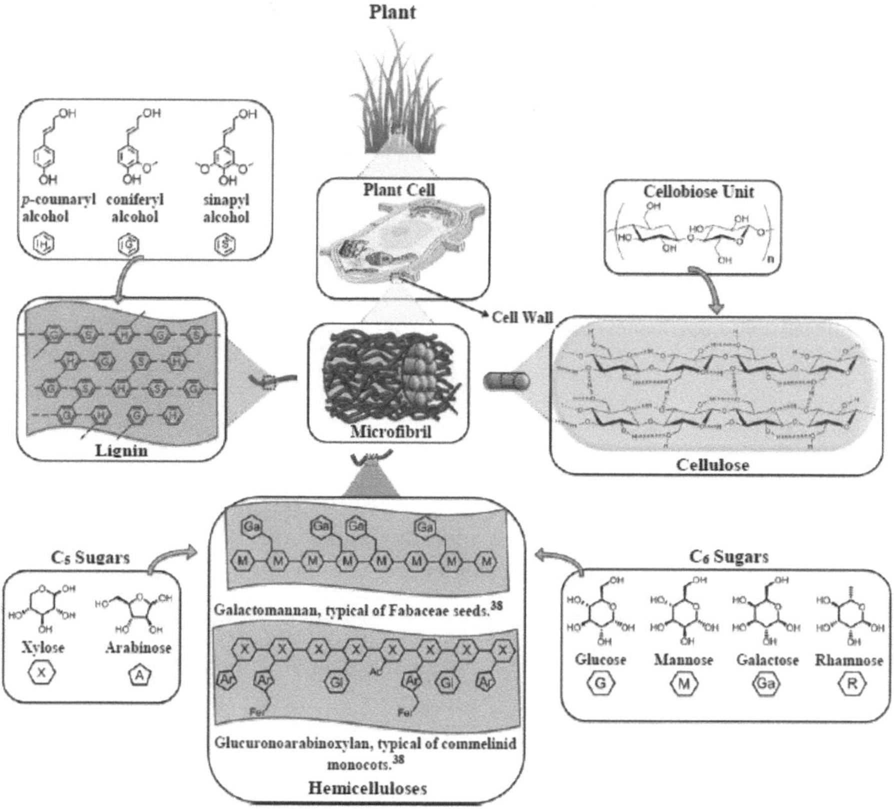

Fig. 2 The main components and structure of lignocellulose.
From F.H. Isikgor, C.R. Becer, Lignocellulosic biomass: a sustainable platform for production of bio-based chemicals and polymers.

walls to form a complex network of bonds that provide structural strength by linking cellulose fibers into microfibrils and cross-linking with lignin [2].

Lignin: They are three-dimensional polymer of phenylpropanoid units. It functions as the cellular glue which provides compressive strength to the plant tissue and the individual fibers, stiffness to the cell wall, and resistance against insects and pathogens.

Lignocellulosic biomass consists of 35%–50% cellulose, 20%–35% hemicellulose, and 10%–25% lignin. Proteins, oils, and ash make up the remaining fraction [2].

The purpose of lignocellulosic feedstock biorefinery is to refine lignocellulosic biomass into intermediate outputs to be processed into a range of products and bio-energy. Lignocellulosic processing consists of following steps [1,3]:

1. Pretreatment of biomass source
2. Fractionation
3. Classical chemical methods

4. Fermentation
5. Ionic liquid-phase reaction
6. Enzymatic transformation

10.2.1.1 Pretreatment of biomass source

During the pretreatment step, native lignocellulosic biomass is converted into cellulose form where hydrolysis is effectively achieved. Pretreatment of lignocellulosic biomass yields materials that can be characterized by cellulose substrate redistribution either as an increase in the accessible cellulose surface area and/or the solubilization of lignin [1].

Pretreatment approaches can be classified into four categories:

1. Physical
2. Chemical
3. Physiochemical
4. Biological

Physical pretreatment techniques include mechanical comminution and ultrasound. This step is of typically lower performance and higher cost.

Chemical pretreatment consists of a high degree of selectivity for the biomass component as they degrade but also involve relatively harsh reaction conditions. These conditions may not be ideal in a biorefinery setup due to possible effects on downstream biological processing. Some of the chemical pretreatment processes include:

- Ozonolysis
- Dilute and concentrated acid
- Alkaline and oxidative H_2O_2 delignification
- Organosolv

Physiochemical pretreatment is a combination of physical and chemical steps. In this stage, milder chemical conditions are used but under more extreme operational conditions. This adds to the cost of implementing these techniques in a biorefinery scheme.

Some of the different physiochemical pretreatment techniques are as follows:

- Liquid hot water such as hydrothermolysis, uncatalyzed solvolysis aquasolv
- Steam explosion such as autohydrolysis with and without chemical addition
- Ammonia fiber explosion (AFEX)
- Carbon dioxide explosion

The last step is the biological pretreatment route which offers significant advantages such as low chemical and energy use, mild operational conditions, and ease of integration into a consolidating bioprocessing setup [1].

10.2.1.2 Fractionation

Lignocellulose fractionation

In this stage, the pretreated biomass feedstock is separated into its core components (cellulose, hemicelluloses, and lignin) such that each individual component may be processed or functionalized. Implementation of cost-effective methods to release soluble sugars from lignocellulosic biomass at an industrial scale is a significant challenge [1].

Acid-based fractionation

It has been shown that cellulose solvents such as concentrated phosphoric acid can completely dissolve cellulose fibers and disrupt the hydrogen bonds that hold together crystalline cellulose, which in turn increase the accessibility of cellulose to cellulase enzymes. Cellulose solvent-based lignocellulosic fractionation (CSLF) technique uses a cellulose solvent, volatile organic solvent, and water to separate lignocellulose at relatively low temperatures and pressures [1].

The process fractionated lignocellulosic biomass components are based on the difference in solubility of cellulose; hemicellulose; and lignin in the cellulose solvent, organic solvent, and water, respectively.

The goal of CSLF technique is the:

- Decrystallization of cellulose fibers
- Partial removal of lignin and hemicellulose from the cellulose fraction
- Relatively modest reaction conditions

Ionic liquid-based fractionation

Ionic liquids (ILs) are organic salts that exist as liquids at low temperatures, often well below 100°C. They have tuneable physiochemical properties, very low vapor pressures, good thermal stability, and unlimited combination of anions and cations that can be used to synthesize ILs.

Recent research has shown that both cellulose and lignin can be dissolved in a variety of ILs, and, perhaps more importantly, easily regenerated from these solutions. The regenerated cellulose microstructure can range from amorphous to crystalline based on the regeneration process [1].

This regenerated cellulose has the same degree of polymerization and polydispersity as the original cellulose. The regeneration of the solute is achieved by precipitation in the presence of an antisolvent (water or methanol) due to preferential solute displacement. This antisolvent can then be stripped from the IL and recovered for reuse.

Regenerated cellulose exhibited a significant improvement in hydrolysis rate and glucose formation over untreated cellulose. The improvement in hydrolysis rates is attributed to the slight decrease in the degree of polymerization of both cellulose and hemicellulose, the decrease in the crystallinity of the cellulose, and an increase in accessibility due to lignin separation [1].

10.2.1.3 Classical chemical methodology

Classical chemical methodology is a chemical approach to build range of block polymers from the biomass sources that will play a significant role in the development of future biorefineries. These methods to derive polymers from biomass involve the conversion of biological fatty acids to polymer building blocks.

Cargill, in conjunction with the Kansas Polymer Research Centre, has chemically converted triglyceride carbon-carbon double bonds to alcohol and methoxy groups, leading to a bio-derived polyol (BiOH).

This methodology can be applied to derive other industrial chemicals:

- Levulinic acid is industrially prepared from biotechnology-based routes, using wood processing and agricultural wastes via conversion of hexose sugars in acid
- Lignin can be converted to xylene, benzene, toluene, or other aromatic compounds

Classical chemical methodology includes:

1. Catalysis
2. Condensation polymerization

Catalysis

Catalysts are used to transform biomass into range of high-value chemicals. There are processes that involve use of catalyst. One such process involves the use of Fischer-Tropsch chemistry on pyrolyzed biomass. For example, catalysts, including Fischer-Tropsch synthesis, have been used in the production of biofuels from palm. Additionally, platinum catalyzed aqueous-phase reforming of glycerol has demonstrated high yields of hydrogen fuel, with low CO content, for use in fuel cells.

Biocatalysis is also widely used in the production of high-value products such as fine chemicals and pharmaceutical products. In this process, the biocatalysts undergo selectively catalyzed reactions depending on the desired products. They also help reduce energy consumption and waste generation while synthesizing products that are not possible via chemical reactions as well as development of specialty chemicals [1].

Condensation polymerization

Condensation polymerization is used to polymerize bio-derived monomers using immobilized enzyme catalysts. Polycondensation of sorbitol or glycerol with diacids is performed using commercially available lipase catalyst. This reaction can be either chemically or biologically derived. By using this approach it has been demonstrated to reduce reaction temperatures and energy consumption, relative to traditional chemical polymerization processing, while also controlling branching during polymerization [1].

10.2.1.4 Fermentation

Fermentation is a widely used technique to produce industrial chemicals. An increasing push to integrate into biorefinery to move this technology forward is becoming a reality. One example is the production of succinic acid through fermentation. Salts of succinic acid can be derived through the fermentation of glucose, which fixes carbon dioxide. Since most microorganisms used in industrial fermentation are not tolerant of acidic conditions, the process is neutralized. These salts then undergo more conventional chemical processing such as separation and recovery where they are separated from the microorganisms and then dissolved in an acid solution to form succinic acid [1].

Other example includes production of 1,3-propanediol through the fermentation of glycerol. Corporation such as Genencor is developing low-cost route of producing

1,3-propanediol by modifying natural routes. 1,3-propanediol is a key building block in production of polypropylene terephthalate [1].

Another example is the production of polymerized esters. They are typically used in adhesives and coatings. Lactic acid, which can be chemically converted to several important chemicals including methyl lactate, lactide, and polylactic acid, can be produced through fermentation. The hydroxyl and carboxyl groups present in lactic acid make it amenable for conversion into a wide range of value-added products [1].

10.2.1.5 Ionic liquid-phase reaction

Ionic liquid-phase reaction in biomass processing has the potential for functional additives to be added directly in solution via dissolution or dispersion before or after cellulose dissolution. For this reason, incorporating this technology into biorefinery will help in reducing the number of processing steps and the corresponding cost and power requirements. The ionic liquid-phase has also been used to stabilize catalysts during reaction. Additionally, polymerization reactions involving cellulose and ILs have been reported in recent literature. The IL has served both as a means to prepare a cellulose-derived initiator for atom-transfer radical polymerization, as a medium for cellulose polymerization reactions, and as a polymerizable composite with cellulose in radical polymerization [1].

10.2.1.6 Direct biological conversion

Extraction

Extraction is a useful tool to synthesize value-added products within biological species, both plant matter and microorganisms. There has been significant development in efficient extraction technologies which needs to be implemented into biorefineries early on to allow for extraction before further processing.

For example, commodity chemicals have been directly extracted from biomass via conventional chemical extraction techniques. One of the most notable chemicals derived in this manner is ferulic acid, extracted in high percentage yields from corn fiber. Ferulic acid is a chemical feedstock used in the production of fine chemicals such as vanillin and guaiacol.

DuPont has been able to extract a monomer from tulips, α-methylenebutyrolactone or tulipalin A, which polymerizes in a manner similar to methyl methacrylate and also has favorable durability and refractive index [1].

Enzymatic transformation

Direct biological conversion is a significant factor producing polymers completely within microbial cells. Family of polyhydroxyalkanoates (PHAs) is a very good example. This technology is incorporated into a biorefinery as a separate processing stream. Carbon-limited feeding strategy in some organisms can be utilized to produce PHA without limiting essential growth nutrients.

There are other areas which have seen increase in the development of enzymatic transformation activities by:

- Genetically modifying plants for direct PHA production.
- Advances in genome sequencing provide the foundation for optimizing PHA synthesis and for developing the ability to produce PHA polymers with a range of properties.

Recent studies have reported that PHAs can be synthesized using a forestry-based biorefinery approach with lignocellulosic process streams, including hemicelluloses hydrolysates, levulinic acid derived from cellulose, and tall oil fatty acids from kraft pulping, being employed as the carbon sources for the bacteria *Burkholderia cepacia*.

As an alternative, use of microbial consortia to produce PHAs from different carbon sources has been researched. For example, industrial waste from methanol-enriched pulp and paper mill foul condensate, fermented municipal primary solids, and biodiesel have been shown to yield PHAs when passed through batch bioreactors containing a mixed microbial consortium from municipal activated sludge [1].

10.2.2 C1–C6 molecules

Fig. 3 gives an overview of the various platform molecules, classified by their number of carbon atoms. It is based on a list produced by Department of Energy in 2004 and its update in 2010 and on some new trends thereafter [4].

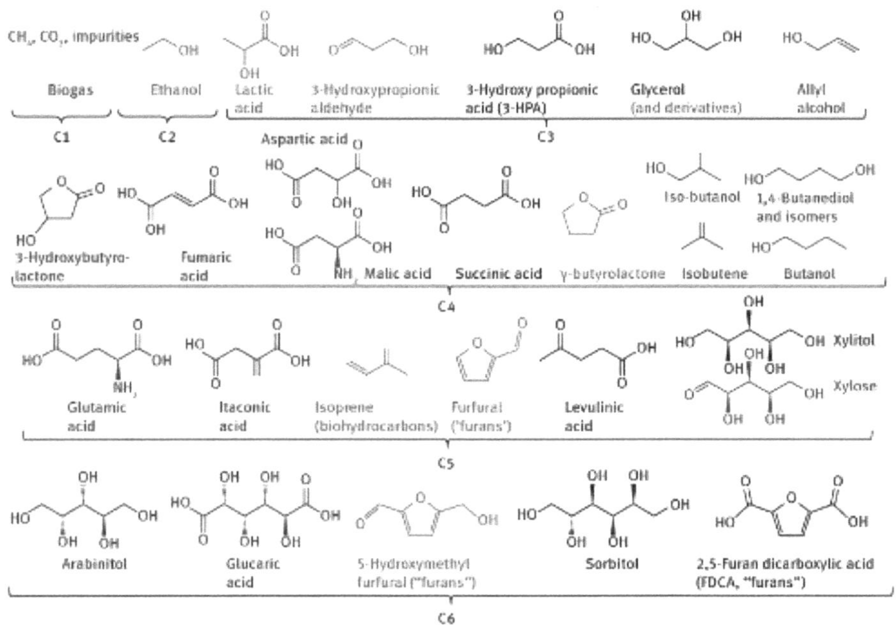

Fig. 3 Platform molecules.
From M. Aresta, A. Dibenedetto, F. Dumeignil, Biomass derived molecules conversion to chemicals, Biorefineries: An Introduction, Walter de Gruyter GmbH, p. 146 (Chapter 6).

10.2.2.1 Biogas

Biogas is a mixture of gases produced from biomass in an anaerobic process. It is composed of methane (50%–80%), carbon dioxide (15%–40%), nitrogen (4%), and hydrogen sulfide (1%) with traces of hydrogen, oxygen, and carbon monoxide.

Anaerobic digestion consists of four steps [5]:

1. Hydrolysis: This is the an anaerobic digestion phase where complex organic molecules are broken down into simpler organic molecules.
2. Acidogenesis: This is where simple organic molecules are converted into fatty acids.
3. Acetogenesis: This is where fatty acids are converted into acetate.
4. Methanogenesis: This is where molecules that have been converted into a suitable food source for methanogenic microorganisms that are converted into methane by those organisms.

A typical biogas plant is shown in Fig. 4 [6]. It consists of five individual units:

1. Liquid manure storage and feeding system for solid biomass
2. Disinfection unit if necessary
3. Pit storage digesters with stirrers and foil covering
4. Storage tank or lagoon
5. Motor or cogeneration unit

Liquid manure storage and feeding system

At this step, following are the list of activities:

- Slurry and small amount of solid biomass are stored temporarily in the liquid manure storage and if necessary they can be diluted or mixed.
- Codigestion plants feeding liquid material and a high amount of solid biomass need special feeding systems depending on the solid materials.
- The liquid material is pumped from the liquid store into the digester. Recommended that the feeding takes place continuously.

Disinfection unit

At this step, animal by-products or food wastes that are used gets disinfected. The recommended time is for 1 h at 70°C before or after the digestion process.

Pit storage digesters with stirrers and foil covering

- The heated digester is the main item of the plant. The digester must be water and gas proof and opaque. Special mixers be installed that will ensure a well-mixed and homogeneous material in the digester so that bacteria and substrate are in close contact to get a high gas yield. An over and under-pressure protection system and foil covering should be integrated in the digester.
- The storage digester should be covered with a foil to utilize further produced biogas.

Storage tank or lagoon

- When the substrate is digested, it is pumped to the storage tank and can be perfectly used as a fertilizer for farmland.

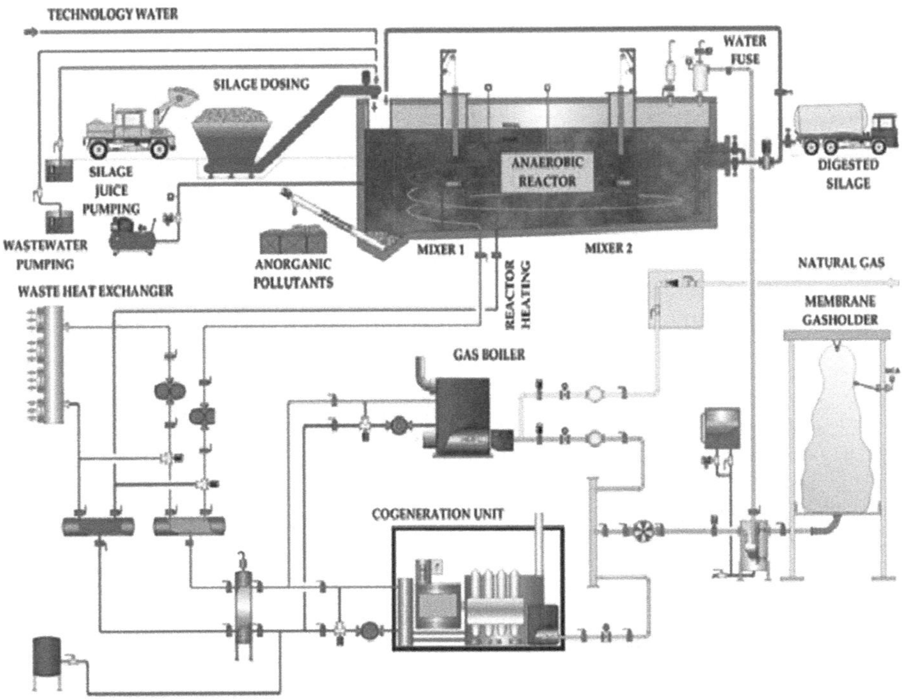

Fig. 4 A typical biogas production plant.
From https://www.intechopen.com/books/advances-in-silage-production-and-utilization/maize-silage-as-substrate-for-biogas-production.

- The produced biogas is cooled and cleaned from biogas ingredients such as hydrogen sulfide. Cooling and cleaning is highly recommended because hydrogen sulfide when interacts with water can produce highly corrosive sulfuric acid.

Motor or cogeneration unit

The last unit is the motor or cogeneration unit. Here, the gas is finally used to produce electricity and heat constantly.

There are two important process parameters that indicate technical and biological stability of the biogas plant [7]:

- Hydraulic retention time (HRT)
- Volume load or organic loading rate

1. Hydraulic retention time

   HRT is the average time the added material remains in the fermenter before being pumped in the storage tank. It is calculated by the utilizable volume of the digester and the amount of organic material loaded daily. The recommended retention time is typically 35–80 days, depending on the substrate and plant components.

2. Volume load or organic loading rate

   Volume load also known as the organic loading rate is weight of the organic dry solids loaded per digester volume and unit of time. The organic loading rate is important for the plant components especially, mixer and for the bacteria. Organic loading rate greater than 4.0 kg DS/m^3 day could damage mixers and pumps. The bacteria could also be stressed by too much feeding. Consequently, no biogas production will take place and the digestion process stops completely.

10.2.2.2 Syngas

Syngas is a gas that contains hydrogen (H_2) and carbon monoxide (CO) as main combustible components and some trace amounts of carbon dioxide (CO_2), water (H_2O), and nitrogen (N_2). Fig. 5 [8] shows a typical biomass gasification plant.

The conversion of biomass into a combustible gas is a two-stage process:

1. In the first stage, pyrolysis takes place at temperatures below 600°C. This is when volatile components contained within the biomass such as organic compounds, hydrogen, carbon monoxide, tars, and water vapor are released. Pyrolysis also leaves a solid residue called char.
2. In the second stage, the char generated during the first stage is reacted with steam or burnt in a restricted quantity of air or oxygen to produce further combustible gas.

Mixture of carbon monoxide, carbon dioxide, methane, hydrogen, and various hydrocarbons are produced by the gasification process. These can be used directly in gas turbines, and boilers, or used as precursors for synthesizing a wide range of other chemicals. These gases can also be further treated to result in high purity gases that can be burned directly for cooking or heat supply, or used in secondary conversion devices, such as internal combustion engines or gas turbines, for producing electricity or shaft power [8].

10.2.2.3 Ethanol

Ethanol is the most important C2 platform molecule produced from biomass. Ethanol is significantly used as biofuel; either directly blended with gasoline or after conversion of ethyl-tert-butyl ether (ETBE) using iso-butene as a reactant. Ethanol is also used as a solvent, a disinfectant, and as a platform molecule to synthesize chemicals.

Ethanol is produced by a fermentation process using mono-, di-, and polysaccharides as the feedstock. Monosaccharides such as glyceraldehyde, xylose, ribose, glucose, and fructose consist of single sugars bound together. One molecule of disaccharide results from a chemical reaction in which a new bond is formed between two monosaccharides after removal of one molecule. Polysaccharides are chains of bonded monosaccharides molecules, which result from dehydration synthesis. Polysaccharides composed of similar subunits known as monomers. Starch and cellulose are polymers composed of monomers of glucose. Polysaccharides must be decomposed in disaccharides and/or monosaccharides through hydrolysis before fermentation to ethanol [9].

Fig. 6 shows a basic flowchart of biomass-to-ethanol production process. The biomass is pretreated to avoid early fermentation and prevent bacterial contamination

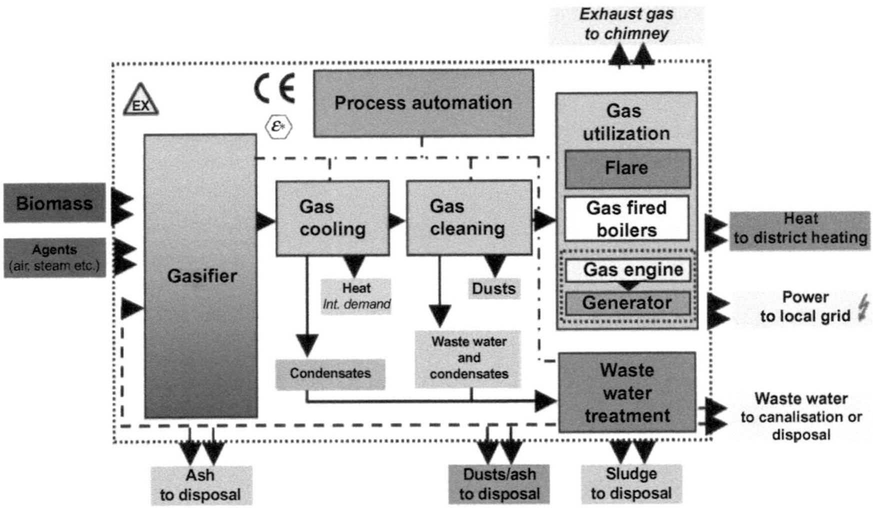

Fig. 5 A typical biomass gasification plant for processing syngas.
From S. Zafar, Gasification of municipal wastes. http:/www.bioenergyconsult.com/tag/syngas/.

during transportation and storage. During this step, simple sugars are made available in proportions depending on the biomass and the pretreatment process. A large portion of fibers may remain for saccharification through hydrolysis reactions to produce simple sugars, which are then fermented [9].

Batch fermentation involves adding hydrolysate, yeasts, nutrients, and other ingredients at the beginning of the step. Batch process involves adding one or more inputs as fermentation progresses. In a continuous process, ingredients are constantly added and the products are removed from the fermentation vessels [10]. The fermentation reactions occur at 25–30°C and last 6–72 h depending on the composition of the hydrolysate, the type, the density, and the activity of the yeasts.

Fermentation of sucrose is performed using commercial yeast such as *Saccharomyces cerevisiae*. Chemical reaction is composed of enzymatic hydrolysis of sucrose followed by fermentation of simple sugars. First, invertase catalyzes the hydrolysis of sucrose to convert it into glucose and fructose.

$$\underset{\text{(Sucrose)}}{C_{12}H_{22}O_{11}} + H_2O \xrightarrow{\text{Invertase}} \underset{\text{(Glucose)}}{C_6H_{12}O_6} + \underset{\text{(Fructose)}}{C_6H_{12}O_6}$$

Then, another enzyme, zymase, also present in the yeast, converts the glucose and the fructose into ethanol and CO_2.

$$\underset{\text{(Fructose)}}{C_6H_{12}O_6} \xrightarrow{\text{Zymase}} \underset{\text{(Ethanol)}}{2CH_3CH_2OH} + 2CO_2$$

The remaining flow from the distillation column, known as vinasse or stillage, can be valorized to produce coproducts, which may include process steam and electricity, products for feeding animals, and other valuable by-products.

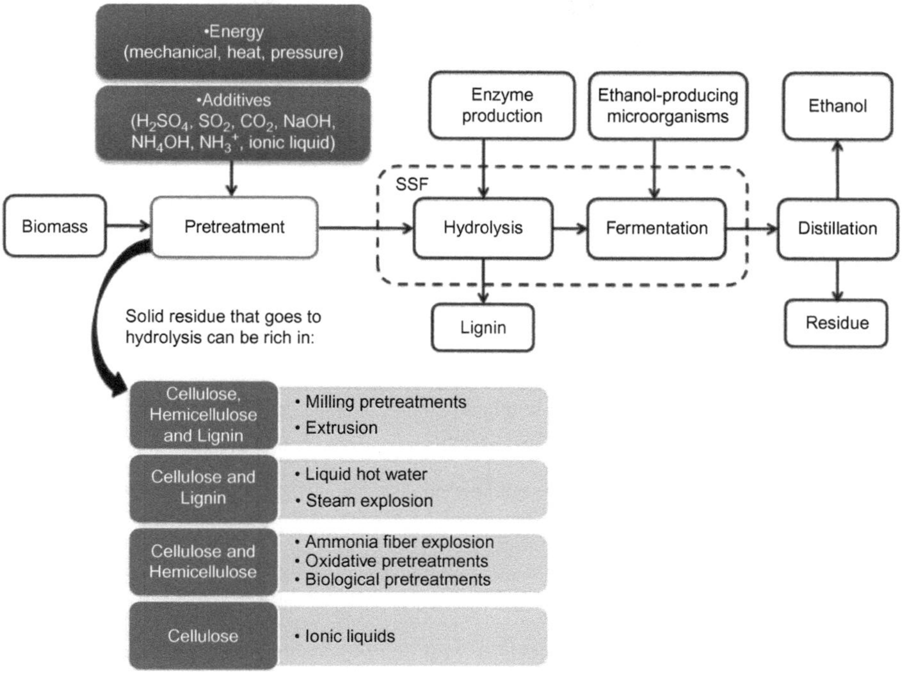

Fig. 6 Flowchart of biomass-to-ethanol production process.
From https://www.intechopen.com/books/sustainable-degradation-of-lignocellulosic-biomass-techniques-applications-and-commercialization/sugarcane-and-woody-biomass-pretreatments-for-ethanol-production.

10.2.3 Oleaginous biomass

10.2.3.1 Lipids

High viscosity of oils causes operational problems in the diesel engine when it is used as fuel. They are therefore converted to biodiesel, bringing their kinematic viscosity closer to fossil-based fuel [11]. Biodiesel is made up of monoalkyl esters of long chain fatty acids derived from triglycerides feedstocks such as vegetable oils, animal fats, or waste oils [11].

Transesterification is the reaction where triglycerides are transformed into biodiesel [12,13]. During this transesterification reaction, triacylglycerol reacts with an alcohol in the presence of a catalyst to form alkyl esters of the fatty acids. 50% to 200% access alcohol is required to obtain high yields of alkyl esters (~99.7%) [14]. The typical alcohol used in transesterification process is methanol. When methanol is used in the reaction, the derived biodiesel product is composed of fatty acid methyl esters [11].

A schematic of biodiesel production from low free fatty acids (FFA) containing feedstock via base catalyst transesterification is shown in Fig. 7. Oil, catalyst, and

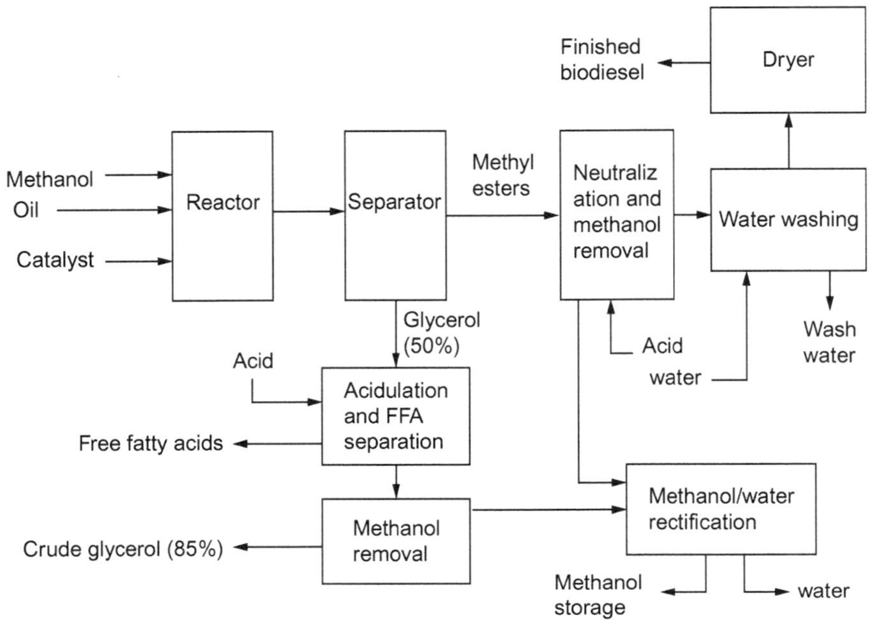

Fig. 7 A schematic of biodiesel production from low FFA containing feedstock via base catalyst transesterification.
From E. Sari, Green Diesel Production via Catalytic Hydrogenation/Decarboxylation of Triglycerides and Fatty Acids of Vegetable Oil and Brown Grease (Doctoral dissertation), Wayne State University, 2013.

alcohol are mixed in a reactor for 1 h at 60°C. After the reaction is completed, glycerol is separated from the fatty acid methyl esters. Fatty acid methyl esters proceed in a neutralizer with acid to deactivate any residual catalyst and then methanol is removed. Any remaining catalyst, soap, salts, methanol, or free glycerol is removed from fatty acid methyl ester during water washing step. At the end, biodiesel product is obtained after the drying process [11,15].

Quality of the biodiesel depends on:

1. Quality of the feedstock
2. Materials used in the process
3. Transesterification process

Efficiency of transesterification process depends on moisture content in alcohol and feedstock. It is recommended that the moisture content should be less than 0.5% of FFA. Moisture in alcohol can lead to a reaction with alkyl esters, triglycerides, diglycerides, and monoglycerides further forming FFA. These fatty acids would then need to be pretreated before transesterification process. Otherwise, during transesterification, they can form soap and causes separation problems in the biodiesel production process [11,16,17].

10.3 Production of second-generation bioethanol

10.3.1 Fermentation

There are several technologies available for the conversion of lignocellulosics to fuel ethanol. Fermentation can follow the hydrolysis or can be simultaneous to it. The main hydrolysis fermentation processes are as follows:

1. Separate hydrolysis and fermentation (SHF)
2. Simultaneous saccharification and fermentation (SSF)
3. Simultaneous saccharification and cofermentation (SSCF)
4. Consolidated bioprocessing (CBP)

10.3.1.1 Separate hydrolysis and fermentation

In the SHF process, pretreated lignocelluloses are hydrolyzed to monosaccharides and subsequently formed to ethanol in separate reactors shown in Fig. 8 [18].

This mode of operation makes it possible to perform hydrolysis and fermentation at their own optimum conditions. The optimum temperature for cellulase is usually between 45°C and 50°C, whereas that of fermenting microorganisms is between 30°C and 37°C [18].

There are certain disadvantages with this process:

1. Inhibition to cellulase activity by the released sugars, mainly cellobiose; β-glucosidase is, on the other hand, inhibited by released glucose.

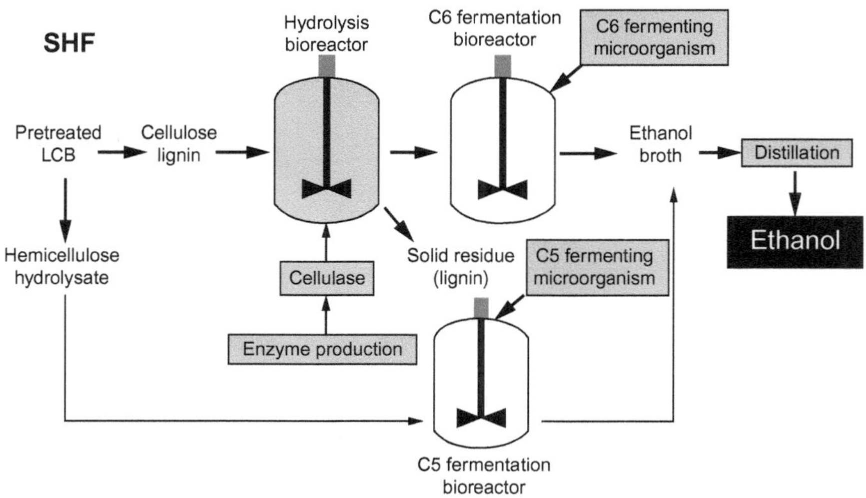

Fig. 8 Schematics for bioethanol production from lignocellulosic biomass using separate hydrolysis and fermentation (SHF).
From M. Aresta, A. Dibenedetto, F. Dumeignil, Process development and metabolic engineering for bioethanol production from lignocellulosic biomass, Biorefinery: From Biomass to Chemicals and Fuels, De Gruyter, 2012, pp. 207–214.

2. Prone to microbial contaminations during the hydrolysis process if cellulase preparations are not sterile.

10.3.1.2 Simultaneous saccharification and fermentation

In a SSF process, the glucose is consumed immediately by the fermenting microorganism present in the culture. Fig. 9 shows the schematics for bioethanol production from lignocellulosic biomass using SSF [18].

There are several advantages of SSF compared to SHF:

- SSF avoids end product inhibition since the inhibition effects of cellobiose and glucose to the enzymes are minimized by keeping a low concentration of these sugars in the media. Consequently, results in higher ethanol yields compared to SHF
- SSF requires lower amounts of enzyme and shows a lower risk of contamination due to the presence of alcohol

There are several issues with SSF scheme:

- There is difference between the optimum temperatures of the hydrolyzing enzymes and fermenting microorganisms
- Impossible to achieve high substrate loads because of difficulties with mechanical mixing and insufficient mass transfer

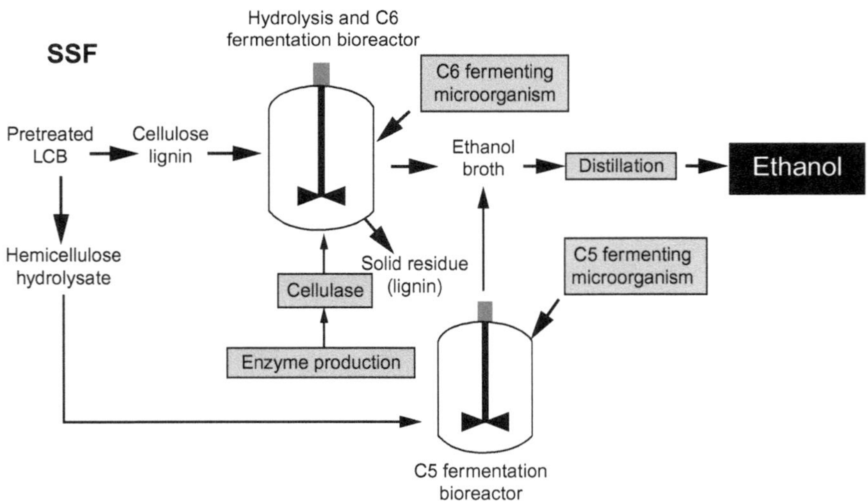

Fig. 9 Schematic for bioethanol production from lignocellulosic biomass using simultaneous saccharification and fermentation (SSF).
From M. Aresta, A. Dibenedetto, F. Dumeignil, Process development and metabolic engineering for bioethanol production from lignocellulosic biomass, Biorefinery: From Biomass to Chemicals and Fuels, De Gruyter, 2012, pp. 207–214.

10.3.1.3 Simultaneous saccharification and cofermentation

SSCF is a variation of SSF in which the fermentation of both five- and six-carbon sugars to ethanol is carried out in the same bioreactor as shown in Fig. 10 [18].

The hemicellulose hydrolyzed during pretreatment and the solid cellulose is not separated after pretreatment. This allows the hemicellulose sugars to convert into ethanol together with the SSF of the cellulose. *S. cerevisiae* recombinant strains capable of fermenting both hexose and pentose sugars are used in this process. Also, the recombinant bacterium *Zymomonas mobilis* and the natural pentose fermenting yeast *Pichia stipites* are used in SSCF. However, yields obtained are lower and a thorough detoxification of the hydrolyzate is necessary [18].

10.3.1.4 Consolidated bioprocessing

CBP is a process in which cellulase production, substrate hydrolysis, and pentose and hexose fermentations are accomplished in a single step by cellulolytic microorganisms shown in Fig. 11 [18].

Some of the CBP are [19] as follows:

- Lower production costs, lower capital investment, and higher conversion efficiency
- Higher hydrolysis rates

10.3.1.5 Very-high-gravity fermentation

Ethanol made by fermentation is the most commonly used fuel in transportation. However, a relative high production cost limits its extensive applications [20]. Features such as low energy and water consumption, and high annual ethanol productivity, make very-high-gravity (VHG) fermentation a promising technology for ethanol industry [21,22].

The concept of high gravity (HG) ethanol fermentation was proposed in the 1980s and successfully applied thereafter, allowing drastic increase of final ethanol concentration

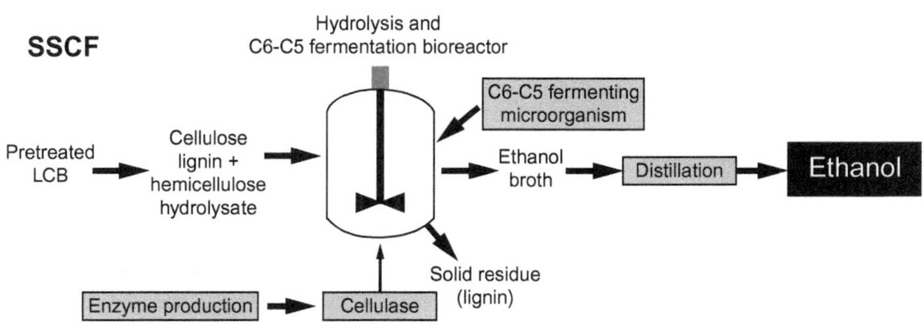

Fig. 10 Schematic for bioethanol production from lignocellulosic biomass using simultaneous saccharification and cofermentation (SSCF).
From M. Aresta, A. Dibenedetto, F. Dumeignil, Process development and metabolic engineering for bioethanol production from lignocellulosic biomass, Biorefinery: From Biomass to Chemicals and Fuels, De Gruyter, 2012, pp. 207–214.

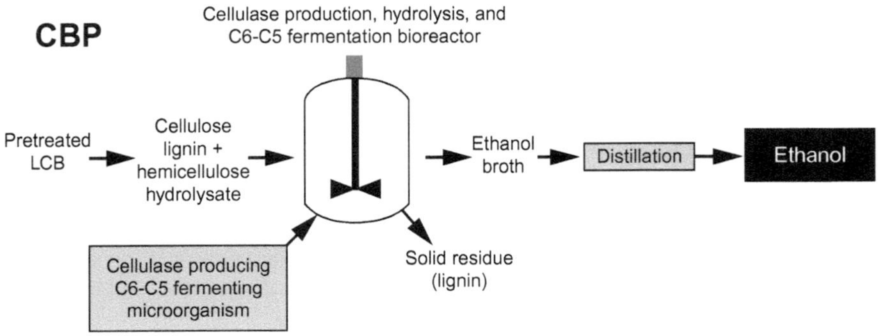

Fig. 11 Schematics for bioethanol production from lignocellulosic biomass using consolidated bioprocessing (CBP).
From M. Aresta, A. Dibenedetto, F. Dumeignil, Process development and metabolic engineering for bioethanol production from lignocellulosic biomass, Biorefinery: From Biomass to Chemicals and Fuels, De Gruyter, 2012, pp. 207–214.

from the previous level of 78% (v/v) [20]. Research has revealed that many strains of *S. cerevisiae* can tolerate far higher concentrations of ethanol much better than previously believed, without any conditioning or genetic modifications [23,24].

Development of high-performance processes that uses starchy materials consists of the fermentation of HG mashes. During HG fermentation, the solid concentration in the medium exceeds 200 g/L, which implies a high substrate load and, consequently, high ethanol concentrations at the end of fermentation [25,26].

Fig. 12 shows yeast cells subjected to high osmolarity stress at initial stages of VHG fermentation [27]. The yeast cells are subjected to high osmolarity stress at

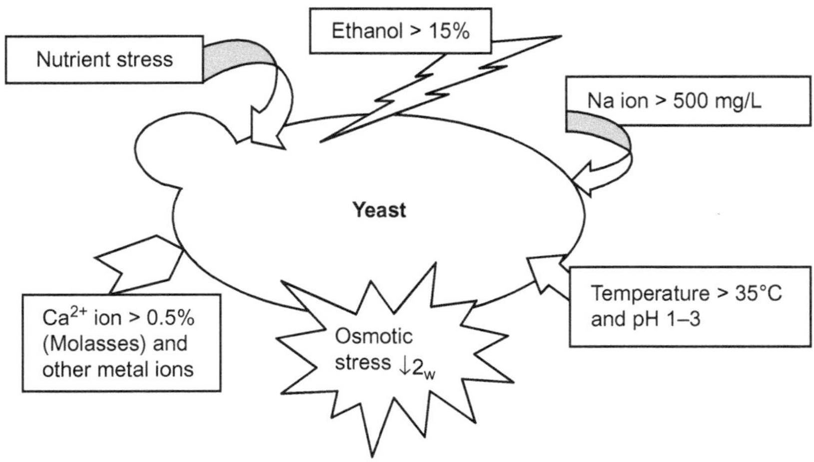

Fig. 12 Stresses experienced by yeast under VHG fermentation conditions.
From P. Puligundla, D. Smogrovicova, V.S.R. Obulam, S. Ko, Very high gravity (VHG) ethanolic brewing and fermentation a research update, J. Ind. Biotechnol. 38 (2011) 1133–1144.

initial stages of fermentation when the sugar level of the medium increases above their normal tolerance limits (>30%, w/v). Therefore sluggish fermentations are often observed, and ultimately the efficiency of ethanol production drops [28]. High osmotic pressure, low water activity (aw), and toxic effects of higher ethanol levels are the key factors [29,30] along with other stress factors, responsible for inhibition of yeast growth and decreased fermentation performance under high substrate concentrations [27].

References

[1] M. Fitzpatrick, P. Champagne, M.F. Cunningham, R.A. Whitney, A biorefinery processing perspective: treatment of lignocellulosic materials for the production of value-added products, Bioresour. Technol. 101 (2010) 8915–8922.
[2] F.H. Isikgor, C.R. Becer, Lignocellulosic biomass: a sustainable platform for production of bio-based chemicals and polymers, Polym. Chem. 6 (2015) 4497–4559.
[3] E.D. Jong, G. Jungmeler, Biorefinery concepts in comparison to petrochemical refineries, in: Industrial Biorefineries and White Biotechnology, Elsevier, Waltham, MA, USA, 2015, p. 13.
[4] M. Aresta, A. Dibenedetto, F. Dumeignil, Biomass derived molecules conversion to chemicals, Biorefineries: An Introduction, Walter de Gruyter GmbH, 2015, p. 146 (Chapter 6).
[5] Processing biomass into biogas. http://articles.extension.org/pages/30313/processing-biomass-into-biogas&gt.
[6] Johnson, Biogas—never ending source of fuel, electricity, and fertilizer, 2003. http://www.pluginindia.com/blogs/biogas-never-ending-source-of-fuel-electricity-fertilizer.
[7] The process and techniques of anaerobic digestion, Sustainable Energy Authority of Ireland. http://www.seai.ie/Renewables/Bioenergy/Bioenergy_Technologies/Anaerobic_Digestion/The_Process_and_Techniques_of_Anaerobic_Digestion/&gt.
[8] S. Zafar, Gasification of municipal wastes, 2015. http://www.bioenergyconsult.com/tag/syngas/.
[9] E. Gnansounou, A. Dauriat, Ethanol fuel from biomass: a review, J. Sci. Ind. Res. 64 (2005) 809–821.
[10] C.E. Wyman, Ethanol fuel, in: C.J. Cleveland (Ed.), Encyclopedia of Energy, Elsevier, Amsterdam, 2004, pp. 541–555.
[11] G. Knothe, J.V. Gerpen, J. Krahl, The Biodiesel Handbook, AOCS Press, Urbana, 2005.
[12] B. Freedman, M.O. Bagby, Heat of combustion of fatty esters and triglycerides, J. Am. Oil Chem. Soc. 66 (11) (1989) 1601.
[13] A. Demirbas, Biofuel sources, biofuel policy, biofuel economy and global biofuel projections, Energy Convers. Manag. 49 (2008) 2106–2116.
[14] Basic research needs: catalysis for energy, U.S. Department of Energy, 2007.
[15] E. Sari, Green Diesel Production Via Catalytic Hydrogenation/Decarboxylation of Triglycerides and Fatty Acids of Vegetable Oil and Brown Grease, (Doctoral dissertation)Wayne State University, 2013.
[16] B. Smith, H.C. Greenwell, A. Whiting, Catalytic upgrading of tri-glycerides and fatty acids to transport biofuels, Energy Environ. Sci. 2 (3) (2009) 262.
[17] B. Freedman, E.H. Pryde, T.L. Mounts, Variables affecting the yields of fatty esters from transesterified vegetable oils, J. Am. Oil Chem. Soc. 61 (10) (1984) 1638.

[18] M. Aresta, A. Dibenedetto, F. Dumeignil, Process development and metabolic engineering for bioethanol production from lignocellulosic biomass, in: Biorefinery: From Biomass to Chemicals and Fuels, De Gruyter, Berlin, 2012, pp. 207–214.
[19] S.-T. Yang, H.E. Enshasy, N. Thongchul, Cellulases: characteristics, sources, production, and applications, in: Bioprocessing Technologies in Biorefinery for Sustainable Production of Fuels, Chemicals, and Polymers, Wiley, Hoboken, NJ, 2013.
[20] F.W. Bai, W.A. Anderson, M. M.-Young, Ethanol fermentation technologies from sugar and starch feedstocks, Biotechnol. Adv. 26 (2008) 89–105.
[21] W. Ingledew, Y.-H. Lin, Biofuels and bioenergy-ethanol from starch-based feedstocks, in: M. Moo-Young (Ed.), Comprehensive Biotechnology, Elsevier, 2011, , pp. 37–49.
[22] Y.-H. Lin, C.-G. Liu, Process design for a very high gravity ethanol formation, Energy Procedia 61 (2014) 2725–2728.
[23] G.P. Casey, W.M. Ingledew, Ethanol tolerance in yeasts, in: Critical Reviews in Biotechnology, vol. 13, CRC Press, Boca Raton, FL, 1986, p. 219.
[24] K.C. Thomas, W.M. Ingledew, Production of 21% (v/v) ethanol by fermentation of very high gravity (VHG) wheat mashes, J. Ind. Microbiol. 10 (1992) 61.
[25] C.A. Cardona, O.J. Sanchez, L.F. Gutierrez, Ethanolic fermentation technologies, in: Process Synthesis for Fuel Ethanol Production, CRC Press, Boca Raton, FL, 2010, p. 167.
[26] K.C. Thomas, S.H. Hynes, A.M. Jones, W.M. Ingledew, Production of fuel alcohol from wheat by VHG technology: effect of sugar concentration and fermentation temperature, Appl. Biochem. Biotechnol. 43 (1993) 211.
[27] P. Puligundla, D. Smogrovicova, V.S.R. Obulam, S. Ko, Very high gravity (VHG) ethanolic brewing and fermentation a research update, J. Ind. Biotechnol. 38 (2011) 1133–1144.
[28] C.J. Panchal, G.G. Stewart, The effect of osmotic pressure on the production and excretion of ethanol and glycerol by a brewing yeast strain, J. Inst. Brew. 86 (1980) 207–210.
[29] M. Jakobsen, J.U. Piper, Performance and osmotolerance of different strains of lager yeast in high gravity fermentation, MBAA Tech. Q. 26 (1989) 56.
[30] G.G. Stewart, T. Stewart, C.J. Panchal, I. Russell, Factors that influence the ethanol tolerance of brewer brewer that influence the ethanol tolerance of nations, MBAA Tech. Q. 25 (1988) 47.

Material and energy balance for conversion of biomass materials

11

11.1 Introduction

Material and energy balances are defined as the material or energy quantities passing through processing operations that obey the law of conservation of mass and energy, and are independent of the type of processes—batch, semicontinuous, and continuous.

These balances are very critical to any industrial operation as they directly impact processing as well as yield of the products. Material balances are performed during the preliminary phase of new process development, when the new plant is commissioned and finally, when the process is optimized and production is streamlined. In essence, material balances should be performed when there is any change in the process [1].

Similarly, energy balances are as important as material balances. Any changes to the process can be translated to increase in the energy cost. By performing energy balances, industries can identify possible ways to reduce energy consumption and losses. Energy balances can be performed at different stages of the production process from raw materials to the finished product and identifying potential root cause [1].

11.2 Basic principles

Fig. 1 presents fundamental understanding of material and energy balances using schematics. Each box is considered as a unit operation. According to law of conservation of mass and energy, the mass and energy going into the box must balance with the mass and energy coming out.

The law of conservation of mass leads to what is called a mass or a material balance.

$$\text{Mass in} = \text{mass out} + \text{mass stored}$$

In a given process, law of conservation of mass leads to:

$$\text{Raw materials} = \text{products} + \text{wastes} + \text{stored}$$

$$\sum m_R = \sum m_P + \sum m_W + \sum m_S$$

where Σ denotes the sum of all terms.

Technology and Applications of Polymers Derived from Biomass. https://doi.org/10.1016/B978-0-323-51115-5.00011-6

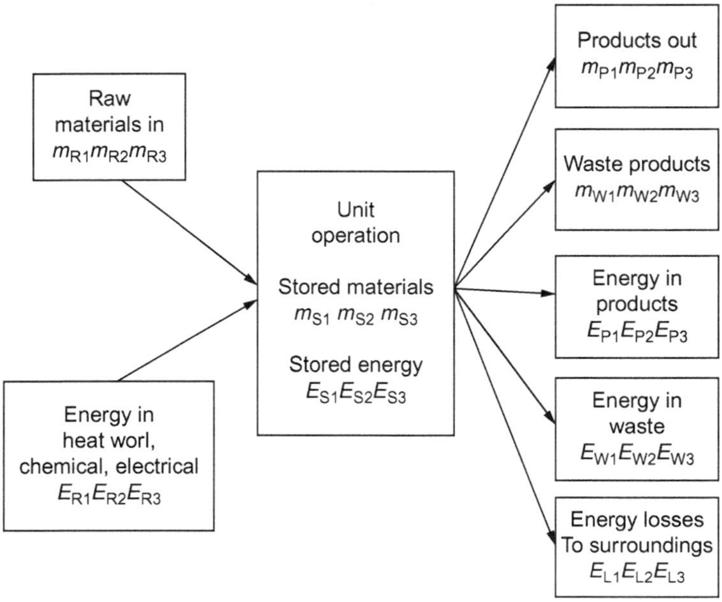

Fig. 1 Basic principles of material and energy balances.
From Bureau of Energy Efficiency, Material and energy balance, The National Certification Examinations for Energy Managers and Energy Auditors Guidebook, pp. 82–107 (Chapter 4).

Total raw materials, $\sum m_R = \sum m_{R1} + \sum m_{R2} + \sum m_{R3}$

Total products, $\sum m_P = \sum m_{P1} + \sum m_{P2} + \sum m_{P2}$

Total waste, $\sum m_W = \sum m_{W1} + \sum m_{W2} + \sum m_{W3}$

Total stored, $\sum m_S = \sum m_{S1} + \sum m_{S2} + \sum m_{S3}$

If there are no chemical changes occurring in the plant, the law of conservation of mass will apply also to each component, so that for component A:

m_A in entering materials $= m_A$ in the exit materials $+ m_A$ stored in the plant

Sometimes, there are some unknown losses that need to be identified. The material balance in such a case will be:

Raw materials = products + waste + stored products + losses

The energy balance is similar to mass balance. The energy coming into a unit operation can be balanced with the energy coming out and the energy stored.

Energy in = energy out + energy stored

$$\text{Total energy, } \sum E_R = \sum E_P + \sum E_W + \sum E_S + \sum E_L$$

Where,

$$\text{Total energy in, } \sum E_R = \sum E_{R1} + \sum E_{R2} + \sum E_{R3}$$

$$\text{Total energy leaving with the product, } \sum E_P = \sum E_{P1} + \sum E_{P2} + \sum E_{P3}$$

$$\text{Total energy leaving with the waste, } \sum E_W = \sum E_{W1} + \sum E_{W2} + \sum E_{W3}$$

$$\text{Total energy lost, } \sum E_L = \sum E_{L1} + \sum E_{L2} + \sum E_{L3}$$

$$\text{Total energy stored, } \sum E_S = \sum E_{S1} + \sum E_{S2} + \sum E_{S3}$$

11.3 The Sankey diagram

A Sankey diagram, originally developed by H.R. Sankey is a visualization chart to depict pictographically from one set of values to another. In this diagram, material and energy balances are used to study different stages of a complex process or the complete process from raw materials to the finished product. The Sankey diagram supports the systematic derivation of improvement measures. In order to fully understand the Sankey diagram, some assumptions have to be applied [2]:

1. Variables are related to a specific period of time
2. Considered values are extensive
3. The arrow width is proportional to its value
4. No stocks, what goes in a process has to go out as well

In order to fully understand the application of the Sankey diagram, the following sections will review four examples.

11.3.1 Example 1: CO_2 reduction by using biomass from the pulp industry

The first example is to derive a Sankey diagram for reducing carbon dioxide by using biomass from the pulp industry. Fig. 2A and B shows Sankey diagrams for two different processing situations: (A) lignin precipitation and biomass gasification, and (B) black liquor gasification. The numbers shown in parenthesis are the conversion efficiency of each product based on the total biomass fuel input while the heat arrows represent district heat. The heat arrows also include some extra LP steam, which can be generated for use in a paper mill in case of an integrated plant [3].

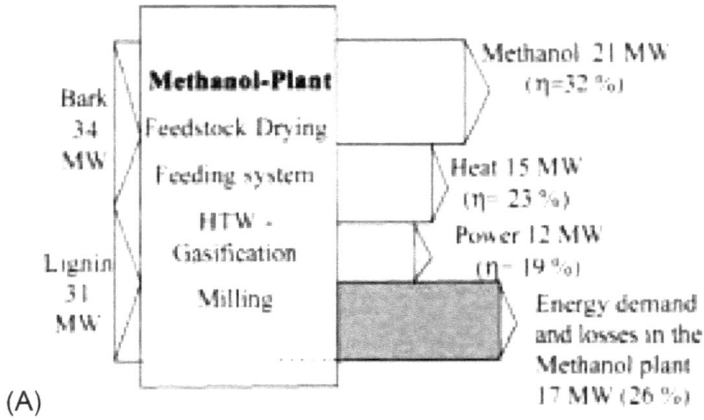

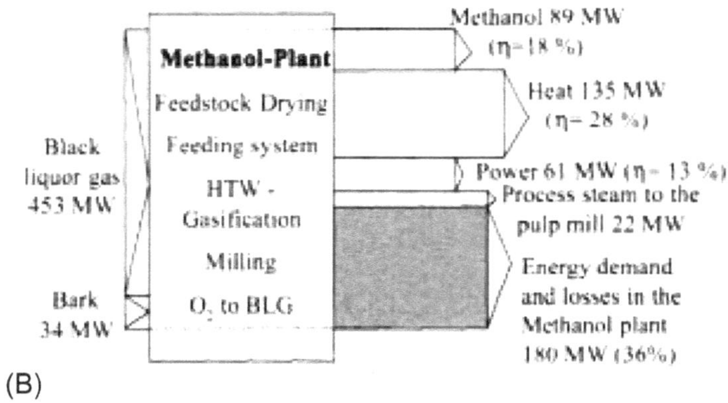

Fig. 2 Sankey diagram examples for (A) lignin preparation and biomass gasification, and (B) black liquor gasification developed using ASPEN PLUS.
From A. Isaksson, V. Martin, J. Yan, CO_2 Reduction by using biomass from the pulp industry—system simulation for cogeneration of methanol, power and heat, in: Greenhouse Gas Control Technologies: Proceedings of the 5th International, CSIRO Publishing, 2001, p. 1280.

11.3.2 Example 2: Efficiency of Rosenheim wood gasifier

In this diagram, it is shown that the city is building or already running a wood gasification plant. Instead of just using the heat from directly burning wood, they decided to work with a wood gas carburetor and use the wood gas to run a gas motor. Overall loss of energy in the system is only 23%. The green box at the bottom displays the avoided fossil GHG emissions per tons of wood for both technologies. Flows are in MWh, but only some selected arrows are labeled [4].

11.3.3 Example 3: Biomass 2-in-1 flipped Sankey diagram

These are actually two diagrams that are flipped over at a vertical center line. The left half is the first Sankey diagram that goes right-to-left orientation and shows the use of cropland to produce animal feed, corn ethanol, and exports. Some cropland sits idle and is not used productively. The right half is a second Sankey diagram and shows a different use of the cropland with major crops and outputs for the maximum ethanol production scenario [4].

11.3.4 Example 4: Bio-char slow pyrolysis process

The Sankey diagram shows how the energy flows. 16,000 MJ of energy is contained in a ton of stover; more than a fourth of it can be approved of as syngas from the pyrolysis process. The main objective however seems to be carbon dioxide capture, and bio-char presents a viable alternative to other carbon reduction technologies [4,5].

11.4 Application benefits of Sankey diagram

The application benefits of a Sankey diagram are [2] as follows:

1. The Sankey diagram represents visually various energy inputs, outputs, losses, and energy stored, so that overall energy improvements can be done in accordance with priority or preference.
2. The Sankey diagram is a visual aid that can illustrate which energy flows represent useful energy and what flows are responsible for waste or emissions.
3. The Sankey diagram can help in locating dominant contributions of an energy component to an overall flow of energy.
4. Apart from energy, a Sankey diagram can be used to visualize material balance, cost transfers, or energy between processes.

11.5 Creating Sankey diagram

The first step in making a Sankey diagram is to identify energy inputs, outputs, and conversion stages. The second step is to identify the amounts of primary energy sources such as natural gas, oil, and electricity and measured them before and after the conversion stage. These measurements must be converted to the same energy units [6,7].

Once all the data is collected and converted, a Sankey diagram can be generated using appropriate software. There are a number of commercial programs available, including e!Sankey [8]. There is also a free program that uses Excel to draw Sankey diagrams. The U.S. Department of Energy's Process Heating Assessment and Survey Tool (PHAST) is a program that helps improve thermal efficiency of heating equipment by performing a heat balance that identifies major areas of energy use under various operating conditions [9]. Sankey diagrams can be generated and what-if scenarios can be generated for testing various options for reducing energy use.

RETScreen is a clean energy management program developed by Natural Resources Canada, also generates Sankey diagrams during its analysis of potential renewable energy, energy efficiency, and cogeneration projects [10].

11.6 Material balances

In order to establish material balances, the first step is to determine (a) materials in, (b) materials out, and (c) materials stored. For each category, it has to be clearly stated whether the materials are to be treated as a gross mass balance, or whether various constituents should be treated separately, listing all the constituents [11,12].

Material balances are based on three types of processes depending on how materials are transferred into and out of the system.

1. Batch
2. Continuous
3. Semibatch

1. *Batch process*: In this, the entire feed of material is added instantaneously to the system at the beginning of the process, and all the contents of the system including the products are removed later, at the end of the process.
2. *Continuous process*: In this case, the materials enter and leave the system as continuous streams.
3. *Semibatch*: In this case, there are two options: (a) feed may be added at once and the products removed continuously, or (b) feed may be added continuously and the products be removed at once.

11.7 Energy balances

There are many forms of energy such as heat, kinetic energy, chemical energy, and potential energy. There are situations where some energy is insignificant [13,14]. Energy balances can be calculated on the basis of external energy used per kilogram of product, or raw material processed, or on dry solids or some key component.

11.7.1 Heat balances

Heat energy is the most common energy form and the heat balance is represented as shown in Fig. 3. In this heat balance, heating and cooling are considered as two individual operations. The total heat conserved is represented by an enthalpy (H). Enthalpy balances can be written for the various items of equipment or process stages, or for the whole plant. It has to be assumed that that no appreciable heat is converted to other forms of energy such as work [13,14].

Enthalpy is always referred to some reference level, so that the quantities are relative to this level. Energy balances are then made considering the various

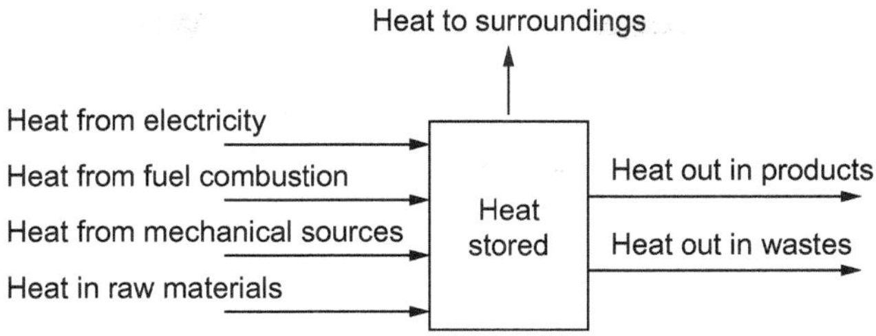

Fig. 3 Heat balance.
From http://www.nzifst.org.nz/unitoperations/matlenerg3.htm; R.L. Earle, M.D. Earle, Material and energy balances, Unit Operations in Food Processing, Pergamon Press, 1983, pp. 12–21.

quantities of materials involved, their specific heats, and their changes in temperature or state (typically latent heats arising from phase changes are encountered) [13,14]. Latent heat is typically used in situations where heat absorbed or evolved by some reactions is insignificant compared with other forms of energy.

Latent heat is the heat required to change, at constant temperature, the physical state of materials from solid to liquid, liquid to gas, or solid to gas. The units of latent heat are J/kg and total latent heat change is calculated by multiplying the mass of the material, which changes its phase by the latent heat. Once the factors that are significant in the overall energy balance are determined, the simplified heat balance can then be used [11,13,14].

11.8 Other forms of energy

Other forms of energy include electrical energy which drives the motor power. There are always losses from the motors due to heating, friction, and windage; the motor efficiency, which can normally be obtained from the motor manufacturer, expresses the proportion of the electrical input energy, which emerges usefully at the motor shaft and so is available.

In situations where flow has to be considered, for example, fluids in pumping, of solids in solids handling, the energy input is largely mechanical. The flow situations can be analyzed by recognizing the conservation of total energy whether as energy of motion, or potential energy such as pressure energy, or energy lost in friction. Similarly, chemical energy released in combustion can be calculated from the heats of combustion of the fuels and their rates of consumption. Eventually energy emerges in the form of heat and its quantity can be estimated by summing the various sources [11].

11.9 Guidelines for material and energy balance

Material and energy balances are important for the following reasons:

- Identify and quantify known and unknown losses
- Monitoring the improvements
- Evaluating cost benefits

Because of the inefficient use of raw materials and energy in production processes, it is important to identify guidelines for material and energy balances. This ensures identifying the losses as well as improving efficiency of the process [11].

Following are the guidelines for material and energy balance:

1. In a complex production process, it is better to first draft the overall material and energy balance.
2. Choose simple discrete subsystem when splitting up the total system.
3. Choose the material and energy balance in such a way that the number of streams entering and leaving is the smallest possible.
4. Always choose recycle material and energy stream within the envelope.
5. The measurement units may include time factor or production linkages.
6. Consider a full batch as the reference in case of batch operations.
7. Include start-up and cleaning operation consumptions of material and energy resources.
8. Calculate the gas volumes at standard conditions.
9. In case of shutdown losses, averaging over long periods should be performed.
10. For each stream, where applicable, indicate energy quality (pressure, temperature, enthalpy, kcal/h, kW, Amps, Volts, etc.).
11. While preparing material and energy balances, precision of analytical data, flow and energy measurements have to be accurate.

References

[1] Bureau of Energy Efficiency, Material and energy balance, The National Certification Examinations for Energy Managers and Energy Auditors Guidebook, Bureau of Energy Efficiency, National Productivity Council of India, http://www.em-ea.org/gbook11.asp.
[2] M.O. Abdullah, Energy resources, supply, and demand, in: Applied Energy: An Introduction, CRC Press, Boca Raton, FL, 2013, p. 100.
[3] A. Isaksson, V. Martin, J. Yan, CO_2 reduction by using biomass from the pulp industry—system simulation for cogeneration of methanol, power and heat, in: Greenhouse Gas Control Technologies: Proceedings of the 5th International, CSIRO Publishing, Collingwood, VIC, Australia, 2001, p. 1280.
[4] http://www.sankey-diagrams.com/tag/biomass/.
[5] K.G. Roberts, B.A. Gloy, S. Joseph, N.R. Scott, J. Lehmann, Life cycle assessment of biochar systems: estimating the energetic, economic, and climate change potential, Environ. Sci. Technol. 44 (2) (2010) 827–833.
[6] K.A. McGuire, Data visualization, The Analytic Hospitality Executive, Wiley Publishers, Hoboken, NJ, 2016.
[7] R. Brath, D. Jonker, Flows, in: Graph Analysis and Visualization: Discovering Business Opportunity in Linked Data, Wiley, 2015, pp. 315–357.
[8] https://www.ifu.com/e-sankey/.

[9] Process heating assessment and survey tool, Office of Energy Efficiency and Renewable Energy. https://www.energy.gov/eere/amo/articles/process-heating-assessment-and-survey-tool.
[10] http://www.nrcan.gc.ca/energy/software-tools/7465.
[11] N.K. Jha, Green Design for Manufacturing for Sustainability, CRC Press, Boca Raton, FL, pp. 25-32.
[12] W.C. Lyons, G.J. Plisga, M. Lorenz, General engineering and science, in: Standard Handbook of Petroleum and Natural Gas Engineering, Elsevier, 2016, p. 115.
[13] http://www.nzifst.org.nz/unitoperations/matlenerg3.htm.
[14] R.L. Earle, M.D. Earle, Material and energy balances, in: Unit Operations in Food Processing, Pergamon Press, Manchester, UK, 1983, pp. 12–21.

Applications of biomass-derived polymers and chemicals

12

12.1 Introduction

Renewable resources like plants and bacteria are the sources of large number of polymeric matrices. The utilization of biomass for development of composites provides a route to utilize huge raw materials available in nature with the advantage of reduction in the volume of waste material, reduction of released atmospheric carbon dioxide, and complete biodegradability.

The use of biomass-derived matrices results in composite material obtained from completely renewable resources and may be categorized as green composites. The biodegradable polymers can be classified depending on their origin from agropolymers, microbial, and chemically synthesized from agro-based products [1]. The growth of biodegradable polymers is driven by the following factors:

1. Consumer preference toward environmental-friendly products
2. Use of renewable and bio-based raw material
3. Biodegradability
4. Government policies toward green procurement

The two most important commercial, biodegradable polymers are polylactic acid (PLA) and starch-based polymers, accounting for about 47% and 41%, respectively, of total biodegradable polymers consumption. Starch sources vary worldwide, but include corn, potatoes, cassava, and sugar beets. In Europe, starch-based biodegradable polymers are the major type consumed, accounting for 62% of the market, due to Europe's large, starch-based capacity and their use in many applications. This is followed by PLA, with 24% and other biodegradable polymer types with 14%.

Some of the biomass-derived polymers and their manufacturers are listed in Table 1 [2].

Biomass-derived polymers and chemicals have found their end use in a variety of commercial applications. Some of the common biopolymers include bio-polyamide, PLA, polyhydroxyalkanoate (PHA), succinic acid, polybutylene succinate (PBS), polyethylene furanoate (PEF), and polybutylene adipate terephthalate (PBAT). These biopolymers have found their use in applications such as automotive, packaging, horticulture, medical devices/biomedical, pharmaceutical/hygiene, and filtration/purification.

In this chapter, commercial applications of different biomass-based polymers and chemical will be discussed.

Technology and Applications of Polymers Derived from Biomass. https://doi.org/10.1016/B978-0-323-51115-5.00012-8

Table 1 **Biomass-derived polymers and their manufacturers [2]**

Biomass-derived polymer	Manufacturer
Cellulose acetate	• Rotuba • Solvay • Eastman chemicals
Bio-polyamide	• DuPont • Solvay • Li Peng Enterprise Co. • Formosa Group • Shenma Industrial Co. Ltd.
Polybutyrate adipate terephthalate (PBAT)	• Lackner Ventures & Consulting GmbH • JinHui Zhaolong
Polybutylene succinate (PBS)	• Myriant • Mitsubishi Chemical Co. • Showa Denko K. K. • Ire Chemical Ltd. • SK Chemicals Co. Ltd. • Gyeonggi-do
Polyhydroxyalkanoates (PHA)	• Metabolix Inc. • Kaneka Corporation • Meredian Holdings Group, Inc. • Shenzhen Ecomann Biotechnology Co. Ltd. • BioMatera • Biomer • Bio-On-Sri • Newlight Technologies, LLC • PHB Industrial S.A.
Polylactic acid (PLA)	• NatureWorks LLC • Corbion Purac • Nexeo Solutions • Plastics Color Corporation • PolyOne Corporation • S&E Speciality Polymers, LLC • PM Plastics • LTI Color Compounders, LLC • Klockner Pentaplast • Hybrid Plastics, Inc.
Polyethylene furanoate (PEF)	• Avantium
Polycaprolactone (PCL)	• Perstorp
Succinic acid	• Faurecia
Levulinic acid	• GF Biochemicals
Ethanol	• ADM • POET • Valero Energy Corp. • Green Plains renewable Energy • Flit Hills Resources LP

From Sorona EP Thermoplastic Polymer, Product datasheet, DuPont. www2.dupont.com/Plastics/en_US/Products/Sorona/sorona_ep.html.

12.2 Automotive

Automotive industry is going through a major change with the focus is on reducing fuel consumption and emissions by reducing a vehicle's weight. Plastics have been a part of automotive components but there has been a concern toward plastics recyclability. For this reason, automotive industry is making a big leap toward using biomass-derived polymers and chemicals. Some of the bio-based plastics such as bio-based polyamides, bio-based polyesters are already successfully being used by leading automotive brands around the world today with the aim of reducing their products' environmental impact [3].

Manufacturers have turned to bio-based or partly bio-based durable bioplastics to produce sturdy dashboard components as well as solid interior and exterior features. Components made from biomass-derived polymers can provide a standard of safety that is of ultimate importance in the transportation sector.

Some of the main biomass-based polymers for automotive applications are as follows:

- Bio-polyamide
- Polylactic acid
- Polyhydroxyalkanoate (PHA)
- Polycaprolactone (PCL)
- Succinic acid

12.2.1 Bio-polyamide

Structures of the key raw materials for some partially or fully bio-sourced polyamides are shown in Fig. 1. Both 11-aminoundecanoic acid and sebacic acid are products from chemical conversion of ricinoleic acid, the major fatty acid component of castor oil [4].

Fig. 2 shows (A) cable gland and (B) RislanHT connectors made from bio-polyamide. These are graded with an IP 66 protection rating and made without halogen or cadmium. Units exhibit high impact strength and operate in temperatures ranging from −45°C to 105°C [5]. Fig. 2C shows a radiator end tank made for Ford cars [4–6].

12.2.2 Polylactic acid

PLA products have a reduced carbon footprint and a high bio-content and hence lesser impact to the environment [7]. From a functional requirement point of view, high heat PLA offers heat resistance up to 140 h, high gloss and excellent colorability, dimensional stability and will withstand the demanding conditions in today's engine compartment. This makes PLA an excellent alternative to most polyesters (PC, PET, and PBT) as well as polystyrenes, polyolefins, and polyamides [8].

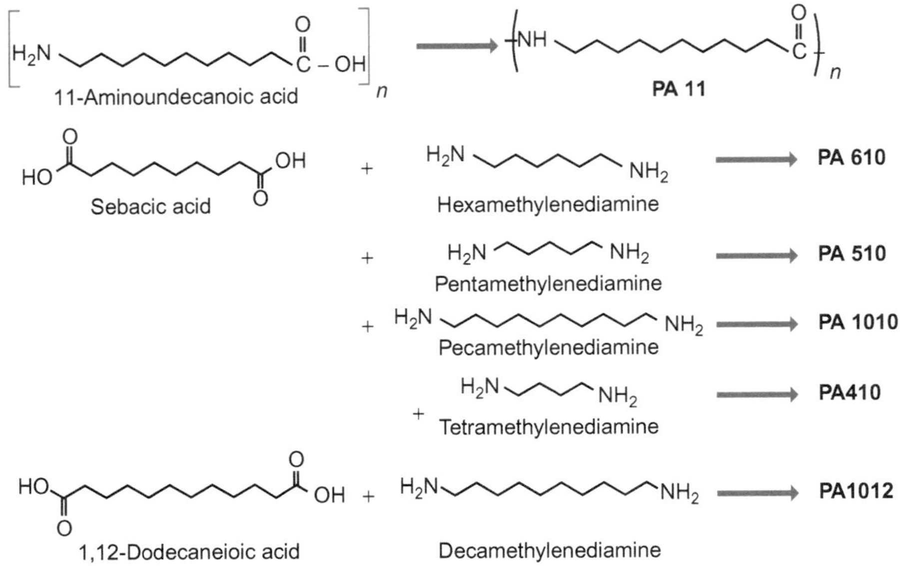

Fig. 1 Structures of the key raw materials for partially or fully bio-based polyamides. From R. Humphreys, Bio-polyamides: where do they come from? Polymer Innovation Blog. https://polymerinnovationblog.com/bio-polyamides-where-do-they-come-from/.

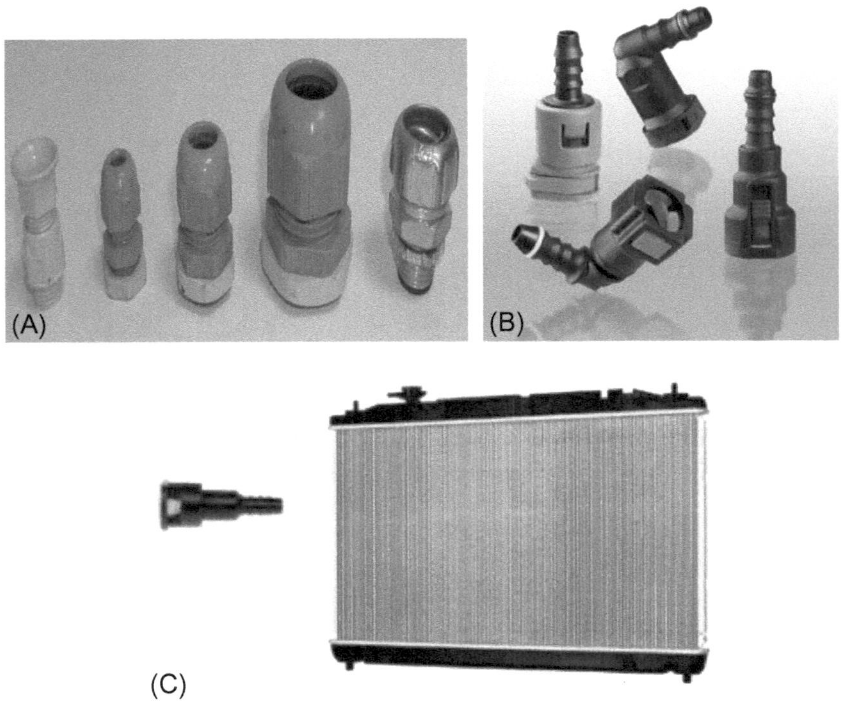

Fig. 2 (A) Bio-polyamide cable gland, (B) RilsanHT connectors, and (C) radiator end tank. From (A) https://en.wikipedia.org/wiki/Cable_gland#/media/File:Presse-etoupe.jpg; (B) R. Humphreys, Bio-Polyamides: Where Do They Come From?, Polymer Innovation Blog; https://polymerinnovationblog.com/bio-polyamides-where-do-they-come-from/; (C) http://www.lawbc.com/share/bcs2013/Molecules%20to%20Market/lee-presentation.pdf

12.2.3 Polyhydroxyalkanoates (PHAs)

BIO-ON has developed an exclusive process for the production of a broad family of PHAs from agricultural waste including molasses and sugarcane and sugar beet syrups. These PHAs are linear polyesters produced in nature by bacterial fermentation of sugar. The various PHAs produced in this way can potentially replace some families of traditional plastics such as LDPE, HDPE, PP, PVC, PS, and PET.

Fig. 3 shows example of car interior made by MINERV PHA, a high-performance PHA biopolymer replacing materials such as PET, PP, and PE [8]. MINERV PHA is endowed with optimal thermal properties with processing range from −10°C to 180°C. This biopolymer is typically processed using injection molding and extrusion.

Fig. 4 shows biodegradability curve for MINERV PHA [8]. The MINERV PHA increases its biodegradability factor in bacteriologically impure water. This type of polymer biodegradation is the future of biodegradability. The natural dissolution of a biopolymer in bacteriologically impure water (e.g., river water) in a few days is a rare and very difficult result to obtain. MINERV-PHA is able to achieve 100% biodegradability in 10 days in normal river water.

12.2.4 Polycaprolactone (PCL)

APS Elastomers, a leader in the polyurethane casting industry specializing in suspension and drive train components for automotive, light truck, and off-road industries have developed PCL-based thermoplastic polyurethane (TPU) grades with good processability and compression set at elevated temperatures. One example is suspension and drive train components made from Zythane 3000 PCL series of TPE as shown in Fig. 5 [9].

Fig. 3 Car interior made by a high performance PHA biopolymer, MINERV-PHA. From http://www.bio-on.it/what.php?lin=inglese.

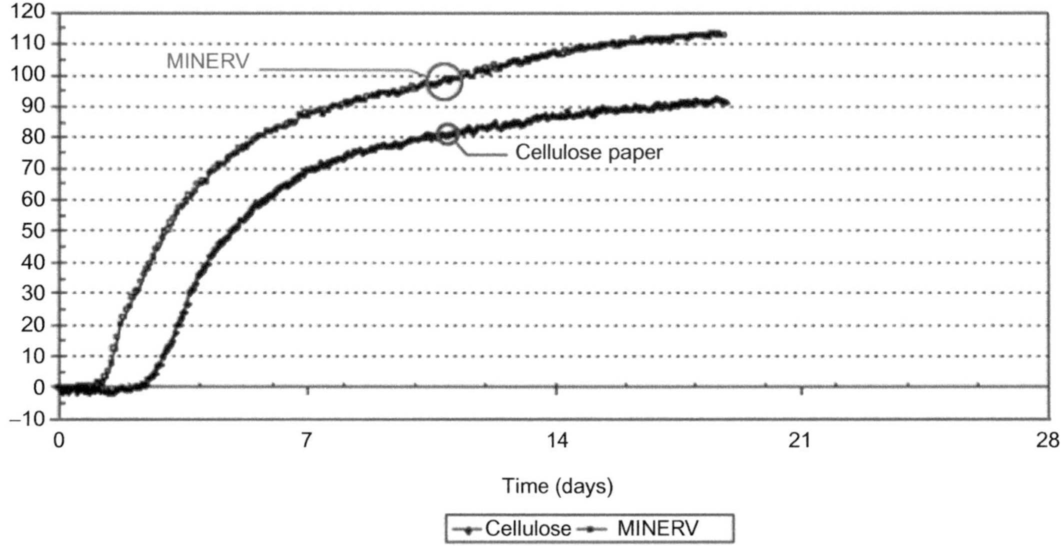

Fig. 4 Biodegradability curve for MINERV PHA.
From http://www.bio-on.it/what.php?lin=inglese.

Fig. 5 Suspension and drive train components made from polycaprolactone-based TPU, Zythane 3000.
From https://en.wikipedia.org/wiki/Polycaprolactone#/media/File:Shapelock_bike_light.png.

Some of the important properties include:

- Working conditions range from −40°C to 130°C.
- High wear resistance against oils, grease, water, and extreme temperatures.

12.2.5 Succinic acid

BioAmber developed a process to convert renewable feedstock into 100% bio-based succinic acid [10].

Faurecia, a leading automotive supplier have collaborated with BioAmber to develop of high-performance natural fiber composite, a blend between bio-PBS and succinic acid [11].

12.3 Packaging

Biomass-derived polymers offer a large range of packaging applications. Some of the packaging options include bags for compost, agricultural foils, horticultures, nursery products, toys, and textiles. Other areas include packaging for contact articles including disposable cutlery, drinking cups, salad cups, plates, overwrap and lamination film, straws, stirrers, lids and cups, plates and containers.

Some of the main biomass-based polymers used for packaging applications are as follows:

- Cellulose acetate (CA)
- Polyethylene furanoate (PEF)
- Polyhydroxyalkanoate (PHA)
- Polycaprolactone (PCL)
- Succinic acid

12.3.1 Cellulose acetate

Cellulose acetate films are used in the packaging for a range of products such as food, perfume, clothing, and confectionary. Some of the key features of cellulose acetate include [12]:

- High transparency and transmittance
- Exceptionally low haze
- Near zero birefringence
- Nonoriented
- UV stable
- Good tensile strength and elongation
- Relatively low tear strength
- High stiffness
- Good dead fold
- High moisture vapor transmission rate
- Extremely low water permeability
- Dimensionally stable
- Hard wearing surface
- Thermoformable
- Resistant to oils, grease, and fat
- Dry touch
- Permanently high surface energy
- Receptive to ink, adhesive, and foil blocking
- Printable
- Light weight
- Good electrical insulator
- Easy to cut
- Easy to write on

Celanese developed Clarifoil packaging for different markets such as:

- Luxury consumer goods
- Pharmaceutical and medical
- Electronics
- Construction
- Footwear
- Anticounterfeit

These films can be bend around corners without any stress whitening and retain tension even if the board expands and contracts in store. An example is shown in Fig. 6 [13].

12.3.2 Polyethylene furanoate (PEF)

PEF has very good barrier properties. This makes it a good candidate for packaging applications. The oxygen barrier of PEF is six times greater than PET, while carbon dioxide and water barrier of PEF are three and two times greater than PET, respectively [14]. In comparison to PET, PEF has higher glass transition temperature,

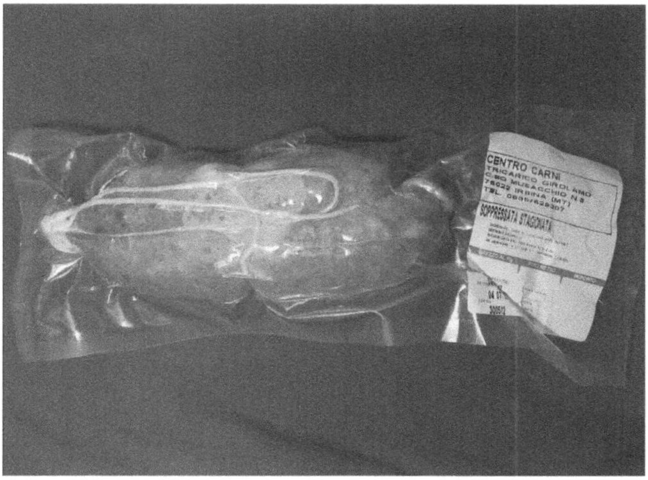

Fig. 6 Celanese clarifoil film used for packaging.
From https://upload.wikimedia.org/wikipedia/commons/thumb/7/7e/Soppressata_sottovuoto.JPG/1200px-Soppressata_sottovuoto.JPG.

Fig. 7 Polyethylene furanoate (PEF) packaging applications.
From https://en.wikipedia.org/wiki/Polyethylene_terephthalate#/media/File:Plastic_bottle.jpg; https://en.wikipedia.org/wiki/Plastic_wrap#/media/File:Pvc-Film.jpg; https://en.wikipedia.org/wiki/File:Kodak-Max-400-35mm-Film.jpg.

low melting point, and a higher modulus. Some of the examples of PEF packaging are shown in Fig. 7 [15–17].

12.3.3 Bio-polyamide

As shown in Fig. 8, MF Folien GmbH has developed bio-polyamide film based on Royal DSM's bio-based EcoPaXX polyamide 410 [17]. These bio-PA films are used for flexible packaging applications and come in three different thicknesses, 30, 40, and

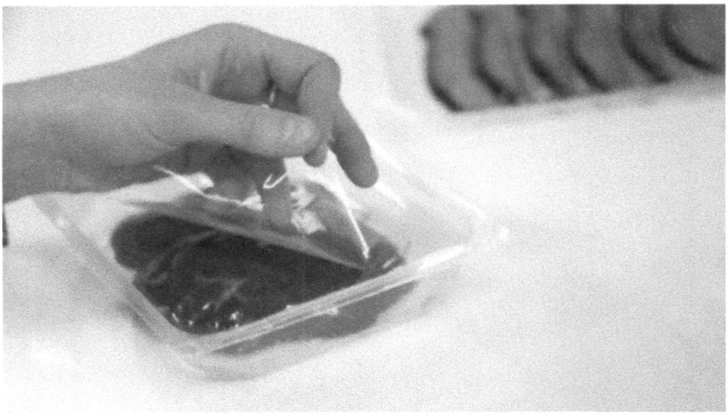

Fig. 8 MF Folien GmbH bio-polyamide film based on Royal DSM's bio-based EcoPaXX polyamide 410.
From http://www.bakeryandsnacks.com/Processing-Packaging/MF-Folien-introduces-first-film-made-from-DSM-s-EcoPaXX.

50 μm, respectively. Some of other properties of bio-polyamide in comparison with polyamide 6 include:

- Low moisture transmission rate
- Higher oxygen barrier

These exceptional barrier properties result from one of its building blocks that are derived from castor oil obtained from plants that grow in tropical regions.

12.3.4 Polybutyrate adipate terephthalate

PBAT is a biodegradable random copolymer. Jinhui Zhaolong High Technology Co., Ltd, Taiyuan City, China has developed a high-performance film, Ecoworld PBAT [18]. This PBAT is completely biodegradable and has the following properties:

- Tear resistance is 120% greater than polyethylene
- Impact strength is 130% greater than polyethylene

Ecoworld PBAT can be decomposed into carbon dioxide and water in 180 days by microorganisms under compostable conditions. The entire process not only avoids toxic gas but also can generate organic fertilizer. JinHui Zhaolong's PBAT production uses one-step polymerization technology [18].

Some of the typical applications of PBAT are plastic films, compostable bags, and bottles as shown in Fig. 9 [19, 20].

12.3.5 Polybutylene succinate (PBS)

The French company ARD (Agro Industry Research and Development) realized the first worldwide production of PBS with 100% bio-based succinic acid using wheat as a feedstock. Natureplast formulated PBS for the processing of food packaging. Films,

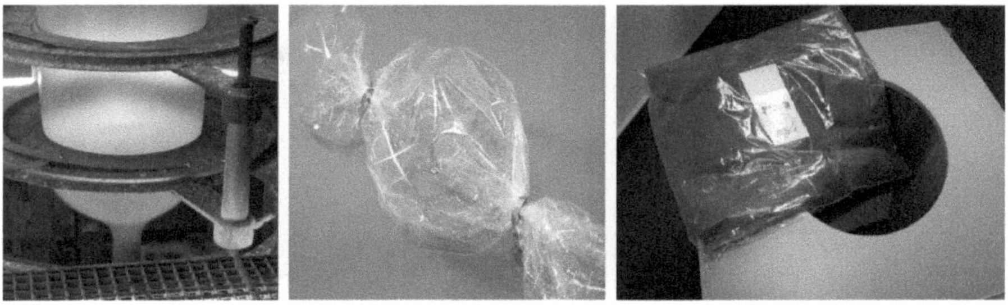

Fig. 9 PBAT applications.
From https://en.wikipedia.org/wiki/Plastic_film#/media/File:Film_extrusion.jpg; https://en.wikipedia.org/wiki/Plastic_film#/media/File:Sweets_packaging_made_of_PL-Blend_Bio-Flex.jpg; https://commons.wikimedia.org/wiki/File:Biodegradable_%22poo%22_bags_(4733211858).jpg.

trays, and cups have been produced by Leygatech, Velfor, and the CTCPA for the packaging of cheese, meat, fish, and vegetarian dishes.

Ecoterah from Precision Color Graphics is a line of paper-based, multilayer compostable bags and pouches used for pet food and snacks. Fig. 10 shows different applications of PBS [21–24].

Fig. 10 Different applications of polybutylene succinate.
From https://en.wikipedia.org/wiki/Plastic_wrap#/media/File:B%C3%BCro_Obst_(27067584185).jpg; https://en.wikipedia.org/wiki/Packaging_and_labeling#/media/File:Food_packages_(1).jpg; https://en.wikipedia.org/wiki/Biodegradable_plastic#/media/File:BiodegradablePlasticUtensils2.jpg; http://www.wikiwand.com/de/Biokunststoff.

Fig. 11 Ricotta cheese packaged in polybutylene succinate (PBS).
From http://www.dairyreporter.com/Library/Packaging-Packing-Materials-Containers/Green-PBS-potential-investigated-in-EU-project.

Succipack, a European FP7 project developed 100% bio-based PBS. Fig. 11 shows ricotta cheese packaged in PBS [25].

12.3.6 Polyhydroxyalkanoates

PHA is a linear polyester derived from microbial fermentation of plant sugar. PHA is an ideal candidate for packaging applications based on the following properties:

- Excellent barrier properties to water and grease
- A-PHA-modified PLA film is both tough and ductile, creating a clear film that provides an excellent surface for printing and messaging, and is suitable for a broad range of packaging applications.

Metabolix's version of PHA popularly known as Mirel is produced by similar microbial fermentation process of corn. Because of its unique properties, it is used in food and medical packaging applications. Mirel also has good heat resistance, and thereby used for microwaveable food packaging applications.

Fig. 12 shows some examples of PHAs [26–30].

12.3.7 Polylactic acid

PLA is aliphatic polyester derived from renewable resources such as corn or sugar cane. PLA is often used in food packaging because of the following properties:

- Excellent barrier properties
- Transparency and clarity

NatureWorks PLA known as Ingeo is used in food and beverage packaging ranging from yogurt to strawberries to coffee. Following are some of its properties:

- Provides excellent gloss, transparency, and clarity
- Easy to shape, print, and emboss

Fig. 12 Different applications of polyhydroxy alkanoate (PHA) [26].
From https://en.wikipedia.org/wiki/Bioplastic#/media/File:Shampoo_Bottle_made_of_PLA-Blend_Bio-Flex.jpg; Danimer Scientific; https://en.wikipedia.org/wiki/Biodegradable_polymer#/media/File:Biodegradable_cutlery.jpg; https://en.wikipedia.org/wiki/Bioplastic#/media/File:Flower_Wrapping_made_of_PLA-Blend_Bio-Flex.jpg.

- Increased rigidity
- Excellent barrier properties

Unitika Limited and FKuR Kunststoff GmbH produce modified version of Ingeo PLA under trade names of Bio-Flex and Terramac. Terramac offers its PLA with higher heat distortion/deflection temperature than that of Ingeo.

Fig. 13 shows some examples of PLA [29,31–37].

12.4 Horticulture

Production of agricultural mulches, seeding strips, and tapes made from biodegradable polymers are another strong potential market. The tape biodegrades in the soil as the seeds germinate and take root. Plastic mulch films are then used to give the new seedlings a head start in the spring; the mulch helps reduce evaporation and

Fig. 13 Different applications of polylactic acid (PLA).
From https://www.plasticstoday.com/content/greening-industry-injection-and-blowmolding-packaging-may-not-be-first-volume-market-bioplastics/7331554586239; http://www.ptonline.com/articles/enhanced-resins-additives-at-biopolymers-conference; http://www.bakeryandsnacks.com/Processing-Packaging/AMR-research-on-PLA-food-packaging-2020; http://makingsociety.com/2013/08/materialmatters-series-n1-pla-plastic/; https://www.greenerpackage.com/bioplastics/pla_filmsheet_masterbatch_increases_impact_strength; http://www.sustainableisgood.com/blog/natureworks_pla/; http://www.wikiwand.com/en/Bioplastic; Food and Beverage Packaging, NatureWorks LLC, http://www.natureworksllc.com/Ingeo-in-Use/Food-and-Beverage-Packaging.

(Continued)

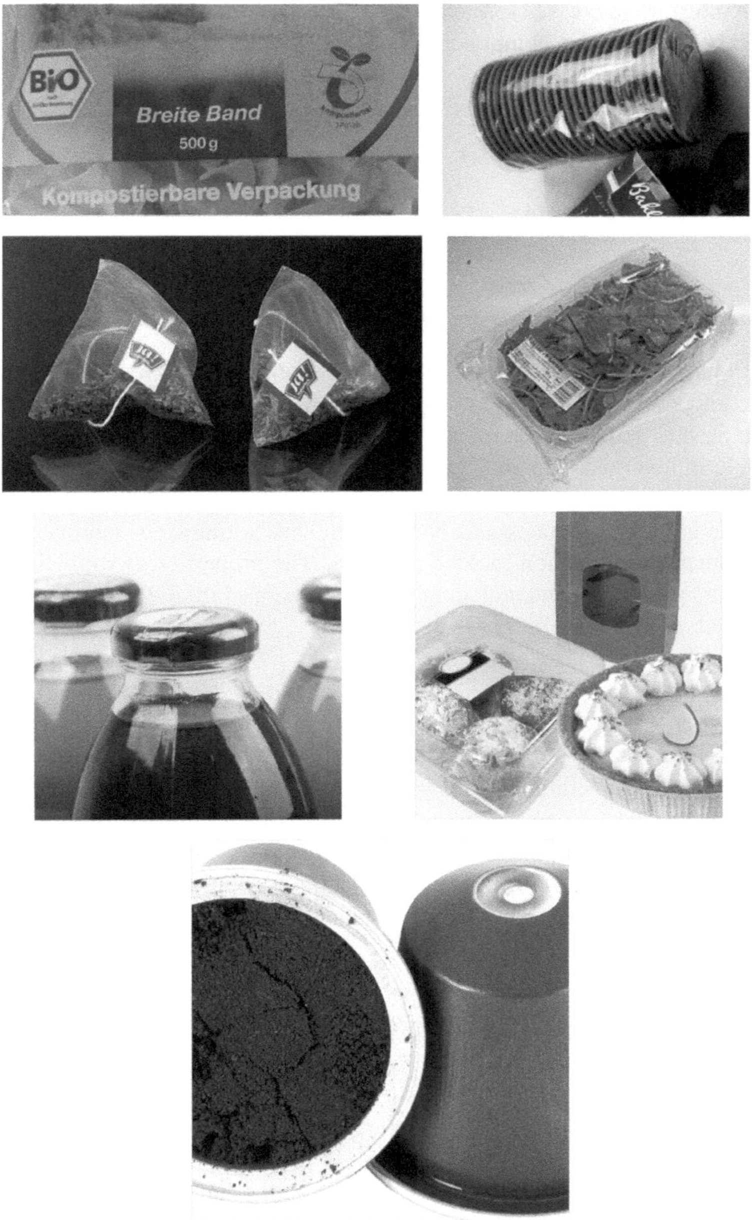

Fig. 13—continued

conserve moisture, increases soil temperature, keeps control on the weeds. Compostable seed belts and active component capsulations made out of bioplastics have also proven to be beneficial.

Some of the main biomass-based polymers used for packaging applications are as follows:

- Polybutylene succinate (PBS)
- Polyhydroxyalkanoate (PHA)
- Levulinic acid

12.4.1 Polybutylene succinate (PBS)

Showa Denko K.K., Tokyo commercialized Bionolle in the 1990s, developing automotive compounds. Bionolle is an aliphatic polyester resin that has the versatility of common plastics and is stable under ordinary conditions. However, Bionolle becomes degradable in compost, wet soil, fresh water, seawater, and activated sludge where microorganisms are present.

Bionolle has a good processability for films, sheets, filaments, laminates, molded foam products, and injection-molded products. The properties of Bionolle makes it useful in film processing and mulching film applications [38]. Fig. 14 shows examples of PBS used as mulching films [38,39].

12.4.2 Polyhydroxyalkanoates

Ball Horticultural Co.'s introduced its new SoilWrap bio-based plantable, compostable plant container to eliminate 100% of the environmental impact of the packaging. The wraparound plastic sleeve eliminates the use of a bottom panel on the pot, which allows plant roots to make immediate contact with the soil upon planting, encouraging faster growth. While SoilWrap degradation begins as soon as soil is placed in the pot, the process accelerates only after the pot is planted, allowing plenty of time for use in

Fig. 14 Application of polybutylene succinate as mulching films.
From http://www.showa-denko.com/news/bionolle-the-pioneer-in-biodegradable/; http://shivampolymers.tradeindia.com/mulching-films.html.

Fig. 15 Horticulture application of polyhydroxyalkanoates (PHA) [41].
From A.M. Mohan, Plantable nursery pot biodegrades in soil, GreenerPackage.com, https://www.greenerpackage.com/compost_biodegrade/plantable_nursery_pot_biodegrades_soil; Biodegradable Landscape Stakes, http://www.erosioncontrol-products.com/landscapestakes.html.

the greenhouse and garden center. Once in the ground, the pot is said to completely biodegrade within 12 months [40].

PHA is also used as a low-maintenance alternative to metal staples, pins, and stakes. These stakes have also been used for erosion control grass, sod, straw, coir, mats, and even light geotextiles (Fig. 15) [42].

12.4.3 Levulinic acid

Levulinic acid is a natural compound made from biomass and it uses sugar beet coproducts as the raw material. It is a platform chemical product that can be used to produce other chemical substances or to replace the synthetic alternatives.

Levulinic acid major agricultural applications include production of:

1. Delta-amino levulinic acid (DALA)
2. Methyl tetrahydrofuran (MTHF)
3. 2-Methylindolee-3-acetic acid

Delta-amino levulinic acid (DALA) is a naturally occurring chemical with a large variety of functions. It is used as an herbicide on lawns and certain grain crops. DALA is a suitable replacement for traditional herbicides and pesticides.

Some of the suppliers of levulinic acid include:

- GFBiochemicals
- Biofine
- DuPont
- Ouyi Pharmaceutical Co Ltd.
- Hebei Langfang Triple Well Chemicals Co Ltd.
- Shijiazhuang Pharmaceutical Group

12.5 Medical devices/biomedical

Nontoxic biodegradable polymer sutures are being used each year by surgeons in life-saving heart operations and other procedures. Easily sterilized, the sutures remain strong and intact until the surrounding tissues have healed. The sutures dissolve and are readily metabolized in the body without leaving any trace.

Both pliable braided sutures as well as wiry monofilament sutures are available, depending on (a) the type of surgery, (b) the knots to be used, (c) the tensile strength needed, and (d) the potential for infection. Plastic sutures are largely made from lactic or glycolic acid and account for about 95% of total market, with various pins, implants, and dental devices making up much of the remainder [41,43,44].

Tepha develops medical devices derived from resorbable, biodegradable, PHA polymers. Fig. 16 shows resorbable mesh made from PHA [45].

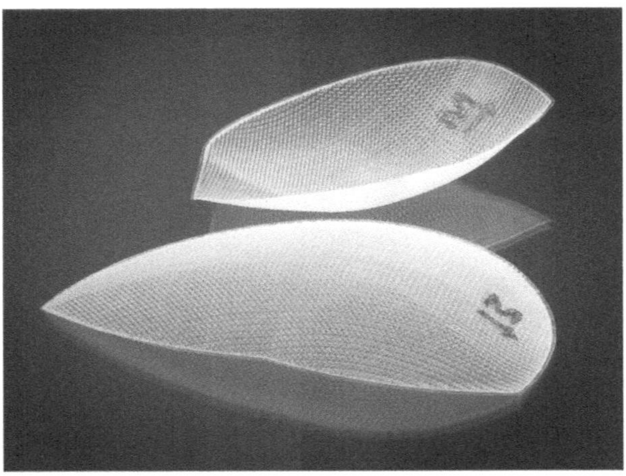

Fig. 16 PHA resorbable mesh developed by Tepha Inc.
From J. Mowbray, Biodegradable polyester license amended, Tepha Inc. https://www.tevonews.com/fibres-yarns-news/341-amendment-for-biodegradable-polyester-license.

12.6 Pharmaceutical/hygiene

Biomass-derived polymers and chemical have found their use in pharmaceutical and hygiene-based applications. The list of polymers is not limited to:

(a) Ethanol
(b) Polyhydroxyalkanoate (PHA)
(c) Levulinic acid
(d) Succinic acid

12.6.1 Ethanol

Ethanol is a very important ingredient in pharma industry. Cargill uses Pharma Agricultural Alcohol to manufacture products such as anesthetics, antiseptics, drugs, liniments, lotions, and also for cosmetics such as fragrances, deodorants, and shampoos [46]. Fig. 17 shows some examples of ethanol-based product manufactured by Cargill [47,48].

12.6.2 Polyhydroxyalkanoate

PHA based complexes have been developed by Bio-on that can be used for the formulation of cosmetics and personal care products. In addition to the biodegradability, this complex acts as an innovative carrier and dispenser of specific active substances, the bioactivity of which is prolonged after each application.

Bio-on has discovered that MinervPHA Bio Cosmetics type C1 shown in Fig. 18 is capable of binding active molecules and antioxidants, such as coenzyme Q10, vitamins, proteins, and active substances in general, transporting them naturally to parts of the body where cosmetics products are normally applied. After having located the active substance where it must act, MinervPHA is naturally washed off or absorbed without a trace [49].

The product range is divided into three macro-families:

1. Scrubs
2. Texturing
3. Active ingredients

12.6.3 Levulinic acid

Levulinic acid is used in antiinflammatory medication such as indomethacin, antiallergy agents, mineral supplements like calcium levulinate, and transdermal patches as shown in Fig. 19 [50,51]. The levulinic acid derivative DALA is used in pharmaceuticals for diagnosis and photodynamic therapy of cancer.

Levulinic acid and its derivatives are used in organic and natural cosmetic compositions for perfuming, skin conditioning, and pH-regulating purposes. This compound gives inherent fresh odor, prevent wrinkles, and stabilize formulations and emulsions.

Fig. 17 Ethanol-based products manufactured by Cargill.
From https://en.wikipedia.org/wiki/Ethanol#/media/File:Ethyl_alcohol_usp_grade.jpg; https://en.wikipedia.org/wiki/Ethanol#/media/File:Sao_Paulo_ethanol_pump_04_2008_74_zoom.jpg; https://en.wikipedia.org/wiki/Ethanol#/media/File:Ethanol_Flasche.jpg; http://lovettcivics8students.wikispaces.com/senate+bill+number+3303.

Fig. 18 MinervPHA Bio Cosmetics Type C1.
From http://www.bio-on.it/production.php?lin=inglese.

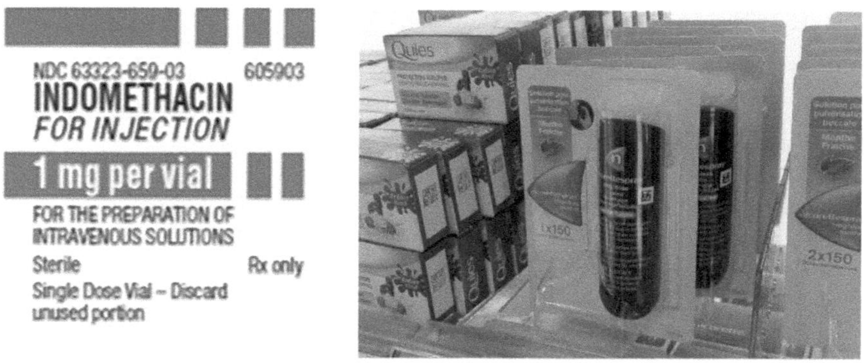

Fig. 19 Different applications of levulinic acid.
From http://www.wikidoc.org/images/b/b8/Indomethacin_injection_drug_lable01.png; https://en.wikipedia.org/wiki/Nicorette#/media/File:Nicorette_QuickMist_Spray.jpg.

Fig. 20 Baltic Amber, a medicinal compound made from succinic acid.
From https://gailfaithedwards.com/2015/10/29/the-medicinal-use-preparation-and-application-of-baltic-amber-gail-faith-edwards/. Used with permission from Belle Glade Hilmer taken at the Kaliningrad Amber Museum, Kaliningrad, Russia.

12.6.4 Succinic acid

Succinic acid, also known as amber acid and 1,4-butanedioic acid, is produced naturally as a product of cellular metabolism and is produced synthetically from maleic anhydride [52].

Baltic amber, a medicinal compound (Fig. 20) is an active pharmaceutical ingredient that contains high concentrations of succinic acid. Succinic acid is a colorless crystalline solid with a melting point of 185–187°C; soluble in water; slightly dissolved in ethanol, ether, acetone, and glycerin. Baltic amber is classified as warming, stimulating, magnetic, slightly sweet and bitter, aromatic, and neutral in nature [53].

12.7 Filtration/purification

Toray has been using separation membranes for water treatment to produce biomass-derived polymers. Fig. 21 shows Toray's setup to produce biomass-derived polymer. This approach drastically improves fermentation efficiency and reduction in refining energy [54].

Similarly, Sartorius has been using cellulose acetate membranes in their filters to purify all aqueous solutions and oils with a pH 4–8. Fig. 22 shows examples of the devices with cellulose acetate membranes.

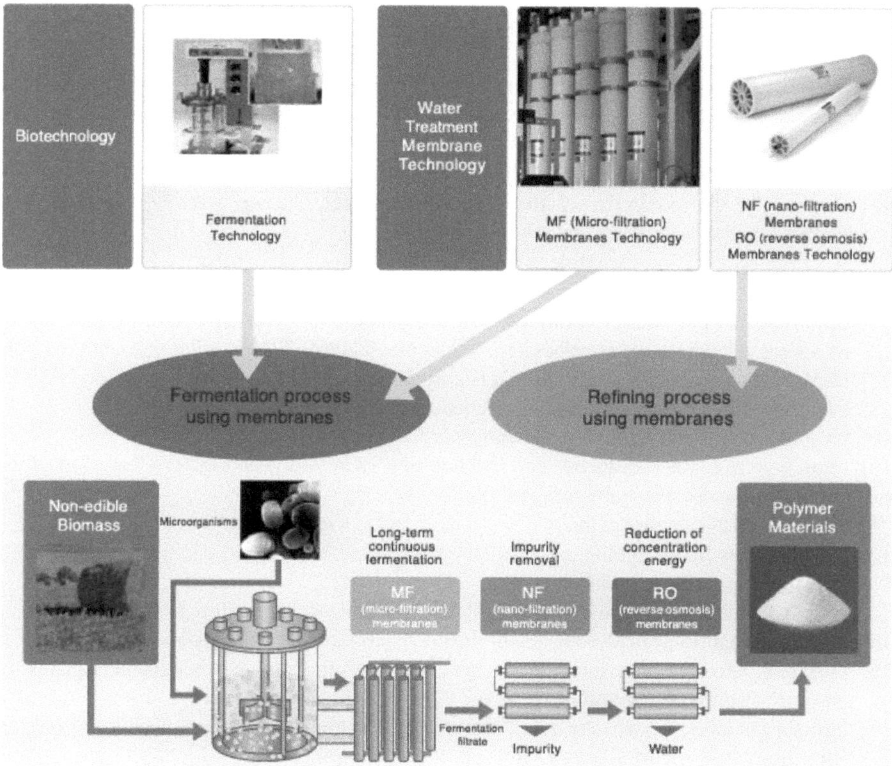

Fig. 21 Biomass-derived polymer membranes for purification.
From http://toray.co.th/en/technology/policy/green/gre_002.html.

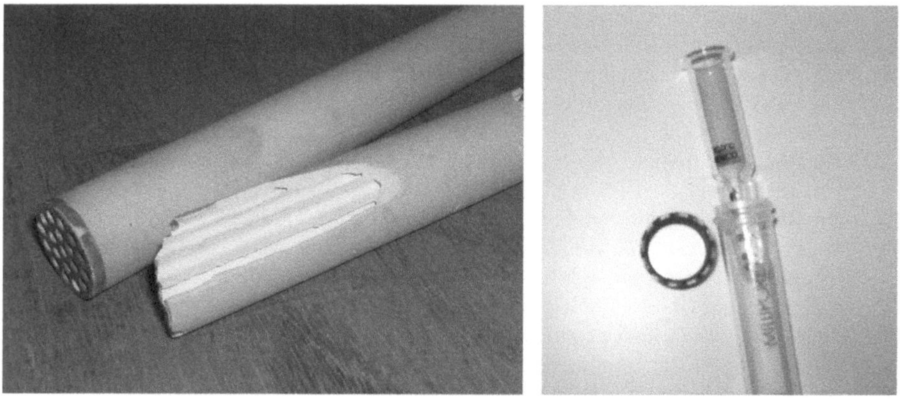

Fig. 22 Purification devices with cellulose acetate membranes.
From https://en.wikipedia.org/wiki/Synthetic_membrane#/media/File:Ceramic_membrane_(crossflow_filtration).jpg; https://en.wikipedia.org/wiki/Ultrafiltration#/media/File:Ultra_ filtration.JPG.

References

[1] S. Kalia, Recent developments in surface modification of natural fibers for their use in biocomposites, in: Biodegradable Green Composites, Wiley, Weinheim, Germany, 2016, p. 83.

[2] http://cen.acs.org/articles/92/i43/Biobased-Polymers.html.

[3] Sorona® EP thermoplastic polymer, Product datasheet, DuPont. www2.dupont.com/Plastics/en_US/Products/Sorona/sorona_ep.html.

[4] R. Humphreys, Bio-polyamides: where do they come from? Polymer Innovation Blog. https://polymerinnovationblog.com/bio-polyamides-where-do-they-come-from/.

[5] http://www.directindustry.com/prod/pma/product-19870-712743.html.

[6] http://www.lawbc.com/share/bcs2013/Molecules%20to%20Market/lee-presentation.pdf.

[7] https://www.total-corbion.com/markets/automotive/.

[8] http://www.plastics.gl/exhibit/latest-generation-of-pla/.

[9] http://www.bio-on.it/what.php?lin=inglese.

[10] https://www.gupta-verlag.com/news/technology/18343/polycaprolactone-based-tpu-for-suspension-and-drive-train-components.

[11] https://www.bio-amber.com/.

[12] http://www.themoldingblog.com/2012/11/13/faurecia-bets-on-bio-pbs-for-auto-interiors/.

[13] https://www.celanese.com/clarifoil.

[14] http://www.flexpackmag.com/articles/86036-cellulose-acetate-films-from-clarifoil.

[15] http://makingoftomorrow.com/green-plastics-from-citrus-fruit-peels-and-sugar/.

[16] http://www.foodengineeringmag.com/articles/89902-fresh-juice-processor-automates-pasteurization-with-continuous-process.

[17] http://www.bakeryandsnacks.com/Processing-Packaging/MF-Folien-introduces-first-film-made-from-DSM-s-EcoPaXX.

[18] http://www.jinhuizhaolong.com/en/chanpinmulu/index.html.

[19] https://www.packworld.com/article/food/refrigerated/pef-names-eastman-kodak-packaging-leader-year.

[20] http://greenchemicalsblog.blogspot.com/2013_10_01_archive.html.

[21] http://hbfilmmachine.com/product-2-5-pla-pbat-cling-film-line-en/134736.

[22] https://plasticsengineeringblog.com/2014/09/18/waiting-for-bio-pbs/.

[23] https://www.fmanet.org/blog/2013/11/21/bioplastics-9-lives.

[24] http://www.ptonline.com/articles/conrerence-report-bioplastics-are-breaking-out-of-their-green-niche.

[25] http://thaikk.co.th/index.php?route=information/information&information_id=128&sidebar=true.

[26] http://danimerscientific.com/pha-danimer-biodegradable-plastics/.

[27] http://www.dairyreporter.com/Library/Packaging-Packing-Materials-Containers/Green-PBS-potential-investigated-in-EU-project.

[28] http://www.ptonline.com/articles/processing-biopolymers-for-rigid-shee-thermoforming.

[29] https://www.plasticstoday.com/content/greening-industry-injection-and-blowmolding-packaging-may-not-be-first-volume-market-bioplastics/7331554586239.

[30] https://pt.slideshare.net/gerlach_insegment/insegment-metabolix-applications-slideshow-1017?smtNoRedir=1.

[31] http://www.bakeryandsnacks.com/Processing-Packaging/Metabolix-to-develop-PHA-latex-coatings-for-kraft-paper.

[32] http://www.ptonline.com/articles/enhanced-resins-additives-at-biopolymers-conference.

[33] http://www.bakeryandsnacks.com/Processing-Packaging/AMR-research-on-PLA-food-packaging-2020.
[34] http://makingsociety.com/2013/08/materialmatters-series-n1-pla-plastic/.
[35] https://www.greenerpackage.com/bioplastics/pla_filmsheet_masterbatch_increases_impact_strength.
[36] http://www.sustainableisgood.com/blog/natureworks_pla/.
[37] http://www.wikiwand.com/en/Bioplastic.
[38] Food and beverage packaging, NatureWorks LLC. http://www.natureworksllc.com/Ingeo-in-Use/Food-and-Beverage-Packaging.
[39] http://www.showa-denko.com/news/bionolle-the-pioneer-in-biodegradable/.
[40] http://shivampolymers.tradeindia.com/mulching-films.html.
[41] Biodegradable landscape stakes. http://www.erosioncontrol-products.com/landscapestakes.html.
[42] A.M. Mohan, Plantable nursery pot biodegrades in soil, GreenerPackage.com. https://www.greenerpackage.com/compost_biodegrade/plantable_nursery_pot_biodegrades_soil.
[43] S. Fox, A new breed of medical screws dissolve in body and promote bone growth. http://www.popsci.com/science/article/2010-03/biodegradable-medical-screws-disolve-body-and-promote-bone-growth.
[44] Euroscrew® TCP, Teknimed. http://www.medicalexpo.com/prod/teknimed/product-70291-499892.html.
[45] Product review—Insorb absorbable skin stapler. http://exploreplasticsurgery.com/category/medical-devices/.
[46] http://license.umn.edu/technologies/99107_bioplastic-for-use-in-orthopedic-devices-orthopedic-implants-and-implantable-medical-devices.
[47] https://www.cargill.com/pharmaceutical/pharmaceutical-ethanol/96-percent-pharmaceutical-agricultural-ethanol.
[48] https://www.cargill.com/personal-care/personal-care/formulations.
[49] https://www.cargill.com/personal-care/ethanol-for-cosmetics/96-pct.-cosmetic-alcohol-fortior.
[50] http://www.bio-on.it/production.php?lin=inglese.
[51] http://www.dd-database.org/other-drugs/indomethacin.html.
[52] https://www.nicotinetogo.com/STEP_1_Habitrol_21mg_Nicotine_Patch_28_box_p/habstep1bulk.htm.
[53] https://www.vitaoto.com/succinic-acid-amber-butanedioic-120-tabs-500-mg-useful-for-all-organism-dietary-supplement-antioxidant.html.
[54] https://gailfaithedwards.com/2015/10/29/the-medicinal-use-preparation-and-application-of-baltic-amber-gail-faith-edwards/.

Further reading

[1] http://www.foodengineeringmag.com/articles/89902-fresh-juice-processor-automates-pasteurization-with-continuous-process.
[2] http://www.bakeryandsnacks.com/Processing-Packaging/MF-Folien-introduces-first-film-made-from-DSM-s-EcoPaXX.
[3] http://www.jinhuizhaolong.com/en/chanpinmulu/index.html.
[4] J. Mowbray, Biodegradable polyester license amended, Tepha Inc. https://www.tevonews.com/fibres-yarns-news/341-amendment-for-biodegradable-polyester-license.

[5] http://toray.co.th/en/technology/policy/green/gre_002.html.
[6] https://www.sartorius.com/sartorius/en/EUR/Products/Laboratory/Microbiology-Products/Minisart%3Csup%3E%C2%AE%3C-sup%3E-NML-Syringe-Filter-16555——————Q,-0-45-%C2%B5m-Surfactant-free-Cellulose-Acetate/p/16555——————Q.
[7] http://www.sigmaaldrich.com/catalog/product/sigma/cls8162?lang=en&region=US.

Environmental impact of biomass conversion

13

13.1 Introduction

Nonsustainable extraction of wood from forests and fossil-fuel combustion has led to global environmental problems such as deforestation and climate change. This has led to the realization of the need to explore efficient and alternative environmentally sound energy options.

Biomass feedstock consumption is expected to be three times of the current biomass by the year 2030. By that time, forest biomass feedstock such as wood, logging residues, trees, shrubs, sawdust, and bark would be extensively used for industrial and electrical applications. Increased forest harvesting and plantation would supplement the residue supply. Biomass generated from plantations will be used to convert into biofuel to be used in transportation industry [1]. Excessive extraction and consumption of biomass has led to forest degradation and deforestation, loss of biodiversity, and soil degradation. In addition, pollution from emissions of greenhouse gases (GHG) caused by the combustion of wood and indoor air pollution have led to health risks [2]. There are multiple issues with deforestation. It not only leads to soil erosion but also poses risks of floods and impacts to the woodlands for agriculture and livestock. There has been some additional concerns which range from nonsuitability of forest soils for agricultural purposes, health problems due to smoke caused by burning of fuelwood, and loss in soil fertility due to use of agricultural residues [2].

Combustion of fuelwood and other biomass fuels leads to carbon dioxide emissions. It is difficult to estimate percentage emission from fuelwood use is from nonsustainable source. In addition to carbon dioxide emissions, combustion of fuelwood and agro-residues leads to emission of products of incomplete combustion [2].

13.2 Life cycle assessment

Life cycle assessment (LCA) evaluates [3–5]:

1. All stages of a product's life from the perspective that they are interdependent, meaning that one operation leads to the next.
2. Enables the estimation of the cumulative environmental impacts resulting from all stages in the product life cycle.
3. Impacts not considered in more traditional analyses (e.g., raw material extraction, material transportation, ultimate product disposal, etc.).

Fig. 1 illustrates the possible life cycle stages that can be considered in an LCA and the typical inputs/outputs measured [4].

Technology and Applications of Polymers Derived from Biomass. https://doi.org/10.1016/B978-0-323-51115-5.00013-X

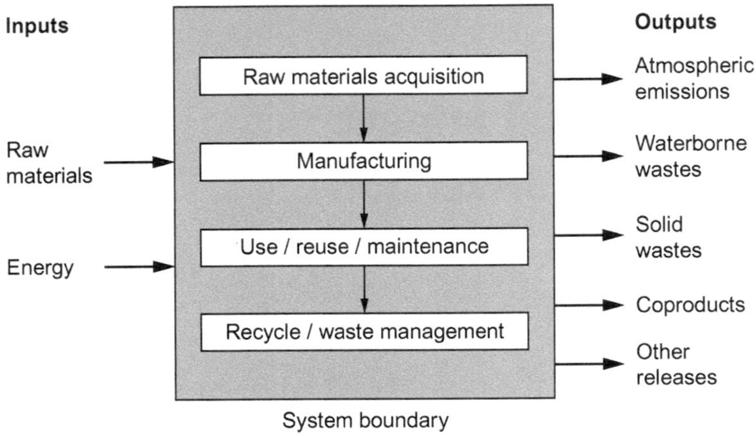

Fig. 1 Different stages of life cycle assessment (LCA).
From W. Klopffer, B. Grahl, Life Cycle Assessment (LCA): A Guide to Best Practice, Wiley, 2014.

When evaluating environmental as well as potential impacts associated with a product, process, or service, LCA can be used by:

- Compile an inventory of relevant energy and material inputs and environmental releases.
- Evaluate the potential environmental impacts associated with identified inputs and releases.
- Interpret the results to help decision-makers make a more informed decision.

Further, a systematic and phased approach is required to fully understand LCA. As shown in Fig. 2, there are four components to the LCA framework [3–5]:

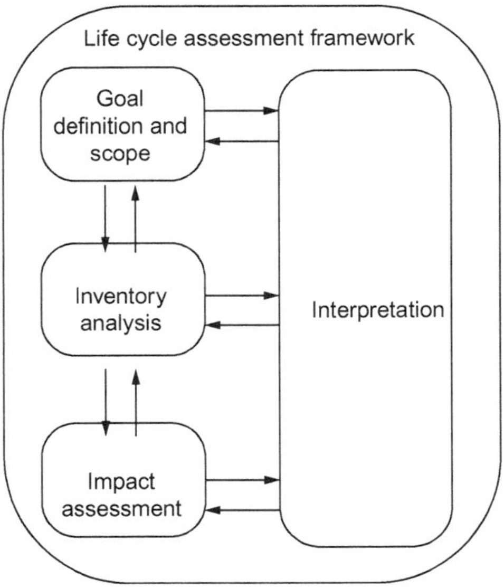

Fig. 2 Life cycle assessment (LCA) process.
From W. Klopffer, B. Grahl, Life Cycle Assessment (LCA): A Guide to Best Practice, Wiley, 2014; M.A. Curran, Life cycle assessment: principles and practice, EPA/600/R-06/060, U.S. Environmental Protection Agency, Cincinnati, OH, 2006.

(1) *Goal definition and scoping*: Define and describe the product, process, or activity. Establish the context in which the assessment is to be made and identify the boundaries and environmental effects to be reviewed for the assessment.
(2) *Inventory analysis*: Identify and quantify energy, water, and materials usage and environmental releases (e.g., air emissions, solid waste disposal, and waste water discharges).
(3) *Impact assessment*: Assess the potential human and ecological effects of energy, water, and material usage and the environmental releases identified in the inventory analysis.
(4) *Interpretation*: Evaluate the results of the inventory analysis and impact assessment to select the preferred product, process, or service with a clear understanding of the uncertainty and the assumptions used to generate the results.

13.3 Benefits of LCA

LCA is an important tool that can help decision-makers select the product or process that result in the least impact to the environment. This ability to track and document shifts in environmental impacts can help decision-makers fully characterize the environmental impacts associated with product or process alternatives. In order to get full advantage of this tool, designers and analysts will be able to [4,5]:

(1) Develop a systematic evaluation of the environmental consequences associated with a given product.
(2) Analyze the environmental trade-offs associated with one or more specific products/processes to help gain stakeholder acceptance for a planned action.
(3) Quantify environmental releases to air, water, and land in relation to each life cycle stage and/or major contributing process.
(4) Assist in identifying significant shifts in environmental impacts between life cycle stages and environmental media.
(5) Assess the human and ecological effects of material consumption and environmental releases to the local community, region, and world.
(6) Compare the health and ecological impacts between two or more rival products/processes or identify the impacts of a specific product or process.
(7) Identify impacts to one or more specific environmental areas of concern.

13.4 Biomass feedstocks and carbon emission

Extreme diversity of biomass feedstocks produces challenges in their characterization. Currently, there are three types of feedstocks which are handled in their unique way [6]:

1. *Organic feedstocks*: These types of feedstocks that can be utilized with conversion processes and end up in landfills. Feedstock include forest products wastes, agricultural residues, organic fractions of municipal solid wastes, paper, cardboard, food waste, green waste, and other waste.
2. *Nonbiodegradable organic feedstocks*: These are regular plastics and cannot be converted by biochemical processes.
3. *Bio-based materials*: Chemical, physical, or biological pretreatment is required.

The major categories of biomass feedstock are as follows:

(1) Forest biomass

Combustion of forest biomass to produce energy results in GHG emissions. Carbon dioxide and other GHG released into the atmosphere get simultaneously reabsorbed by growing forests. This makes the emissions generated from combustion of forest biomass as carbon neutral. Carbon neutrality ultimately depends on reforestation and maintenance of the overall stock of forests [6]. Table 1 shows common carbon emissions from burning of forest biomass. Outdoor burning releases carbon monoxide, particulate matter, and volatile organic compounds into the air [6,7].

(2) Biorenewable resources

Biorenewable resources are organic materials of biological origin that are generally classified as:

- Wastes are material that has been discarded. They are converted into process heat, electricity, fuels, or chemicals as coproduct. Some materials qualify as biorenewable such as agricultural residues, yard waste, municipal solid waste, and food processing waste.
- Energy crops are plants grown specifically for production of bio-based products [8].

Fig. 3 shows carbon emission from agricultural resources in 2013. Agricultural activities contribute directly to emissions of GHG through a variety of processes. Methane (CH_4) and nitrous oxide (N_2O) are the two primary GHG emitted by agricultural activities. Methane emissions from enteric fermentation and manure management represent 25.9% and 9.6% of total CH_4 emissions from anthropogenic activities, respectively. Of all domestic animal types, beef and dairy cattle were by far the largest emitters of CH_4. Rice cultivation and field burning of agricultural residues were minor sources of CH_4. Agricultural soil management activities such as fertilizer application and other cropping practices were the largest source of US N_2O emissions, accounting for 74.2%. Manure management and field burning of agricultural residues were also small sources of N_2O emissions [8].

(3) Food crops

GHG emissions from different crop management practices in South Africa in 2012 are shown in Table 2. Nitrous oxide emissions that result from the application of synthetic fertilizer are significantly higher with incorporation of crop residues into the soil [9].

Table 1 **Forest biomass and carbon emissions**

Pollutant	Forest biomass (LB pollutant/LB/MMBtu)
NO_x	0.220
CO	0.600
SO_2	0.025
VOC	0.017
PM	0.570
CO_2	206.94
HCl	1.900E−02
Hg	3.500E−06
Mn	1.600E−03

From A. Demirbas, Biomass feedstocks, Biofuels: Securing the Planet's Future Energy Needs, Springer, 2009, pp. 54–55 (Chapter 2); ORCAA assessment 2/3/2010, Ecology 2/24/2010.

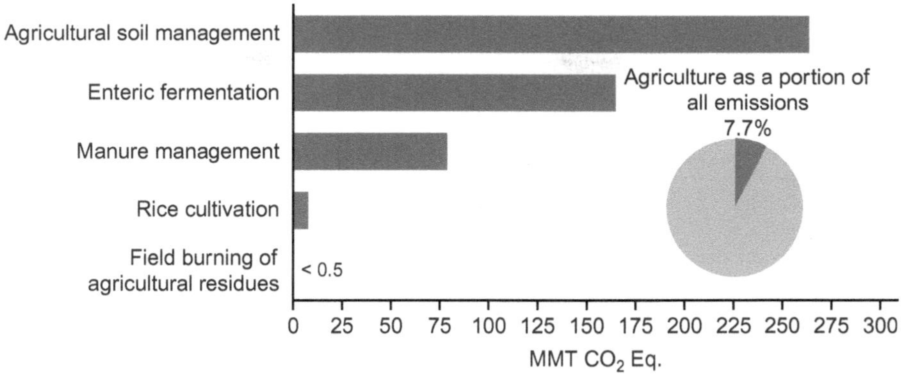

Fig. 3 Carbon emission from agricultural sources in 2013.
From Inventory of U.S. greenhouse gas emissions and sinks: 1990–2013. http://www.epa.gov/ttn/chief/net/2014inventory.html.

Table 2 Greenhouse gas emissions from different crop management practices in South Africa in 2012

Crop	Synthetic fertilizer (%)	Crop residue (%)	Lime (%)	Total GHGs (%)
Cereals	72.6	72.0	57.2	68.0
Legumes and oilseeds	6.7	4.6	22.9	11.2
Vegetables	8.7	5.7	2.9	6.6
Other field crops	12.0	17.7	17.1	14.3

From M. Tongwane, T. Mdlambuzi, M. Moeletsi, M. Tsubo, V. Mliswa, L. Grootboom, Greenhouse gas emissions from different crop production and management practices in South Africa, Environ. Dev. 19 (2016) 23–35.

(4) Sugar crops

Sugarcane is one of the leading food producing crops providing about 75% of the sugar for human consumption [9]. Projected sugar biomass will significantly increase and will provide up to 20% of all worldwide energy usage. GHG emission resulting from climate change, soil management activities, and sugarcane production process in agricultural sector has been studied by researchers.

Researchers from Center for Agricultural Policy and Trade Studies, North Dakota State University developed models to study impact of ton of CO_2e sugar produced by US sugar industry on the carbon emission. Table 3 shows the calculations for GHG emission under the various scenarios. The US beet sugar industry emits 1.16 tons of CO_2e per ton of refined sugar, while the cane sugar industry emits 0.57 tons of CO_2e per ton of refined sugar [10].

(5) Landfills

Wastes generated from a processing operation are treated in waste disposal sites. Solid wastes disposal sites are used to treat and/or dispose of solid wastes including landfills, land treatment units, waste piles, and composting operations. Solid wastes at the disposal sites undergo biological transformation to produce carbon dioxide (CO_2) under aerobic

Table 3 CO_2 emissions from sugar processing for US sugar consumption

		Model 1	Model 2	Model 3	Model 4
	CO_2e (tons)/ton of sugar	**CO_2e emissions (1000 metric tons)**			
US beet	1.16	4699	3905	2908	1402
US cane	0.570	1836	1786	1769	1718

From Food and Agriculture Organization and IFA: global estimation of gaseous emission of NH_3, NO and N_2O from agricultural land, FAO, Rome, 2001, Accessed 12 September 2009.

conditions and a mixture of methane (CH_4), and carbon dioxide (CO_2) under anaerobic conditions [11].

Among all the solid waste disposal sites, landfills are the most commonly used. They have predominantly contributed to climate change primarily due to their CH_4 production. Other solid waste disposal sites may also be important when considering CO_2 emissions. Composting operations and land treatment units are specifically operated for the aerobic degradation of organic waste materials, and therefore can have significant CO_2 emissions.

Landfill gas collection systems at active solid waste disposal sites measure the actual biogas composition. The average composition of the biogas is 41% CH_4, 34% CO_2, 22% nitrogen N_2, and 3% oxygen O_2 [12]. It is common to assume that approximately 50% of the carbon degraded within the landfill will be converted to methane, and the remaining 50% will be converted to carbon dioxide, with only trace amount of carbon being released as carbon monoxide.

13.5 Estimating carbon dioxide (CO_2) emissions

There are two main methods for estimating carbon dioxide emissions:

1. Continuous emissions monitoring system (CEMS) method
2. Fuel analysis method

13.5.1 Continuous emissions monitoring system

Continuous emissions monitoring system (CEMS) keeps a continuous and uninterrupted record of emission over an extended time. The monitoring is based on the principles of photometric measurements to measure the concentration of pollutants in the gas stream. One the pollutant concentration is measured, emission rates are obtained by multiplying the pollutant concentration by the volumetric stack gas flow rate [13,14].

Every active disposal site will have the site-specific CEMS data. This data is preferred for estimating NO_x, CO, CO_2, SO_2, and total hydrocarbon emissions because it provides a detailed record of emissions over time. In situations where CEMS data is not available, alternative methods are available to estimate emissions of these pollutants and provide only short-term emissions data.

Instrument out of calibration can be a problem for CEMS to accurately capture the data and any uncaptured data can create long-term incomplete data sets. It is the important that the source owner properly operates, calibrates, and validates the monitoring equipment and the corresponding emission data.

CEMS data can be manually estimated to predict annual measurements by:

- Calculating hourly emissions from concentration measurements
- Calculating gas stack flow rate
- Calculating emission factors from heat input

13.5.1.1 Calculating hourly emissions from concentration measurements

Although CEMS can report real-time hourly emissions automatically, it may be necessary to manually estimate predicted annual emissions from hourly concentration data. This section describes how to calculate emissions from raw CEMS concentration data [13].

Hourly emissions can be based on concentration measurements using the following equation [13]:

$$E_x = \frac{(C \times M_W \times Q \times 60)}{(V \times 10^6)}$$

where

E_x = hourly emissions in lb/h of pollutant x.
C = pollutant concentration in ppmvd.
M_W = molecular weight of the pollutant (lb/lb mol).
Q = stack gas volumetric flow rate in dscfm.
V = volume occupied by 1 mol of ideal gas at standard temperature and pressure.

13.5.1.2 Calculating stack gas flow rate

When direct measurements of stack gas flow rates are not available, Q can be calculated using fuel factors (F factors) according to EPA Method 19 [13] as shown as follows:

$$Q = F_d \times \frac{20.9}{(20.9 - \%O_2)} \times \frac{H_{in}}{60}$$

where

F_d = fuel factor, dry basis (from EPA Method 19) in dscf/MMBtu.
$\%O_2$ = measured oxygen concentration, dry basis expressed as a percentage.
H_{in} = heat input rate in MMBtu/h.

The F factor is the ratio of the gas volume of the products of combustion to the heat content of the fuel. F_d includes all components of combustion less water. F_d can be calculated from fuel analysis results using the following equation:

$$F_d = \frac{10^6[3.64(\%H) + 1.53(\%C) + 0.57(\%S) + 0.14(\%N) - 0.46(\%O)]}{HHV}$$

where

H, C, S, N, and O = concentrations of hydrogen, carbon, sulfur, nitrogen, and oxygen in the fuel expressed as a percentage as determined by a fuel analysis.
HHV = higher heating value of the fuel in Btu/lb.

13.5.1.3 Calculating emission factors from heat input

Sometimes it is desirable to calculate emissions in terms of pounds of pollutant per unit of heat combusted. For regulatory purposes, heat input is calculated based on the HHV of the fuel as measured by analysis. The heat input in terms of MMBtu/hr is calculated using:

$$H_{in} = \frac{(Q_f \times HHV)}{10^6}$$

where

H_{in} = heat input rate in MMBtu/h.
Q_f = mass fuel flow rate in lb/h.
HHV = higher heating value in Btu/lb.

An emission factor relating emissions to the heat input rate for the boiler is expressed as:

$$EF_x = \frac{E_x}{H_{in}}$$

where

EF_x = emission factor in lb/MMBtu of pollutant x.
E_x = emissions of pollutant x in lb/h.

13.5.2 Predictive emissions monitoring

Predictive emissions monitoring (PEM) is based on developing a correlation between pollutant emission rates and process parameters. The evaluation is done in two steps [13]:

First, a correlation test must first be performed to develop this relationship.
Emissions can then be estimated or predicted using process parameters to predict emission rates based on the results of the initial source test.

13.5.3 Fuel analysis method

The fuel analysis method is used for calculating carbon dioxide (CO_2) emissions. It involves determining a carbon content of fuel combusted using either fuel-specific information or default emission factors, and then applying that carbon content to the amount of fuel burned to quantify CO_2 emissions. This approach is appropriate for pollutants such as metals, SO_2 and CO_2 [13,14].

While emission factors are available for most metals, the use of site-specific fuel analysis data provides a more accurate emission estimate. Fuel analysis may also be used to calculate emissions by assuming complete conversion of the carbon in the fuel to CO_2 [13,14].

Fuel analysis method uses one of the three equations to calculate CO_2 emissions for each type of fuel combusted. Two of these equations can also be used to calculate CH_4, and N_2O emissions, using appropriate emission factors. The appropriate equation to use depends on what is known about the characteristics of the fuel being consumed.

(1) Equation 1

This equation is recommended when fuel consumption is known only in mass or volume units, and no information is available about the fuel heat content or carbon content. This equation is the least preferred. It has the most uncertainty because its emission factors are based on default fuel heat content, rather than actual heat content [13,14].

$$\text{Emissions} = \text{Fuel} \times EF_1$$

where

Emissions = mass of CO_2, CH_4, or N_2O emitted.

Fuel = mass or volume of fuel combusted.

EF_1 = CO_2, CH_4, or N_2O emission factor per mass or volume unit.

(2) Equation 2

This equation is recommended when (a) the actual fuel heat content is provided by the fuel supplier or is otherwise known, and/or (b) when the fuel use is provided in energy units. In such cases, the fuel use in energy units can be multiplied directly by the emission factor (EF_2). Equation 2 is a preferable approach over Equation 1 because it uses emission factors that are based on energy units as opposed to mass or volume units [13,14].

$$\text{Emissions} = \text{fuel} \times \text{HHV} \times EF_2$$

where

Emissions = mass of CO_2, CH_4, or N_2O emitted.

Fuel = mass or volume of fuel combusted.

HHV = fuel heat content or higher heating value, in units of energy per mass or volume of fuel.

EF_2 = CO_2, CH_4, or N_2O emission factor per energy unit.

(3) Equation 3

Equation 3 is recommended to calculate CO_2 emissions when the actual carbon content of the fuel is known. Carbon content is typically expressed as a percentage by mass, which requires fuel use data in mass units. This equation is most preferred for CO_2 calculations because CO_2 emissions are directly related to the fuel's carbon content. Because Equation 3 is only applicable to CO_2 emissions, Equation 1 or 2 should be used in conjunction to calculate CH_4 and N_2O emissions [13,14].

$$\text{Emissions} = \text{fuel} \times \text{CC} \times \left(M_{\text{w}_{\text{CO}_2}} / M_{\text{w}_{\text{carbon}}}\right)$$

where

Emissions = mass of CO_2 emitted.

Fuel = mass or volume of fuel combusted.

CC = fuel carbon content, in units of mass of carbon per mass or volume of fuel.

$$\left(M_{\text{w}_{\text{CO}_2}} / M_{\text{w}_{\text{carbon}}}\right) = 44/12$$

13.5.4 Steps for estimating emissions

Following are the steps to estimating emissions [13–15]:

Step 1: Select the appropriate equation

Based on the information available on the characteristics of the fuel being consumed, select the appropriate equation to use in calculating emissions.

Step 2: Determine the amount of fuel combusted

Each fuel type should be quantified separately. This can be based on fuel receipts, purchase records, or through direct measurement at the combustion device.

Step 3: Determine equation inputs

Based on the available inputs available, selected equation is selected to calculate emissions. As appropriate, determine the fuel carbon content, fuel heat content, and/or emission factors associated with each fuel consumed.

Step 4: Calculate emissions

Use the appropriate equation with the fuel consumption and other equation inputs to calculate the emissions of CO_2, CH_4, and N_2O.

References

[1] T.R. Miles Sr., T.R. Miles Jr., Environmental implications of increased biomass energy use: Final report, National Renewable Energy Laboratory, Portland, OR, 1992, NREL/TP-230-4633, Task No. BP111010.

[2] N.H. Ravindranath, K.U. Rao, Environmental effects of energy from biomass and municipal wastes, Encyclopedia of Life Support Systems (EOLSS), Eolss Publishers, Paris, France, 2005, http://www.eolss.net.

[3] J. Guinee, Handbook on Life Cycle Assessment: Operational Guide to the ISO Standards, Springer, Dordrecht, 2004, pp. 5–9.

[4] W. Klopffer, B. Grahl, Life Cycle Assessment (LCA): A Guide to Best Practice, Wiley, Weinheim, Germany, 2014.

[5] M.A. Curran, Life Cycle Assessment: Principles and Practice, U.S. Environmental Protection Agency, Cincinnati, OH, 2006. EPA/600/R-06/060.

[6] A. Demirbas, Biomass feedstocks, in: Biofuels: Securing the Planet's Future Energy Needs, Springer, London, UK, 2009, pp. 54–55 (Chapter 2).

[7] ORCAA assessment 2/3/2010, Ecology 2/24/2010.

[8] Inventory of U.S. greenhouse gas emissions and sinks: 1990–2013. http://www.epa.gov/ttn/chief/net/2014inventory.html.

[9] M. Tongwane, T. Mdlambuzi, M. Moeletsi, M. Tsubo, V. Mliswa, L. Grootboom, Greenhouse gas emissions from different crop production and management practices in South Africa, Environ. Dev. 19 (2016) 23–35.

[10] FAO, Food and Agriculture Organization and IFA: Global Estimation of Gaseous Emission of NH_3, NO and N_2O from Agricultural Land, FAO, Rome, 2009. Accessed 12 September 2009.

[11] R.D. Taylor, W.W. Koo, Impacts of Greenhouse Gas Emission Regulations on the U.S. Sugar Industry, Center for Agricultural Policy and Trade Studies, North Dakota State University, Fargo, ND, 2010.

[12] U.S. EPA (Environmental Protection Agency), Greenhouse Gas Emission Estimation Methodologies for Biogenic Emissions From Selected Source Categories: Solid Waste Disposal Wastewater Treatment Ethanol Fermentation, Sector Policies and Programs Division, U.S. EPA, Washington, DC, 2010. EP-D-06-118.

[13] U.S. EPA (Environmental Protection Agency), Background Information Document for Updating AP42 Section 2.4 for Estimating Emissions From Municipal Solid Waste Landfills, Office of Research and Development, U.S. EPA, Washington, DC, 2008. EPA-600/R- 08-116.

[14] Overview of Available Methods for Estimating Emissions, Preferred and Alternative Methods for Estimating Air Emissions From Boilers, vol. II, Emission Inventory Improvement Program (EIIP), Eastern Research Group, 2001, pp. 2.3–1 (Chapter 3).

[15] Direct emissions from stationary combustion sources, Greenhouse gas inventory guidance, U.S. Environmental Protection Agency, 2016. www.epa.gov/climateleadership.

Index

Note: Page numbers followed by *f* indicate figures, and *t* indicate tables.

FSC
www.fsc.org
MIX
Papier | Fördert gute Waldnutzung
FSC® C083411

Kolibri 360 GmbH
Ferdinand-Jühlke-Straße 7
99095 Erfurt, Deutschland
produktsicherheit@kolibri360.de